Surface Mount Technology

Surface Mount Technology

Rudolf Strauss, Dipl.Ing., FIM

Butterworth-Heinemann Ltd
Linacre House, Jordan Hill, Oxford OX2 8DP

℞ A member of the Reed Elsevier plc group

OXFORD   LONDON   BOSTON
MUNICH   NEW DELHI   SINGAPORE   SYDNEY
TOKYO   TORONTO   WELLINGTON

First published 1994

British Library Cataloguing in Publication Data
Strauss, Rudolf
    Surface Mount Technology
    I. Title
    621.3815

ISBN 0 7506 1862 0

Library of Congress Cataloguing in Publication Data
Strauss, Rudolf.
    Surface mount technology/Rudolf Strauss.
    p.   cm.
    Includes bibliographical references and index.
    ISBN 0 7506 1862 0
    1. Surface mount technology.   2. Printed circuits–Design
    and construction.   I. Title.
    TK7870.15.S63   93–44368
    621.3815'31–dc20   CIP

Typeset by Vision Typesetting, Manchester
Printed and bound in Great Britain by
Biddles Ltd of Guildford and King's Lynn

Contents

Preface

This book has been written for all those who have to solder surface mounted devices to circuit boards, and it should therefore be of interest to practitioners of soldering in all its various forms. Apart from them, people concerned with inspection and quality control, or with the choice and acquisition of equipment, may find some sections of the book useful.

I have tried to cover all the practicalities of soldering in electronic manufacture in such a way and such language, that it can be read and, I hope, understood by those in direct charge of assembling circuit boards by soldering. Temperatures are given in degrees Centigrade and Fahrenheit; as a rule, operating temperatures are rounded up or down to the nearest round figure, unless they relate to physical constants such as melting or boiling points. Dimensions are given both in metric and imperial units.

Since it is in the first place about soldering, the book covers all aspects of the soldering process itself, which include principally solders, fluxes, soldering heat and solderability. Because it deals with the soldering of SMDs, it describes the dimensions and features of these components as far as they are relevant to soldering. What goes on inside an SMD is of no concern here.

All practical and, as far as is necessary, the theoretical aspects of wavesoldering and of the various methods of reflowsoldering are comprehensively treated. Features of the circuit board and of component placement are considered as far as they are relevant to the soldering process.

Cleaning after soldering is treated in detail. The text is based on the state of the art, which this quickly evolving technology had reached by the middle of 1993. The restrictions on the use of cleaning media and methods, which the book mentions, are those which were in force or anticipated at that point in time.

Practitioners of soldering need to know what constitutes a 'good' joint, and how to correct soldering defects. Therefore, quality control and inspection are discussed in detail, as is corrective soldering. Some readers may find the contents of these sections of the book provocative or controversial. This was difficult to avoid, because I hold strong views in these matters, many of them based on practical experience.

My own interest and involvement in soldering go back a long time. Having studied experimental physics, as it was then called, in Germany in the mid-thirties, I joined and later managed the research department of a leading smelter of alloys of lead, tin and copper, when I came to London in 1939. During the war, which broke out soon after, I was involved in the development of the new technologies and materials which it demanded, many of them related to soldering.

In the mid-1950s, I was closely associated with the invention and introduction by my company of wavesoldering of printed circuit boards, which themselves had been invented during the war by Paul Eisler in London. My involvement with electronic soldering continued until my retirement in the mid-seventies. My engagement in consultancy, lecturing, and writing, still on that same subject, continues.

A large number of friends and former colleagues have given me much help, advice and support in writing this book, and I have drawn on the published work of several of them. My thanks are due to them all. They are too numerous to mention individually, but I must single out Dr Wallace Rubin of Multicore Solders Ltd for guiding me through the maze of the standard specifications of solders and fluxes, and Russ Wood, formerly of Dage (GB) Ltd, and Gordon Littleford, formerly of Kerry Ultrasonics Ltd, who have put me right on the finer points of today's cleaning technology.

Finally, my special thanks are due to Dr Colin Lea of the National Physical Laboratory in Teddington, who never hesitated to let me draw on his and his colleagues' wide fund of scientific knowledge.

Rudolf Strauss

Glossary

As is characteristic of any upwardly mobile technology, its practitioners are continuously coining new technical terms and abbreviations, which are given a more or less agreed meaning. It will be useful to provide a necessarily limited list of them at this point.

SMD	A surface-mounted device
TMD	Throughmounted device: a component with connecting wires or legs, for insertion in the throughplated holes of a circuit board
TPH	Throughplated hole
IC	Integrated circuit: an electronic circuit carried on the surface of a silicon wafer
Chip	The term 'chip' has acquired several meanings, among them the following: an IC; an SMD which carries an IC in its inside; a resistor or ceramic capacitor, encased in a rectangular ceramic body. In this book the term 'chip' will always have this last meaning
ASIC	Application-specific IC
I/O or IO	In/out: the number of leads of an SMD
Melf	Metal electrode face-bonded component: a resistor or a diode, encased in a cylindrical ceramic body
SO	Small-outline: an SMD, with a body made of plastic, with up to about 50 gullwing legs, at a pitch of 1.25 mm/50 mil
SOT	An SO transistor
SOIC	An SO, with an IC inside
VSOIC	Very-small-outline IC: an SOIC with 0.75 mm/30 mil pitch

CC	Chip carrier: a square-bodied, plastic or ceramic SMD with an IC inside
LCCC	Leadless ceramic chip carrier: a CC with a ceramic body and sintered metallic solderable IOs along its four sides
PLCC	Plastic leaded chip carrier: a CC with a body made of plastic, carrying J-shaped legs on all four sides
DIL	Dual-in-line. A TMD containing an integrated circuit with two parallel lines of legs

1 Why SMDs?

The relationship between the manufacturers of electronic components and the assemblers of electronic circuitry resembles that between two different orders of living beings, for example insects and plants: they need one another to be able to exist, and for that reason there are close links between the evolutionary paths of both. The shapes and the dimensions of their bodies, or respectively their functions, must match one another, so that whatever is needed to ensure the survival of either species can be properly performed. Any mistakes or mismatches are punished by extinction.

Here the similarity ends: the evolutionary paths of plants and animals started to go their different ways over three hundred million years ago. The evolutionary periods in the world of electronics are measured in units smaller than decades, sometimes years. Also, bees and flowers cannot talk to one another; the designers and makers of components and the designers and makers of electronic assemblies can and should. Sometimes, in the past, maybe not often enough but more frequently now, they do. Unless this communication develops into a continuing, orderly and purposeful dialogue, extinction of isolated species with insufficient evolutionary mobility continues to be a threat. At the other end of the mobility scale, a few large manufacturing houses have managed to bring three orders of electronic species together into one closely-knit symbiosis: components, component-placement equipment, and electronic assemblies are all designed, made, used and marketed by one single vertically-structured organization.

The particular branch of electrical engineering, which from about 1905 onwards was called 'electronics', could be said to have begun with the transmission of the first Morse signal across the Atlantic by Marconi, on 12 December 1901. Then, as now, one of the principal uses of electronics was the creation and transmission of signals. Then, as now, the basic constructional elements of electronic apparatus were of two kinds, components and the conducting links between them. On the one hand there were active devices such as spark-gaps, later on thermionic valves and passive components like inductive tuning and coupling coils, capacitors, and resistors. On the other hand there was a tangle of wires which connected these devices with one another. Judging by contemporary drawings and photographs, the style of these installations, whether landbased or on board a ship, was that of a rather untidy laboratory. The terminals of the various electronic devices were usually screw connectors.

After the First World War, radio started to develop as a vehicle for the transmission of news and entertainment to the public at large. To begin with, most receivers were assembled by domestic amateurs, who soldered connecting wires to a set of components supplied by their makers, complete with the necessary wiring diagrams. Industrial manufacture of domestic receivers and electronic apparatus in general began in the early twenties.

Soldering was the universal method of joining the connecting wires to the component terminals. An electronic apparatus was a three-dimensional assembly: the valves, coils, and resistances were all fairly large, measuring several inches across and in height, and their terminals were not always close together or in one plane. Soon they began to be assembled together on a common chassis, and the connecting wires were prefabricated as a three-dimensional wiring loom. Teams of skilled operators, mainly girls, handsoldered the wire ends to the component terminals, which themselves were either short wires or soldering lugs. They worked with electric soldering irons and solderwire with a rosin-flux core. Soldering quality on the whole was excellent, because every operator was his (or her) own quality inspector: she would not lift the iron off a joint until she had seen the solder flow into it. Making a wrong connection was the main danger.

The three-dimensional nature of electronic assemblies had two consequences: they did not lend themselves to mechanized mass production (though some attempts were made) and post-assembly inspection was almost impossible. Testing was functional, and the location of faults was a skill, not a science.

Paul Eisler's invention of the printed circuit in 1943 (Section 6.1) changed all that: he replaced the three-dimensional wiring loom with a two-dimensional pattern of thin strips of copper foil, carried on one side of an insulating phenolic paper board. Wherever a conductor had to be soldered to the terminal of a component, a hole was drilled into the board, and surrounded by a ring of conductor foil, the 'land'. The components, which at that time were axial resistors, axial or radial capacitors, sockets for thermionic valves and, increasingly, three-legged transistors, were placed on the other side of the board with their terminal wires pushed through their appropriate holes in the board. Their protruding ends were crimped over the lands, which surrounded the holes, and soldered to them, one by one (Figure 1.1). Again, teams of girls inserted the components, crimped the wire ends and handsoldered them to the lands on the board. Rules were established for what a good joint should look like, and some of these rules persist to this day (Chapter 9).

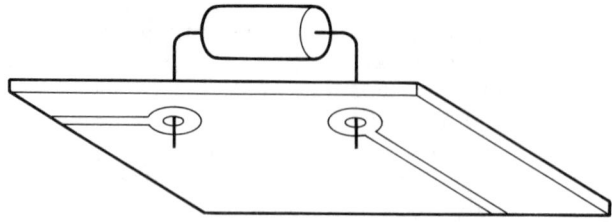

Figure 1.1   *Component, on a single-sided circuit board*

Because all the joints of the assembly were in one plane, soldering all of them in one operation was the obvious next step. This was made possible by the invention of wavesoldering in 1956 (Section 4.1). From then on, the forward march of the printed circuit board became unstoppable, and it soon conquered the world.

However, the assembly itself was still three-dimensional. Though the circuit pattern was in two dimensions, the thermionic valves, later the transistors, and all the resistors and capacitors were sitting on it like houses on a flat piece of ground, with their terminal sticking through it. Surface-mounted resistors stuck to circuit boards had been described in 1952,[1] but the first mention of a device with its terminals in contact with conductors on a circuit board occurs in a British patent in 1960.[2]

In the mid sixties, the growth of hybrid technology provided the incentive to design surface-mounted devices which had no connecting wires.[3] Thick–film circuitry, carried on ceramic wafers, provided a rugged basis for electronic assemblies for use in demanding environments. Because it was impracticable to provide the wafers with holes, the components had to be surface mountable by necessity. To begin with, some of them were simply wired components with their legs cut off, like melfs. Others were already purpose-designed for mounting on ceramics, like chips.

SOs with their angled legs came soon after chips and melfs, followed by PLCCs. They were the direct descendants of the DILs, and the pitch of their legs is still 1.27 mm/50 mil, like that of the DILs. All of them are decidedly three-dimensional, and obviously originally conceived for handsoldering. The component manufacturers issued detailed soldering advice, but left it to the makers and users of wavesoldering machines to cope with the problems caused by the three-dimensional nature of the components (Section 4.1.2).

In spite of these problems, and the initial reluctance of the assembling industry to cope with them, the advance of SMDs was as unstoppable as that of the printed circuit board twenty years earlier. With the arrival of integrated circuits and their multiple functions, the number of component legs – their pincount – began to grow beyond the 68 legs which were manageable with the old 1.27 mm/50 mil pitch, and this signalled the approaching end of the species of inserted components with their legs or wires stuck into holes drilled in a board. Surface-mounting technology (SMT) began to take over. At the same time, components became flatter, and approached the two–dimensionality of the boards on which they sit. The designers of soldering equipment and of SMDs had started to work together. TABs are a typical example of the benefits of this cooperation.

Today, a number of driving forces, which are pushing SMT further forward, can be discerned:

1. **Related to the individual component:** large-scale integration of chips and high switching rates demand short leads of roughly equal lengths. This requirement can only be met with the close-pitch design of QFPs (quad flat packages) and TABs (tape automated bonding packages) (Figure 1.2).
2. **Related to the assembly as a whole:** the number of functions per component, and consequently per assembly, has grown almost exponen-

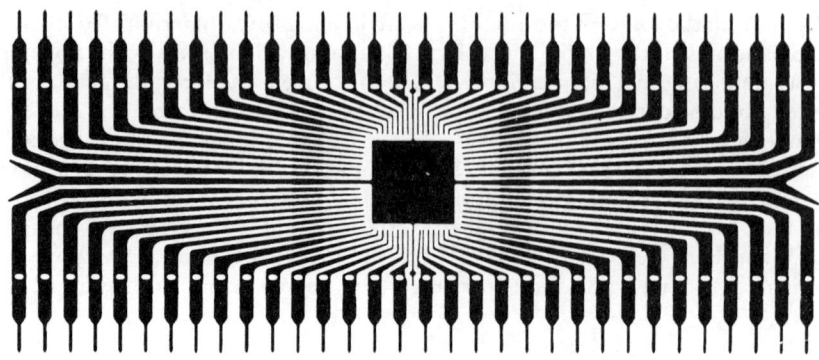

a

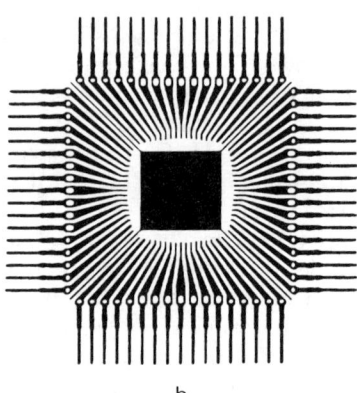

b

Figure 1.2   *DIL and PLCC. (a) 64-lead DIL, 2.54 mm/100 mil pitch; (b) 68-lead PLCC, 1.27 mm/50 mil pitch (Philips)*

tially over the years since the introduction of SM technology. A fine-pitch-technology board is many times smaller than a board with the same functions but populated with inserted components only. Electronic devices like the controls for a camcorder, the circuitry of a car telephone, or a pacemaker, are unthinkable without SMT.

3.   **Related to circuit manufacture:** automatic insertion of fine–pitch wired components, even if they did exist, would pose insurmountable difficulties.

Thus, SMDs have come to stay, and their use continues to grow. They have already overtaken inserted components, which have now reached their plateau (Figure 1.3). The forecast from which these data are quoted detects a steadily increasing growth rate for surface-attached components such as chips directly attached and wirebonded to the circuit board (chip-on–board, COB). Since this

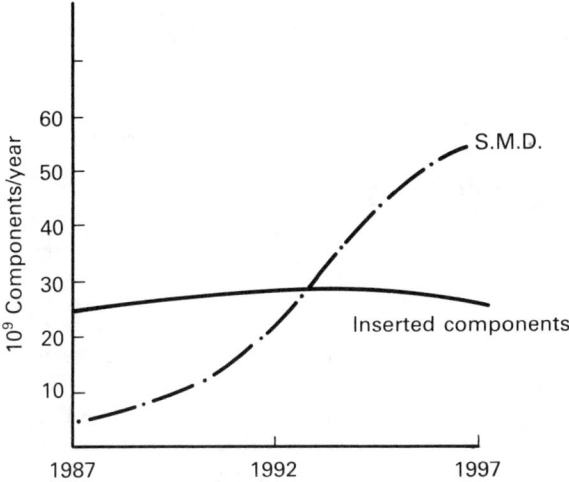

Figure 1.3 *Forecast of electronic component usage (from Surface Mount Technology, BPA Ltd, Dorking RH4 1DF, UK).*

book is devoted entirely to surfacemounting by soft soldering, other surface attachment methods such as wirebonding and using conducting adhesives are outside its scope.

As of now, almost all SMDs are connected to the conductor pattern of the board on which they sit by soldering. Alternative joining techniques, such as wirebonding in the case of COB constructions, or joining with a conductive, adhesive polymer, or even elastomer, have established a toehold in SMD technology, which will almost certainly grow in the future.[4] This present book, however, is confined to soldering as the method of joining components to the conductor pattern on the circuit board.

References

1. Anon. (1952) Adhesive Resistors, *Circuit World (UK)*, Feb. 1952, p. 70.
2. Brit. Patent 853 987, Nov. 1960.
3. Kirby, P. L. and Pagan, I. D. (1987) The Origins of Surface Mounting. *Proc. Europ. Microelectron. Conf.*, Bournemouth, UK, June 1987.
4. The Swedish Institute of Production Engineering Research (1993) *Proc. International Seminar, Recent Achievements in Conductive Adhesive Joining Technology in Electronics Manufacture*, Gothenburg, Sweden, September 1993.

2 The SMD family

Like all electronic components, surface-mounted devices (SMDs) can be classified according to their function.

Passive components include resistors, capacitors and inductors, all of them bipolar devices. Active components have three or more terminals, apart from diodes which have two. Basically, active devices contain switches or gates. These can number from the one of a single transistor with its three leads up to 25 000 for gate arrays with three hundred leads, or 250 000 for memories with about thirty leads.[1]

A detailed discussion of the function and internal construction of SMDs is outside the framework of this book, which is concerned with the problem of soldering SMDs to circuit boards. Therefore, it is the shapes and the solderable surfaces of these SMDs which are our main interest. Components for direct attachment techniques like 'naked' chips for chip on board mounting, and flip chips, which carry solder depots on their bodies in the form of 'solder-bumps', are also outside the scope of this present book. The former because they are connected to the circuit board not by soldering but by a wirebonding technique, the latter because at the time of writing, their industrial application is confined to a few large, vertically-integrated companies.[2]

2.1 Shapes, sizes and construction

SMDs come in three basic shapes: short cylinders, rectangular blocks and flat slabs. Table 2.1 lists them, together with their main characteristics, and sketches their main features. The names by which they have become known have emerged as they were developed, and for this reason they do not follow any logical scheme.

2.1.1 Melfs and chips

Cylindrical SMDs, made solderable by thick–film metallization or metallic caps at either end of their ceramic or glass bodies, are known as melfs (Metal Electrode Faced components). A small proportion of melfs are bipolar diodes with glass bodies, metal endcaps, and markings indicating their polarity. To ensure correct placement, diodes are packed in blistertape. SMDs with small square ceramic

Table 2.1 *The SMD family*

Shape	Nature, name	Number of leads	Pitch mm/inch	Body, size mm/inch	Package	Soldering method
◯⬤⬤	Passive Melf	2	—	Ceramic 5.0 × 2.22 dia./ 0.2 × 0.08 dia.	T,B	W,R
	Mini melf	2	—	3.6 × 1.4 dia./ 0.14 × 0.06 dia.	T,B	W,R
	Micro melf	2	—	2.0 × 1.1 dia./ 0.08 × 0.04 dia.	T,B	W,R
◯⬤⬤	Active SOD	2		Glass 3.5 × 1.6 dia./ 0.14 × 0.64 dia. to 5.0 × 2.3 dia./ 0.2 × 0.09 dia.	T	W,R
▱	Passive chips	2	—	1.25 × 2.0 dia./ 0.05 × 0.08 dia. to 5.7 × 5.0 dia./ 0.23 × 0.2 dia.	T,B	W,R
⬛	Active SOT	3–4	—	Plastic 3.0 × 1.3 dia./ 0.12 × 0.05 dia.	T	W,R
⬛	SO, SOIC	6–28	1.27/ 0.050	4.0 × 5.0 dia./ 16 × 0.2 dia. to 7.6 × 18 dia./ 0.3 × 0.72 dia.	T,B	W,R
▱	VSO	40–56	0.75/ 0.030	up to 11 × 21 dia./ 0.44 × 0.85 dia.	T	W,R
◻	SOJ, PLCC	20–84	1.27/ 0.050	up to 29.4 × 29.4 dia./ 1.16 × 1.16 dia.	T,M	W,R

Table 2.1   *(continued)*

Shape	*Nature, name*	*Number of leads*	*Pitch mm/inch*	*Body, size mm/inch*	*Package*	*Soldering method*
	QFP	44–148	0.5–0.8	up to 25.4 × 25.4 dia. Glass	M	R
	TAB	up to 400	down to 0.35/ 0.014	8, 16, 35 mm Film	Film	Impulse

W: Wavesolderable; R: Reflowsolderable; Impulse: Impulse solderable; T: on tape; M: in magazine; B: in bulk

bodies are called chips. Their endfaces are metallized by thick-film metallic deposit.

Resistors and capacitors, being passive components without polarity, can be mounted either way on their footprints. Therefore they can be picked from bulk feeders. Of the 170×10^9 passive SMDs which were used worldwide in 1990, 57% were resistors, 42% were ceramic condensers, with tantalum and electrolytic condensers forming the rest.[2,3]

Chips have been given a four-digit number code which indicates their approximate size in hundredths of an inch. The first two give their length, the second their width. Thus a 0805 chip is about 2 mm/0.08 in long and 1.25 mm/0.05 in wide. On the other hand, the designation of melfs gives their approximate size in millimetres. An 0102 micromelf has a diameter of 1.1 mm/0.04 in and is 2.2 mm/0.09 in long.

The main feature of melfs and chips is their tendency to get smaller and smaller. At present, a limit seems to have been reached with 0805 chips and 0102 micromelfs. The makers of automatic component placement equipment had to face and solve the formidable task of handling these small components and placing them with the required speed and precision (Chapter 7).

2.1.2   LCCCs

Until recently, there was a class of integrated circuits, housed in flat, square ceramic bodies, in sizes of up to and above 25.4 × 25.4 mm/1 × 1 in. Each of the four sides carried a row of thick-film solderable patches, usually of gold or a gold–platinum alloy, at a pitch of 1.27 mm/50 mil as connectors. These were known as leadless ceramic chip carriers (LCCCs). They had originally been designed to be soldered to the ceramic substrate of hybrid circuits. In the context of printed circuit board assemblies, they became extinct because of the problems caused by the thermal expansion mismatch between them and the FR4 circuit boards to which they had to be soldered (Section 6.2).

2.1.3 SOs, PLCCs, PFQs and TABs

Most other SMDs have moulded plastic bodies, which contain semiconductor devices, and which do not suffer from the thermal expansion mismatch problem. Their solderable connectors have the form of angled legs numbering from three up to 400 or more, depending on the number and complexity of their functions. The devices which they contain may be single transistors, integrated circuits (ICs), or application-specific integrated circuits (ASICs) such as gate arrays, microprocessors, or random access memories (RAMs).

The 'small-outline' SMDs (SOs) are the direct descendants of insertion-technique components with flat push-through legs, spaced at a centre distance (pitch) of 1.27 mm/50 mil. Their bodies can be up to 10 mm/0.4 in high, and they include SOTs (small-outline transistors), and SOICs (small-outline integrated circuits). Their solderable legs are angled downwards ('gullwing' shape), raising the underside of the housing off the board surface ('stand-off', Section 8.1.1). SO legs are relatively thick, from 0.1 mm/4 mil to 0.3 mm/12 mil, and therefore somewhat rigid. This lack of 'compliance', which can cause stresses on the soldered joints in the case of a temperature difference between component and board, can be a problem (Section 3.1). The gullwing legs of 'very-small-outline' SMDs (VSOs) also have gullwing legs, with a pitch of 0.75 mm/30 mil.

PLCCs (plastic leadless chip carriers) are SOs with their J-shaped legs tucked underneath the edge of the component body instead of pointing outwards, in order to save space. The price which the board assembler may have to pay for this gain in real estate is the problem of skipped wavesoldered joints, especially with closely set PLCCs (Section 6.4.1), or of 'wicking' when they are reflowsoldered (Section 9.3). For impulse soldering with a thermode (Section 5.7), PLCCs require a special thermode which heats the sides of the J-legs, since the footprints are not directly accessible. Equally, visual inspection demands special optics for oblique viewing, because the ends of the J-legs are out of the line of sight when viewed vertically.

2.2 High-pincount components

The steadily increasing complexity of ICs, with their need for more and more outer leads, led to the development of flatpacks, quad-flatpacks (QFPs) and ICs for tape-automated bonding (TABs), with more closely spaced and much thinner legs. TABs with 600 leads are already available, and lead numbers of up to 800 are planned for the next few years (Figure 2.1).

The design of a component imposes a limit on its possible maximum pincount. SOs with a minimum pitch of 0.75 mm/25 mil (VSOIC) and a maximum size of 25.4 mm/1 in square could accommodate about 130 leads, which is well below today's maximum demands. QFPs with pitch spacings down to 0.5 mm/20 mil allow pincounts of up to 150–200. At these leadcounts, the gullwing legs are necessarily very thin and narrow (Section 3.3), and are therefore in danger of being deformed during handling, even by the gentlest of pick-and-place machines.

The concept of the TAB device provides an answer. TABs demonstrate the benefits which result from a close cooperation between the designers of electronic components, of placement systems, and of soldering experts at the design stage. The TAB carries a chip which is no longer enclosed in a moulded plastic body, but covered by a thin glass wafer, with its copper leads bonded to a polyimide tape, which has the form of a standard cine-film, from super-8 up to 35 mm or more (Figure 2.2). These tapes are robust, and lend themselves to automatic placement and soldering on purpose-designed equipment, which by now has become well developed and widely available.

The inner ends of the leads are soldered to the pads of the IC by the manufacturer of the TAB by a solder-bump technique (inner-lead bonding, ILB), which is outside the scope of this book. The outer ends, with a pitch of down to 300 μm/12 mil, can be impulse-soldered with a thermode (outer-lead-bonding, OLB). Hot-air thermode (HAT) techniques, which can also work with hot nitrogen, and laser techniques for OLB have been developed.[5] One of the advantages of TABs is the ease with which each can be tested by the manufacturer, using as test pads ends of the outer leads. The footprints to which the TAB leads are to be soldered are given a solder coating of about 15 μm/0.6 mil. For details of the required soldering techniques, see Section 5.7.

TABs are approximately 0.2 mm/8 mil thick. Therefore, they lend themselves to being stacked, one on top of the other, on the same set of footprints. In this way, a large number of possible functions can be crowded into a very small board area, for applications such as camcorders and photographic equipment.[6]

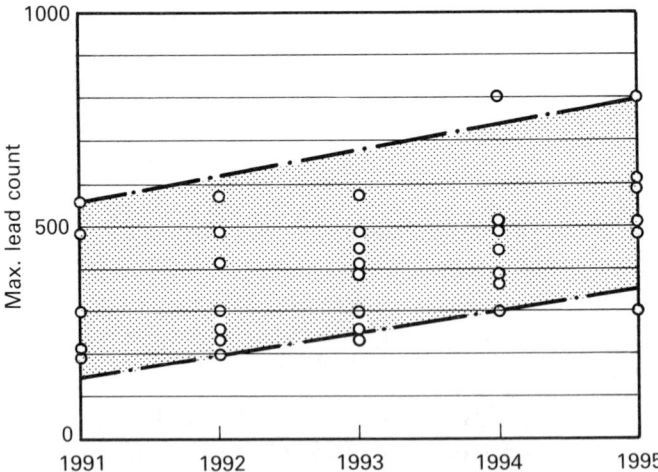

Figure 2.1   *Maximum lead count trends (after E. J. Vardaman*[4]*). General trend: +60 leads/year (courtesy Techsearch International Inc., Austin, Texas, USA)*

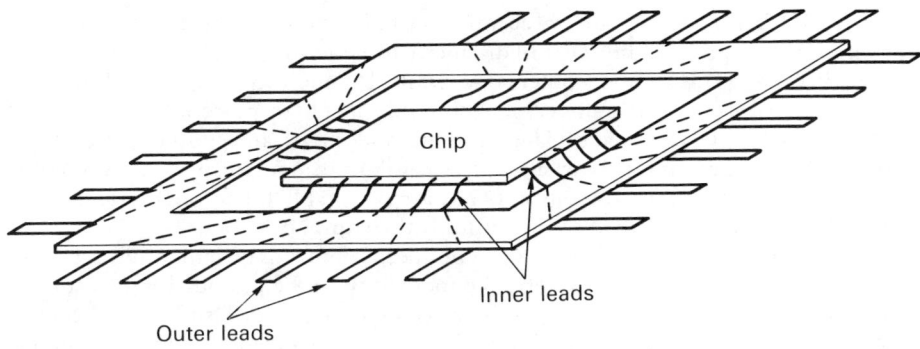

Figure 2.2 *The TAB*

A recent newcomer to the species of high-pincount SMDs is the 'land-grid array' (LGA), which itself is derived from the pin-grid array (PGA), an insertion device suitable for wavesoldering. The pins of PGAs were inserted in a matching pattern of holes in the circuit board, and wavesoldered in the standard manner. LGAs are designed for reflowsoldering. Their pins sit on a matching pattern of footprints, which are disposed underneath the component.[7]

With grid arrays, the leads take the form of straight pins, which are arranged on the underside of the flat component in a two-dimensional pattern (Figure 2.3). Since they are not in single-file rows, their spacing can be wider than that of a row of gullwing legs.

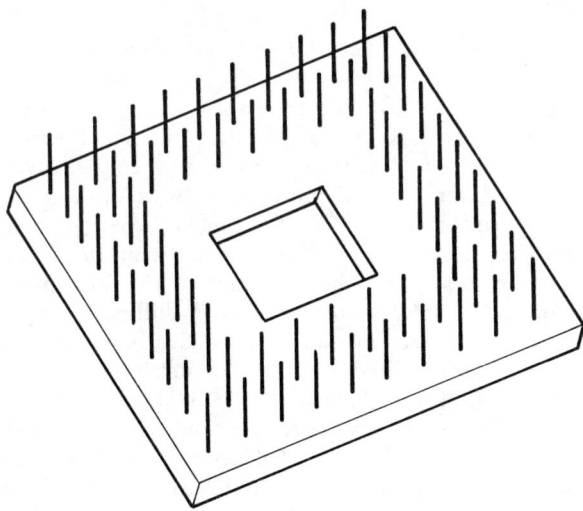

Figure 2.3 *Pin-grid or land-grid array*

An SMD 25 mm/1 in square can accommodate 240 pins, arranged in three rows of three pins abreast, at 1 mm/40 mil distance from one another. The outer row is 1 mm/40 mil inboard from the outer edge, while the innermost row is at a distance of 4 mm/160 mil from the outer edge. This results in a high-pincount component with a manageably wide pitch. The joints are underneath the component body, but even the innermost row is still accessible to an efficient convection reflowsoldering process. Obviously, LGAs are unsuitable for infrared soldering, since all joints sit in the shadow underneath the component body.

Their wide spacing of the joints makes no special demands on either the paste or the printing technique; alternatively, the footprints can be provided with a solid solder depot, and be fluxed prior to the placement of the SMDs (Section 6.3.2). The resulting solder joints are butt joints, which offer the advantage of being sound and mechanically reliable (Section 3.6.3). On the other hand, all joints are hidden from direct view underneath the component, and not accessible to visual inspection. Open joints can only be detected and identified by functional testing.

2.3   The solderable surfaces of SMDs

Obviously, the solderable surfaces and the leads of SMDs must maintain a consistently high solderability, without demanding strong fluxes, and their shelf-life under normal storage conditions should be at least a year. Their design and metallurgical structure are governed by these requirements, and by the nature of the component body and the soldering methods which are likely to be used.

2.3.1   Melfs and chips

Melfs and chips carry solderable, metallic surfaces at two opposing ends of their ceramic bodies.

The ends of the cylindrical melf resistors are normally in the form of small caps, pressed from steel sheet. Next to the steel comes an electroless deposited nickel/phosphorus layer, which in turn carries a thin copper layer, covered by another nickel layer, about 5 μm/0.2 mil thick. It is finally topped by an up to 10 μm/0.4 mil thick tin coating. The first nickel coating ensures a good, diffusion-free bond between the iron and the copper, while the second one acts as a 'diffusion barrier', which prevents the formation of the tin/copper intermetallic 'eta' (η) compound (Section 3.2). Once the η layer diffuses through the outer tin cover to the surface, it oxidizes and spoils the solderability of the component.

The end faces of chips are covered by a sintered thick-film metallic deposit. Such deposits consist of solderable metallic particles, embedded in a frit of low-melting glass. The particles on the surface represent the solderable face of the chip. They consist mostly of 80% to 90% silver, the balance being palladium, which has the function of slowing down the otherwise rapid dissolution of the silver particles during reflowsoldering (Section 3.6.4). Once they have been leached away by the molten solder, only the unsolderable glass frit remains on the surface.

Alternatively, and more recently, a thick–film layer of plain silver is covered with a diffusion barrier of nickel, and a top coat of up to 10 μm of tin for solderability.

2.3.2 Components with legs

SOs and PLCCs

The legs of most SOs and PLCCs are stamped from a 42% nickel/58% iron (Alloy 42) sheet, which has a low coefficient of thermal expansion. They are between 0.2 mm/8 mil and 0.3 mm/12 mil thick, which means that they are sufficiently sturdy not to get bent during normal handling, which ensures their 'coplanarity'. The trade-off is their lack of compliance, when a temperature difference between the component and the circuit board puts a stress on their soldered joints (Section 6.2).

Coplanarity in an SMD means that the ends of all its legs are in one plane, so that each one of them is in firm contact with its footprint when placed on the board. A leg which has been bent upwards may cause an open joint. One or more legs which are bent downwards may lift their neighbours off their footprints.

Normally, coplanarity of multilead components is expected to be within 0.2 mm/8 mil. However, with fine-pitch technology, the thickness of the solderpaste printdown on the footprints may drop below that figure, which means that with today's multilead components, coplanarity must be nearer to 0.1 mm/4 mil.

Many automatic component placement machines are designed to check the coplanarity of multilead components between collecting them from the feeder magazine and putting them down on their footprints (Section 7.3). Naturally, coplanarity only matters with wavesoldering or an infrared, hot air or vapour-phase reflow process, where the component legs rest freely on their footprints. With impulse soldering, a hot soldering tool (thermode) presses them against their footprints and, within reasonable limits, lack of coplanarity becomes irrelevant (Section 5.7).

The solderability of the legs of SOs and PLCCs is ensured by a galvanic, reflowed coating of 60%Sn/40%Pb, which may have an undercoating of 1–5 μm copper to ensure adhesion. Alternatively the topcoat may be hot-tinned with 60/40 tin–lead solder (up to 15 μm/0.6 mil thick). Alternatively, and more recently, legs of SOs and PLCCs are stamped from sheet rolled from an alloy consisting of 98%Cu/2%Fe, which has very good mechanical and electrical properties. The solderable topcoat is the same as with the legs made from Alloy 42.

QFPs and TABs

QFP legs are mostly made from an alloy 89%Cu/9%Ni/2%Sn with a galvanic and reflowed, or hot-tinned, topcoat of 60Sn/40Pb. TAB legs are made either from tinned plain copper or 98%Cu/2%Fe alloy sheet. They are produced by an etching technique, using a photomechanically produced etch resist pattern.

QFP and TAB legs are thin and slender. They may be as thin as 0.1 mm/4 mil, and their width depends on the pitch at which they are spaced. With a pitch of 0.5 mm/20 mil for example, footprints are 0.25 mm/10 mil wide, and the component legs are narrower still. This means that they are easily deformed, causing the above-mentioned problem of coplanarity.

It is one of the several virtues of TABs that they possess built-in coplanarity: the outer ends of their legs are fixed to the frame of the carrier film, from which they are cut free as they are being placed. Furthermore, they are mostly soldered to their footprints with a thermode, which presses them down on the board, and only rarely by a non-contact reflow process.

The metallurgical data of the solderable component surfaces and legs are summarized in Chapter 3, Table 3.7.

2.4   SMD shapes and wavesoldering behaviour

As has been said already, SMDs were originally conceived for hybrid technology in the early seventies, for handsoldering to their ceramic substrate or reflowsoldering on a hotplate. The latter was no problem, considering the single-sided construction of the circuit and the stability and good heat conductivity of the ceramic.

The soldering problems began when SMDs had to be wavesoldered to printed circuit boards. Their design did not turn out to be particularly user-friendly, especially as far as the SOs and PLCCs were concerned. This difficulty is due to the contours which their designers gave them: the ends of their legs, which have to be soldered to the footprints, are too close to the relatively high bodies which enclose their semiconductor circuitry. Thus, the 'angle of aspect', which is formed between the upper edge of the component body and the end of the solderable leg, is between 60° for SOs and 90° for PLCCs (Figure 2.4). The solderwave finds it difficult to penetrate into these corners, because of the surface tension of the molten solder (Shadow effect, Section 4.4.3). These difficulties were the reason for the development of the 'chip-wave' technology.

As chips acquired ever more functions, and the number of their leads increased beyond the 84 legs of the PLCCs, their plastic housings became larger and flatter. QFPs are not much above 2 mm/8 mil in height, and their soldering terminals are more readily accessible to the solderwave.

The even flatter TABs would be even easier to wavesolder if their fine pitch did not make wavesoldering very difficult, if not impossible, because of the close setting of their legs and the consequent danger of bridging. However, recent refinements of wavesoldering under nitrogen have pushed the limit of wavesolderability without bridging back to a pitch of 0.6 mm/24 mil, and efforts are being made to extend this down to 0.4 mm/16 mil.

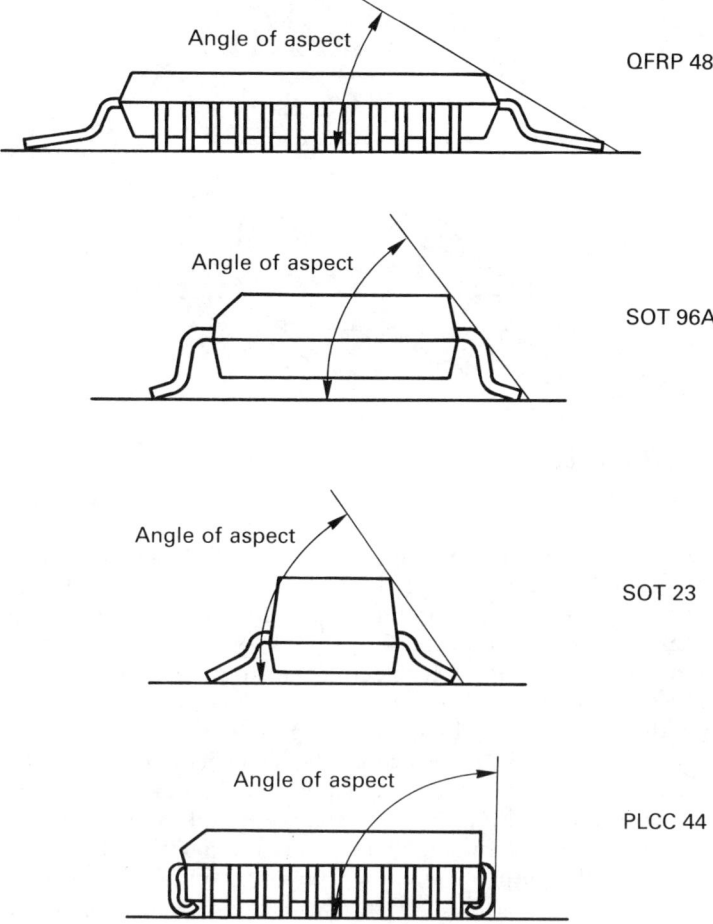

Figure 2.4 *The contours of SOs, PLCCs and QFPs.*

2.5 The popcorn effect

Under some circumstances, the plastic bodies of large, flat SMDs like PLCCs and QFPs can develop cracks if they are soldered in a solderwave, by vapourphase soldering or in an infrared or hot-air convection oven. In all these cases, the whole component is heated to well above the melting point of the solder (183 °C/361 °F). These cracks are caused by the so-called 'popcorn effect', which is due to the following mechanism.

The bodies of SOs, PLCCs and QFPs, being made from plastic, are capable of absorbing moisture if they are manufactured or stored under humid conditions.

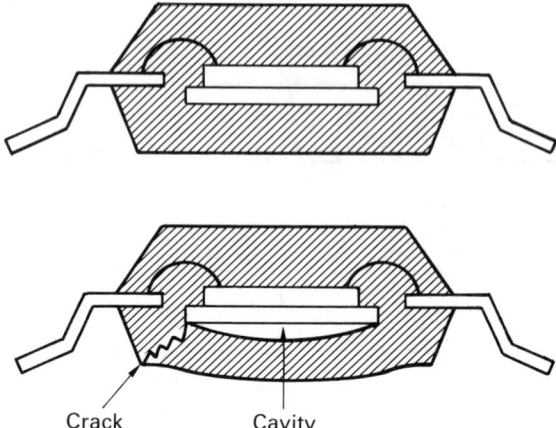

Crack   Cavity

Figure 2.5   *The popcorn effect*

The moisture not only accumulates in the body itself, but can also penetrate into the inner cavity of the SMD. During soldering, it turns to steam. As a consequence, pressure of several atmospheres can build up in the interior of the component, which causes the housing to 'balloon' and crack (Figure 2.5).[8] Naturally, the popcorn effect does not arise with soldering methods where only the leads are heated while the body of the SMD remains cool. These methods include impulse soldering with a thermode, and laser soldering.

   The popcorn effect can be prevented by the manufacturer of the SMD, if the components are moulded in a strictly dry environment, or dried out by a heating process, and packed in moisture-impermeable metal foil. Alternatively, the user can carry out the heat treatment himself (48 hours at 125 °C/260 °F or 100 hours at 100 °C/215 °F). For this purpose, the components must of course be removed from their magazine trays. Afterwards, they are sealed in polythene bags, together with a desiccating compound, and used as soon as possible. It is another virtue of TABs that, since they are enclosed in a glass cover, they do not suffer from this problem.

2.6   References

1.   Reiner, M. (1985) VLSI Packaging, *Hybrid Circuits*, No. 6, pp. 9–13.
2.   BPA Ltd (1989) *Surface Mount Technology, A Critical Analysis*, BPA Ltd, Dorking RH4 1DF, UK, pp. 146–147.
3.   Anon. (1993/4) *Surface Mount Technology (Germany)*, p. 68.
4.   Vardaman, E. J. (1992) New TAB Developments and Applications, *Proc. Nepcon West*, pp. 590–594.

5. Palmer, M.J. (1991) HAT Tool for Fluxless OLB and TAB. *Electronic Components and Technology Conf.*, pp. 507–510.
6. Wolski, G.B. (1993) Impulses from SMT Production. *EPP (Leinfelden, Germany)*, June 1993, pp. 15–21 (in German).
7. Burgess, T., McCall, G. and Poulter, S. (1993) New Trends in IC Packaging, *Electronics Manufacture and Test (UK)*, June 1993, pp. 14–15.
8. Gordon, S.F., Huffman, W.D., Prough, S., Sandkuhle, R. and Yee, M. (1987) Moisture Effect on Susceptibility to Package Cracking in Plastic SMDs. *IPC Techn. Rev.*, **29**, 2, pp. 18–20.

3 Soldering

3.1 The nature of soldering and of the soldered joint

Soldering, together with welding, is one of the oldest techniques of joining two pieces of metal together. Today, we distinguish between three 'metallurgical' joining methods: welding, hard soldering (or brazing) and soft soldering. The term 'metallurgical' implies that at and near to the joint interface, the microstructure has been altered by the joining process: what has happened has made one single piece of metal out of the two joint members, so that electric current can flow and mechanical forces can be transmitted from one to the other.

With both hard and soft soldering, the joint gap is filled with a molten alloy (an alloy is a mixture of two or more pure metals) which has a lower melting point than the joint members themselves, but which is capable of wetting them and, on solidifying, of forming a firm and permanent bond between them. The basis of most hard solders is copper, with additions of zinc, tin and silver. Most hard solders do not begin to melt below 600 °C/1100 °F, which rules them out for making conductive joints in electronic assemblies.

Soft solders are alloys of lead and tin, which begin to melt at 183 °C/361 °F. This comparatively modest temperature makes them suitable for use in the assembly of electronic circuits, provided heat-sensitive components are adequately protected against overheating.

3.1.1 The roles of solder, flux and heat

Soft soldering (from here on to be simply called 'soldering') is based on a surface reaction between the metal which is to be soldered (the substrate) and the molten solder. This reaction is of fundamental importance; unless it can take place, solder and substrate cannot unite, and no joint can be formed.

The reaction itself is 'exothermic', which means that it requires no energy input to proceed, once it has started. Soldering heat is needed to melt the solder, because solid solder can neither react with the substrate (or only very slowly), nor flow into a joint.

The reaction between solder and substrate is of crucial importance for both the process of soldering, and for the resultant soldered joint; only the tin in the tin–lead solder alloy takes part in it. The reaction products are so-called intermetallic compounds, hard and brittle crystals, which form on the interface between the solid substrate and the molten solder. The bulk of them stay where they have formed. They constitute the so-called 'intermetallic layer' or 'diffusion zone', which has a profound effect on the mechanical properties of the soldered joint and on its behaviour during its service life.

Any non-metallic surface layer on the substrate, such as an oxide or sulfide, however thin, or any contamination whatever, prevents this reaction, and by implication prevents soldering. Unless the contamination is removed, the reaction cannot occur. Unfortunately, under normal circumstances all metal surfaces, with the exception of gold and platinum, carry a layer of oxide or sulfide, however clean they look.

The soldering flux has to remove this layer, and must prevent it from forming again during soldering. Naturally, the surface of the molten solder is also one of the surfaces which must be considered here, because an oxide skin would prevent its mobility. Clean solder can flow freely across the clean substrate, and 'tin' it. (The expression 'tinning' derives from the fact that solder is often called 'tin' by the craftsmen who use it, and not from the fact that tin is one of its constituents.)

It is important at this point to make it quite clear that the flux only has to enable the reaction between substrate and molten solder to take place. It does in no way take part in the reaction once it has arranged the encounter between the two reaction partners. Hence it follows that the nature and strength of the bond between solder and substrate do not depend on the nature or quality of the flux. What does depend on the quality of the flux is the quality of the joint which it has helped (or failed to help) to make. For example, if the flux did not remove all of the surface contamination from the joint faces, the solder will not have been able to penetrate fully into the joint gap, and a weak or open joint will result.

Thus there are three basic things which are required to make a soldered joint:

1. Flux, to clean the joint surfaces so that the solder can tin them.
2. Solder, to fill the joint.
3. Heat, to melt the solder, so that it can tin the joint surfaces and fill the joint.

3.1.2 Soldering methods

Handsoldering

The various soldering methods which are used with electronic assemblies differ in the sequence in which solder, flux, and heat are brought to the joint, and in the way in which the soldering heat is brought to the joint or joints.

With handsoldering, the heat source is the top of a soldering iron, which is heated to 300–350 °C/570–660 °F. A small amount of flux may have been applied to the joint members before they are placed together. The assembled joint is heated by placing the tip of the soldering iron on it or close to it. Solder and flux are then applied together, in the form of a hollow solderwire, which carries a core of flux, commonly based on rosin.

The end of the cored wire is placed against the entry into the joint gap. As soon as its temperature has reached about 100 °C/200 °F, the rosin melts and flows out of the solderwire into the joint. Soon afterwards, the joint temperature will have risen above 183 °C/361 °F; the solder begins to melt too, and follows the flux into the joint gap (Figure 3.1). As soon as the joint is satisfactorily filled, the soldering iron is lifted clear, and the joint is allowed to solidify.

Thus, with handsoldering, the sequence of requirements is as follows:

1.   Sometimes, a small amount of flux.
2.   Heat, transmitted by conduction.
3.   Solder, together with the bulk of the flux.

Clearly, this operation requires skill, a sure hand, and an experienced eye. On the other hand, it carries an in-built quality assurance: until the operator has seen the solder flow into a joint and neatly fill it, he – or more frequently she – will not lift the soldering iron and proceed to the next joint. Before the advent of the circuit board in the late forties and of mechanized wavesoldering in the mid fifties, this was the only method for putting electronic assemblies together. Uncounted millions of good and reliable joints were made in this way. Handsoldering is of course still practised daily in the reworking of faulty joints (Section 10.3).

Mechanized versions of handsoldering in the form of soldering robots have become commercially available. These robots apply a soldering iron together with a metered amount of flux–cored solderwire to joints on three-dimensional assemblies, which because of their geometry do not lend themselves to wavesoldering nor to the printing down of solderpaste. Naturally, soldering with a robot demands either a precise spatial reproducibility of the location of the joints, or else complex vision and guidance systems, to target the soldering iron on to the joints.

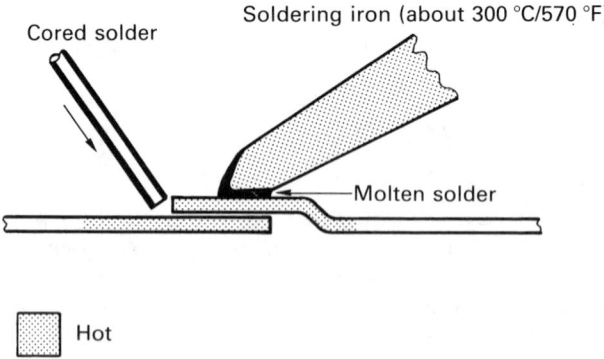

Figure 3.1   *The principle of handsoldering*

Wavesoldering

With wavesoldering (Figure 3.2) the following sequence applies:

1. Flux is applied to all the joints on a board.
2. Preheating the board to about 100 °C/210 °F supplies part of the soldering heat by radiation.
3. Molten solder (250 °C/480 °F) is applied in the form of a wave, which also supplies the bulk of the heat to the joints by conduction.

Reflowsoldering

With reflowsoldering (Figure 3.3), the possible sequences are:

1. A mixture of solder and flux (solderpaste) is applied to every joint before assembly.
2. Heat is applied by radiation, convection or conduction.

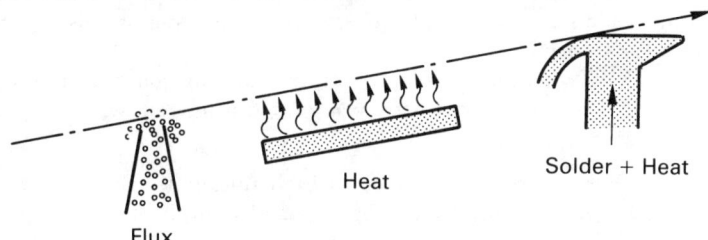

Figure 3.2 *The principle of wavesoldering*

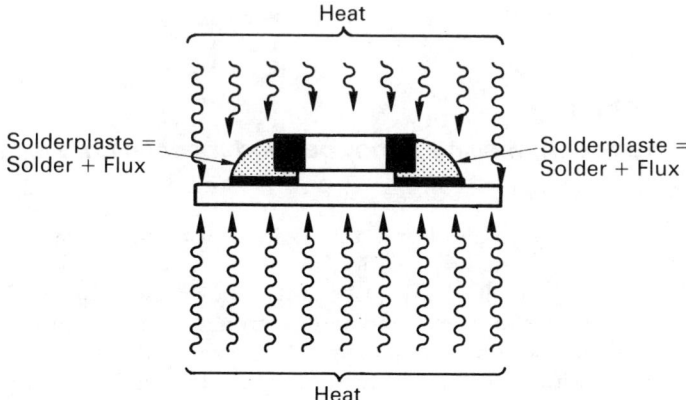

Figure 3.3 *The reflowsoldering principle*

Or

1.   Solder is preplaced in solid form on one of the surfaces of every joint.
2.   Heat is applied by radiation, convection or conduction.

Both wavesoldering and reflowsoldering are highly developed methods of quantity production. Being mechanized and automated, they demand integrated systems of controlling and monitoring their various operating parameters.

3.1.3   Soldering success

Whether a completed soldering operation has been successful or not can be unequivocally decided by answering the following three questions by either 'Yes' or 'No':

1.   Has the solder reached, and remained in, every single place where its presence is required for the functioning of the completed assembly? In other words, are there no open joints?
2.   Has any solder remained in a place where its presence prevents (or endangers) the functioning of the completed assembly? In other words, are there no bridges (or solder balls)?
3.   Is every SMD where it was placed before soldering started? In other words, did any SMDs swin away in the wave, or during reflow, or are there any tombstones?

What matters here is that the answer to each question is in the nature of an objective verdict, not a subjective judgement of compliance with an arbitrary definition of quality.

   This distinction between soldering success and soldering quality has an important bearing on the whole area of quality control. It means that two separate inspectors, including any automatic functional test or opto–electronic inspection method, must arrive at the same judgement.[1] This argument is pursued further in Section 9.3.

3.2   The solder

3.2.1   Constituents, melting behaviour and mechanical properties

Solidification and microstructure

Soft solders are alloys of lead and tin. Lead, a soft, heavy metal, melts at 327 °C/621 °F. Tin, a slightly harder metal of a white colour, melts at 232 °C/450 °F. Lead by itself is hardly ever used for normal soldering in electronics, because it has a high melting point and needs a strong, corrosive flux or a strongly reducing atmosphere in order to tin copper. Tin takes readily to most substrates, and can be used with mild fluxes, but it too is rarely used by itself, for several reasons: it is expensive, and its melting point is inconveniently high for electronic soldering.

The series of tin–lead alloys form a so-called eutectic system: both alloy partners, added to one another, lower the melting point of the resultant mixture; the two descending melting-point curves meet not far from the middle, at the eutectic composition of 63% tin/37% lead, which melts sharply at the eutectic temperature of 183 °C/361 °F. To either side of the eutectic composition, all the tin–lead alloys, which have a tin content between 19.2% and 97.5%, begin to melt at that eutectic temperature when heated from the solid. They also set completely solid at that temperature when cooled down from the molten state.

There is a further feature: to either side of the eutectic, the alloys have no sharp melting *point*, but a melting *range*, which gets wider as the composition moves away from the eutectic. The lower end of the melting range is always 183 °C/361 °F (called the eutectic temperature), but the top end rises towards the melting points of tin and lead respectively (Figure 3.4). Towards both ends of the melting point diagram certain complications arise, which can be disregarded in the present context.

This melting behaviour is reflected in the microstructure of the solidified alloys: seen under the microscope, the eutectic itself forms a finely interlaced pattern of thin layers of tin and lead. Solders of eutectic composition have the lowest melting point within the whole range and they, as well as their close neighbours, solidify with a smooth, bright surface. On the tinny side of the eutectic, small crystals of nearly pure tin are embedded in its microstructure; lead-rich crystals are embedded on the leady side. On heating, the eutectic always melts at the eutectic temperature,

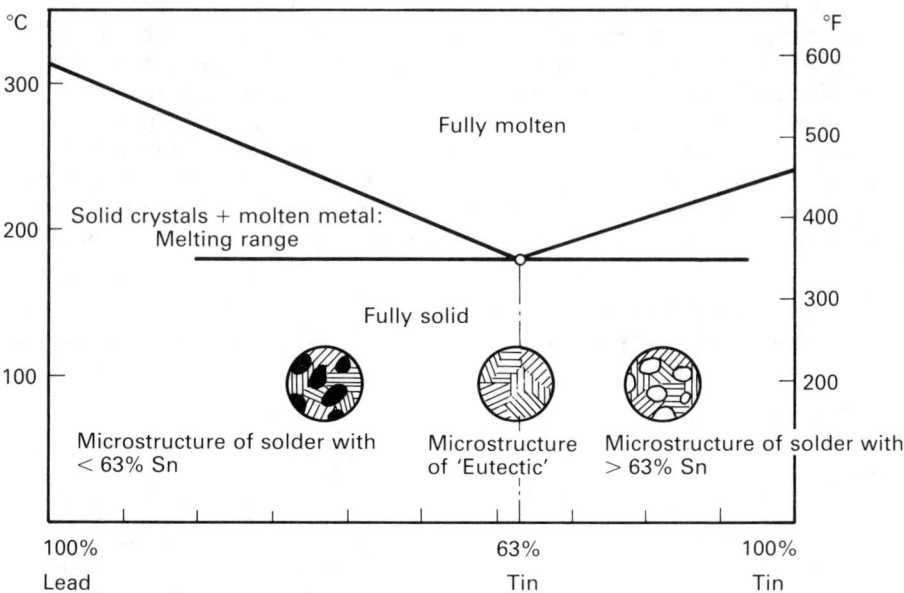

Figure 3.4 *Melting-point diagram and microstructure of the tin–lead alloys*

called the solidus, but the tin- or lead-rich crystals do not melt until the temperature has reached the top end of the melting range, called the liquidus (Figure 3.4).

Mechanical strength

Alloys are almost always mechanically stronger than their individual constituents. The tin/lead alloys confirm this rule, as Table 3.1 shows. This table lists the tensile behaviour at room temperature (20 °C/68 °F) of lead, tin and eutectic solder, with and without a small addition of silver.

As a constructional engineering material, solder is not impressive:

Tensile strength of rolled copper sheet: 12 tons/sq. in
Tensile strength of rolled brass sheet: 21 tons/sq. in
Tensile strength of cast solder 64Sn/37Pb: 4 tons/sq. in

These figures teach us an important lesson: there are only three legitimate functions which a soldered joint should be asked to fulfil. They are the following:

1. To conduct electricity.
2. To conduct heat.
3. To make a liquid- or gas-tight seal.

No soldered joint ought to be required to transmit any constructional loads or forces unless it is mechanically strengthened, e.g. by forming a double–locked seam, as in pressurized cans. The design of a soldered electronic assembly in which joints are used not only as elements of conduction of electricity or heat, but also of construction, should be carefully examined and possibly reconsidered.

This of course begs a question: anchoring reflowsoldered SMDs to a board is undeniably a constructional function. However, as long as the mass of an SMD is below 10 g, say half an ounce, the soldered joints between its leads and the footprints should be well able to hold the SMD where it belongs. If the soldered assembly has to survive extreme accelerations (e.g. in military or rocketry

Table 3.1   *The mechanical strength of solder*[2]

| Metal | Tensile strength at 20 °C/68 °F | | Elongation |
	tons/sq. in	*N/sq. mm*	%
Tin	1.1	17.0	70
Lead	1.0	15.5	45
63% Sn/37% Pb	4.0	62	50
3% Sn/2% Ag/35% Pb	4.2	65	40

At elevated temperatures, tin, lead, and tin/lead solders lose strength progressively: at 100 °C/212 °F about 20 per cent of the room-temperature strength is lost, while at 140 °C/284 °F between 40 and 50 per cent is lost.

hardware) or vibrations, the joints should be relieved of the resulting loads by suitable means such as brackets or encapsulation.

The loads which are placed on joints between SMDs and their footprints by reason of the thermal expansion mismatch between component and board will be dealt with in Section 3.3.5.

As the temperature rises, the strength figures of solders fall off, at first slowly, and then above 100 °C/212 °F rather more quickly, dropping of course to zero at the melting point of the metal concerned. The reason is that, in terms of absolute temperature (absolute zero is located at -273.2 °C $= 0$ K, see Section 5.5.2), at room temperature a solder is already within 35% of its melting point.

3.2.2   Composition of solders for use in electronics

The preferred composition of solders chosen for the soldering of electronic assemblies is at or near the eutectic, for obvious reasons (Figure 3.5). This choice holds good for all forms of solder: ingot solder for wavesoldering machines, solder wire for handsoldering, and solder powder for solderpastes. Solderpastes and solderwire for the handsoldering of SMDs often contain a small addition of silver, which provides several advantages (Section 5.2.3):

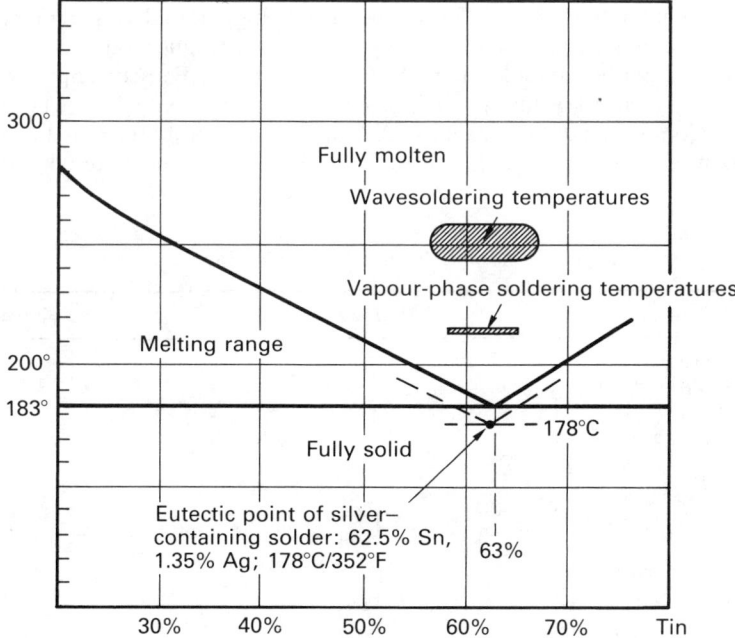

Figure 3.5 *Electronic solders and their melting behaviour*

1. It improves the strength and fatigue resistance of the soldered joints.
2. It reduces the leach-out of silver from silverbased substrates, such as the metallized faces of certain chips.

Tin and lead form a 'binary' eutectic, at near enough 63% Sn (61.9% according to the last critical study,[3] with a melting point of 183.0 °C/361.4 °F. Tin, lead and silver form a 'ternary' eutectic, with a composition of 62.5% Sn and 1.35% Ag, the balance being lead.[4] This would be the metallurgically correct composition of a silver-containing electronic solder. Some standard specifications, and consequently some silver-containing solders and solderpastes which are being marketed, have somewhat higher silver contents. The effect which such an excess of silver may have on the joint is discussed in more detail in Section 5.2.3.

Relevant standard specifications

The standard specifications of the main industrial nations list large numbers of solder alloys to meet the needs of the various joining technologies: Federal Solder standard QQ-S-571 E, for example, lists 27 alloys; the British BS 219 lists 22; while the German DIN 1707 contains 29 solders based on lead and tin. Out of all these, only the very few, which are listed in Table 3.2, are of direct interest for the industrial soldering of electronic assemblies.

It is inadvisable to choose a solder with a tin content below 60%, or an antimony-containing solder. The former needs higher soldering temperatures, which can be an important factor when a soldering technique is pushed to its limits, as is the case with wavesoldering of boards which are densely populated with SMDs, or which carry fine-pitch ICs. Also, lower tin-content solders give less attractive joints, of a dull appearance. Antimonial solders are suspected of a lower wetting power on copper and its alloys, especially brass. The material cost of the

Table 3.2   *Some industrial standards relevant to electronic solders*

Name of alloy	Tin	Lead	Silver	Antimony
US Fed. Spec. QQ-S-571 E				
Sn63Pb37	63%	37%	—	—
Sn62Pb36Ag02	62%	26%	2%	—
UK BS 219				
AP	63%–64%	Rest	—	<0.2%
62S	61.5%–62.5%	Rest	1.8%–2.2%	<0.1%
German DIN 1707				
L–Sn63Pb	63%	Rest	—	—
L–Sn63PbAg	63%	Rest	1.4%	—
L–Sn60PbAg	60%	Rest	3.5%	—

solder forms a minute part to the total cost of soldering; an attempt to save money by buying a cheaper alloy carries an unjustifiable risk.

Many solder vendors offer a special solder quality for wavesoldering, with claims for a drastically reduced dross formation in the machine. Their composition conforms to the relevant standard specifications, but their purity exceeds the requirements laid down in these standards. Several have been manufactured by special processes which aim to improve the behaviour of these solders in the wavemachine still further. With the advent of wavesoldering in an oxygen-free atmosphere, which drastically reduces the formation of dross by removing its cause, these solders may have lost some of their relevance. On the other hand, here too any attempt to save material costs is counterproductive: a cent saved by the purchasing department may cost many dollars spent on rework after soldering.

Alternative solders

Circumstances can arise where solders which are not based on lead and tin become attractive. During the mid eighties, it was thought that the malfunctioning of ICs after soldering was to a large extent due to their exposure to the hot solderwave. It could be shown that wavesoldering at temperatures between $180\,°C/356\,°F$ and $200\,°C/390\,°F$, using the tin–bismuth eutectic (43% Sn, 57% Bi, melting point $139\,°C/282\,°F$) as a solder gave good results and strong, reliable joints, provided a flux was used which was sufficiently active at these low temperatures.[5]

However, the quality of ICs improved at about the same time, and low-temperature wavesoldering never took off.

Recently, driven by environmental factors, interest in lead-free alternative soldering alloys has revived, and evaluation of such solders is being actively pursued. At the time of writing (1993), no concrete results have become known.

Under some exceptional circumstances, it can become attractive to carry out a number of reflow operations one after the other, each at a lower temperature than the preceding one (sequential soldering). Solderpastes with melting points ranging from $139\,°C/282\,°F$ up to $302\,°C/576\,°F$ can be supplied by most paste vendors for sequential soldering (Section 5.2.3).

3.2.3 Solder impurities

Impurity limits

Impurities in solder merit discussion mainly in the context of wavesoldering, because the initially pure solder bath is liable to become contaminated during use. Solderpaste and solderwire are always used in the form and with the degree of purity in which they are received from the vendor. Provided the vendor is experienced and reliable (and nobody should be tempted to buy from any other source) the purity of the solder can be taken as granted, given the present state of the art. Furthermore, the normal buyer is in no position to check the chemical composition of the purchased solder product, and he must of necessity turn to the vendor or to an independent laboratory for a chemical analysis (Section 4.8.5).

The impurity limits which the industrial standards of the USA, the UK and Germany prescribe for the vendor are given in Table 3.3. These figures are obviously based on international agreement, as would be expected. They must be regarded with one proviso, however: general soldering practice, but particularly wavesoldering practice, shows that with some of these impurities, like iron, aluminium and cadmium, limits tolerable in practical production are lower than those allowed in some standards, as Table 3.4 shows. It should also be added that the purity of the solder supplied by the experienced and reliable vendors mentioned above does indeed keep well within these strict limits.

Harmful impurities

What matters most is the effect of an excess of a given impurity on the behaviour of the solder in its molten state. If it is acceptable under that heading, the impurity will certainly not harm the finished joint after it has solidified. The behaviour of a contaminated solder in a wavesoldering machine is the best yardstick of the effect which a given impurity will produce. These damaging effects are conveniently considered under two headings:

1. Impurities which make the molten solder 'gritty'

Copper

If the copper content of a solder rises above the solubility limit of copper at the temperature at which the solder is being used, small, needle-shaped crystals of an intermetallic copper–tin compound (Cu_6Sn_5) form in the melt (we will encounter them again in every solder joint made on a copper substrate; Section 3.3.1). These crystals reduce the mobility of the molten solder, and can lead to incomplete filling of narrow joints. They give the joints a rough, 'sandy' appearance, and increase the formation of dross. In reflowsoldering, the effect of

Table 3.3 *Impurity limits according to US, UK and German standards*

Element	US: QQ-S-57E	UK: BS 219	D: DIN 1707
% Cu	0.08	0.08	0.05
% Al	0.005	0.001	—
% Cd	0.005	0.005	Al + Cd + Zn
			0.002
% Zn	0.005	0.003	—
% Bi	0.1	0.1	0.1
% Sb	0.12	—	0.12
% Fe	0.02	0.02	0.02
% Ni	0.01	—	—
% all others	—	0.08	0.08

Table 3.4   *Impurity limits according to actual practice*[6]

Element	Tolerable practical limit in wavesoldering	Actual analysis of a solder bath after six months' running
% Cu	0.35	0.227[a]
% Al	0.0005	<0.001[b]
% Cd	0.002	<0.001
% Zn	0.001	<0.001
% Bi	0.25	0.015[c]
% Sb	0.1	0.018
% Fe	0.005	0.001
% Ni	0.005	0.001
% Ag	1.35	0.025[d]
% Au	0.5	0.001[d]
% As	0.03	0.01

[a]In a wavesoldering bath, the copper content may rise well above the limit set by the various standard specifications.
[b]A more precise analysis requires a specially prepared sample, not available in this instance.
[c]Bi is sometimes added intentionally in amounts up to 2% in order to give a dull, non-reflective finish to the soldered joints. This makes visual inspection easier, without harming the performance of the solder or the quality of the joints.
[d]In a wavesoldering bath, both Ag and Au may safely rise above the limits set for them under the general heading 'all others', without affecting the performance of the solder or the quality of the joints.

any copper contamination is much less serious, and in any case copper is not a likely impurity in any solder depot which has been reflowed unless this reflow process has been excessively prolonged.

The solubility of copper in molten solder depends on both its temperature and its tin content (Figure 3.6). In a 63% tin solder bath, running at 250 °C/480 °F, the crystals will appear when the copper content exceeds 0.43%. To make sure that none form in some cooler parts of a wavesoldering bath, it is wise to set the upper copper limit for any wavesoldering bath at 0.35%.

In actual wavesoldering practice, copper will hardly ever rise above that point unless something has gone wrong, such as a circuit board having become stuck in the solderwave, or the solder temperature having drifted upwards. Normally, the copper content of a solderbath will settle down between 0.20% and 0.25% after some time, without approaching the solubility limit. As the copper content of the solder rises, its rate of attack on copper slows down, so that the solubility limit is never reached, taking the drag-out of solder on the joints and its replenishment with fresh solder into account. The general use of soldermasks, which leave only the footprints uncovered, has drastically reduced the occurrence of copper contamination.

Up to 0.2% of copper are soluble in solid 63% tin solder before the copper–tin crystals appear. Because such solder, when molten, erodes copper soldering bits

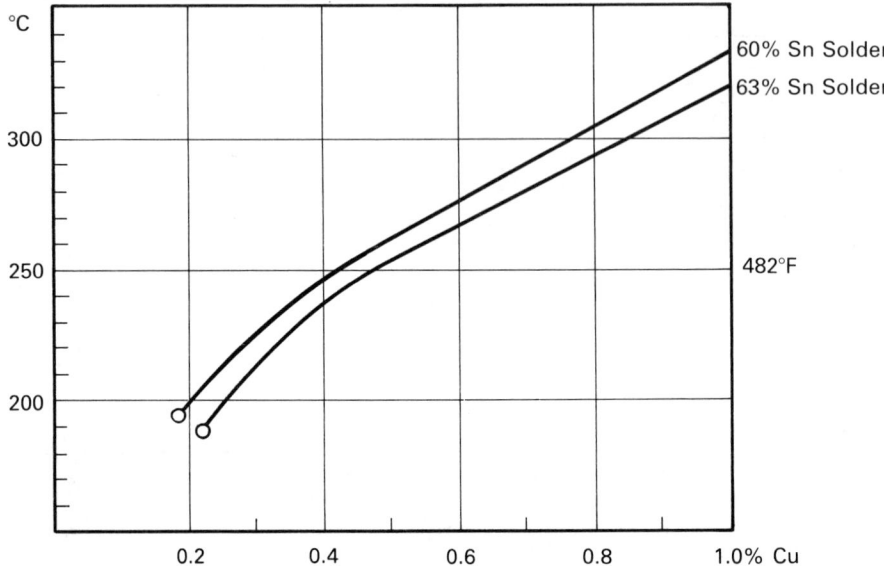

Figure 3.6 *Solubility of copper in non-antimonial solder (author's measurements)*

much more slowly, some vendors market solderwire with this deliberate copper addition. Such solders are of course outside the provisions of the relevant standard specification.

Should the copper contamination have risen above the danger limit, the simplest way of dealing with it is to empty part of the solderbath and replenish it with fresh solder. The drawn–off coppery solder will normally be taken back and credited by the vendor, after deducting a charge for refining.

Iron
Iron is much less soluble in solder than copper: its solubility limit in 63% tin solder at 250 °C/482 °F is 0.017% (author's measurement). Above that percentage, iron forms small, globular crystals of Fe_2Sn. They too give the soldered joints a gritty appearance.

Iron contamination can occur when filling the solderbath of a new, carelessly prepared soldering machine: iron fillings, loose nuts, etc., which have been left behind may start to dissolve in the molten solder. Immediate and excessive formation of dross, as soon as the solder pump is switched on, may be a sign of this kind of trouble. Immediate skimming, repeated at frequent intervals, can be a simple remedy.

Another cause of iron pick–up by the solder may be local attack by the molten solder on a mild steel or cast iron solderbath. This can happen if a sharp inorganic flux, such as zinc chloride, has erroneously been put on the bath surface. In this case too, excessive dross formation is an indicator.

Under normal conditions though, an iron solderbath can be run during the whole useful life of the machine without the slightest iron pick-up in the bath. Fresh solder from any reliable source carries no more than 0.003%–0.005% of iron. If a control analysis shows an iron content of 0.01% or more, it is a sign that something must have gone wrong.

In this case, the bath ought to be emptied and examined for any signs of tinning by the solder. Any affected spots should be cleaned by grinding, and the exposed iron surface must be oxidized with a blowflame, before the container is refilled with clean solder. The contaminated solder is best returned to the vendor.

2. Impurities which produce a skin on molten solder

Impurities in this class cause the formation of a tough oxide skin on the solder surface. In appearance it is similar to the skin formed on hot milk. Even repeated skimming will not remove it. Skinny solder causes bridging, 'icicles', and the formation of solder adhesions on the substrate, often called 'snails' trails'. It is important to realize, however, that there may be another cause for these troubles, related to deficient fluxing (Section 4.2). Before jumping to any conclusion, first check whether there is a skin on the solder surface which cannot be skimmed off.

There is no remedy against skin-forming impurities once one of them has entered the solderbath. The solder container must be emptied, and most carefully cleaned of any remaining solder, before it is refilled with fresh solder.

These considerations do not apply to wavesoldering in a controlled, oxygen-free atmosphere.

Zinc

Zinc causes skinning when present in amounts above 0.001%. It can occasionally get into a solderbath by mistake: immersing any galvanized steel implement in the molten solder is enough to ruin it. The diecasting alloy Mazak is based on zinc, but fortunately it is difficult to tin. A diecast item which has fallen into the bath will not immediately cause any harm if it is retrieved within a reasonable time, say a few hours. Brass too contains zinc, but there is no record of any zinc pick-up by a solder bath, even if brass screw heads or rivets pass regularly through the solderwave and are tinned in the process.

Cadmium

Cadmium acts like zinc, if present in amounts above 0.002%. Before cadmium was outlawed as a poison, cadmium-plated fittings on circuit boards were a frequent cause of cadmium contamination. Today, cadmium can be disregarded as a skin-former in wavesoldering operations.

Before its threat to health was recognized, eutectic lead–tin solders, laced with up to 1 per cent of cadmium, were often used for reflowsoldering or 'sweatsoldering' silver-coated ceramic insulators to condenser lids or similar metal parts. Cadmium in solder slows down the rate of its attack on silver by at least one order of magnitude (author's measurements), because it forms a high-melting intermetallic compound with silver. With this particular type of 'sweatsoldering', the skin-forming tendency of Cd-containing solders was no serious impediment.

Aluminium

Aluminium is the most virulent skin former: above 0.0005% it will cause the skin effect, but there are few circumstances under which it could ever get into a solderbath. Some caps of milk bottles are made of aluminium foil, as is the metal foil in many cigarette packs. Any of those found floating on a solderbath are, in any case, signs of bad works discipline and lack of supervision. Fortunately, aluminium is difficult to tin under normal circumstances, and if removed within the working day the above items will not have caused any harm.

Harmless impurities and deliberate additions

Silver

Silver cannot be regarded as a harmful impurity unless it exceeds the silver content of the ternary tin–lead–silver eutectic (1.35% Ag) by a significant amount ('hyper-eutectic' silver content). Even then, it will not form high-melting intermetallic crystals, which make the molten solder gritty, because it is very soluble in solder above its melting point. Hyper-eutectic silver affects only the microstructure of the solidified solder: with two or more per cent of silver, as specified in some standard specifications (Table 3.2), there is the possibility of large, plate-shaped crystals of Ag_3Sn appearing in soldered joints which have solidified rather slowly, as can occur in vapourphase soldering.[7] Though these crystals are supposed not to be brittle, their presence is undesirable, and it is advisable to use solders with less than 2% Ag unless specifically requested by the customer.

The presence of silver in its correct eutectic proportion confers several advantages: for one, it lowers the melting point of the solder from that of the binary eutectic (183 °C/361 °F) to that of the ternary (178 °C/352 °F). In many difficult soldering situations, that gain of 5 °C/9 °F can be useful. For another, silver has been found to improve the strength and fatigue resistance of soldered joints.[7] Finally, the silver addition slows down the rate of attack of the molten solder on silver-based substrates such as the metallized faces of certain chips (Section 5.2.3).

Gold

Gold dissolves extremely quickly in molten solder,[8] which can hold 16% gold in solution at 250 °C/480 °F. On solidifying, most of the dissolved gold precipitates out in the form of large, brittle crystals of $AuSn_4$, which weaken the joint structure and cause cracks, even at slight loads.[9]

The solubility of gold in solid eutectic solder is 3%, which would therefore be the safe limit for gold contamination in a wave solderbath. In practice, this would mean a gold content of 3000 g/97 oz troy in a solderbath of 100 kg/220 lb, representing a value many times that of the soldering machine. Normally, it will be worthwhile to empty a solderbath when its gold content has risen to 0.2%, and offer it to a metal refiner for sale.

Gold contamination arises from the wavesoldering of circuit boards with partially unprotected gold-plated edge contacts, or gold-plated footprints or component leads. The latter are more frequently encountered with military or space contracts in the US than in Europe.

Reflowsoldering of gold-plated footprints or component leads can result in dangerously embrittled joints. With reflowsoldering, the products of the solder–substrate reaction remain trapped within the joint. Any gold carried on its surfaces turns into the brittle $AuSn_4$ which, especially with narrow fine-pitch impulse-soldered joints, might fill much of the joint gap, with possibly dire long-term results. Hence the frequently encountered prescription to remove the gold from gold-plated leads by briefly dipping them in a small solderbath, which can be disposed of when it has become sufficiently loaded with gold. With gold-plated footprints, the only recourse may be local de-plating, equipment for which is commercially available.

Indium

Indium, like Cd, forms a high-melting intermetallic phase not only with silver, but also with gold. It is therefore sometimes added to eutectic tin–lead solder to prevent the leach-out of gold-plated joint surfaces. A binary alloy of the composition 70% In/30% Pb has a melting range of 165 °C/329 °F to 173 °C/ 343 °F,[10] and makes good, strong joints, not only on gold-plated but also on copper surfaces. It is, however, an expensive alloy. Solderpastes based on indium-containing solderpowder are commercially available.

3.3 The soldered joint

3.3.1 Soldering as a surface reaction between a molten and a solid metal

Preconditions for the reaction

A soldered joint is the result of a reaction between a molten metal, the solder alloy, and a solid metal surface, the substrate. This reaction can only start and proceed if the solid and the liquid can directly touch one another, without any intervening obstacle.

As has been mentioned already, a non–metallic surface film forms on all metallic surfaces, except on gold and platinum, when they are exposed to the normal atmosphere. In perfectly clean air this film will be an oxide, but under many circumstances sulfides can form as well, and some water vapour also may have been adsorbed.

Before they come to be soldered, most substrates will have passed through a number of manufacturing operations. These leave their mark in the form of solid organic or metallic particles, or films such as left-over processing chemicals or fingerprints. All these must be cleaned off before any soldering operation (Section 8.1.1), but no normal precleaning can remove the oxides and sulfides; only a chemical reaction can remove these. Dealing with them is the task of the soldering flux, which is the subject of Section 3.4. For the present, we will assume that that task has been accomplished, and that the molten solder and the substrate touch one another without any intervening non-metallic film.

The solder–substrate reaction

Several things matter here: on the one hand, the surface temperature of the substrate must be above the melting point of the solder; if it is not, the reaction cannot start because the solder will set solid at the interface. The joint members and their immediate environment must receive enough soldering heat to achieve this.

On the other hand, the substrate must of course never melt, and the soldering temperature must always keep well below its melting point. If it approaches it too closely, the solder/substrate reaction will get out of control and the solder will alloy with it and very likely destroy it, instead of reacting with it. This is unlikely to happen in normal soldering practice, except perhaps when soldering with laser energy.

Because the reaction between the tin in the solder and the substrate is exothermic, it occurs spontaneously as soon as the molten solder and the clean substrate touch one another. As a further consequence of its exothermic nature, the reaction will proceed and the solder will cover as large an area of the substrate as its available amount and its surface tension, which tries to prevent its spread, will allow. These matters will be dealt with under the aspect of 'Wetting' (Section 3.4.1). In the present context we are concerned with the events at the interface between the molten solder and the solid substrate, once wetting has been established.

Furthermore, since copper is the most commonly encountered substrate in SMD soldering, we shall from now on deal with the interaction between solder and copper unless specifically stated otherwise.

3.3.2   Structure and characteristics of the soldered joint

The reaction products

The products of the molten-solder/substrate, or more correctly the tin/copper reaction are the so-called 'intermetallic compounds'. They appear as a solid layer on the interface between the two partners, which is commonly known as the 'intermetallic layer', or 'diffusion zone'. The diffusion zone itself consists of two distinct layers of different intermetallic compounds.

Next to the copper comes a thin layer of Cu_3Sn, to which metallurgists have assigned the Greek letter 'epsilon' (ε). This layer is covered with a somewhat thicker one, with the composition Cu_6Sn_5, known as 'eta' (η) (Figure 3.7). At normal soldering temperatures these layers are solid, but above 415 °C/959 °F they begin to dissolve in the molten solder.

Substrates other than copper form their own specific diffusion zones, which are listed in Table 3.5. Though they differ in their composition, they share their crystalline structure with that of the zone formed on copper, with the same consequences for the mechanical properties of the joint itself.

Intermetallic compounds

The intermetallic copper–tin compound η can appear in the soldered joint itself if the molten solder becomes oversaturated with copper during solidification (see Figure 3.6). It forms hexagonal needles, which are sometimes hollow, embedded in the tin–lead crystal structure of the solder. Unless they have become excessively large, due to the soldered joint having solidified very slowly, they do no harm. On the contrary, they may well have a strengthening effect, since η crystals were an intentional constituent of tin- and lead-base bearing metals when these alloys were still a common engineering material. The same applies to the crystals of $NiSn_2$ which form on the interface between the solder and the nickel diffusion barrier on certain component leads (Section 3.6.6).

Other intermetallic compounds are less innocuous interlopers in a soldered joint. Plate-shaped crystals of Ag_3Sn form in solder which is supersaturated with silver (Section 3.2.3). They too can reach considerable size in a slowly solidified joint, but are taken to be harmless because they possess a certain ductility (private communication, International Tin Research Institute). On the other hand, the flaky crystals of $AuSn_4$, which are liable to form in a soldered joint made on gold,

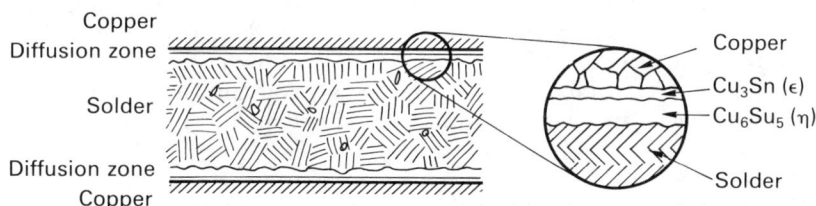

Figure 3.7 *Cross-section through a soldered joint, made with eutectic solder*

Table 3.5 *Substrates and their specific diffusion zones*

Substrates	Intermetallic compounds
Cu	Cu_3Sn (ε) Cu_6Sn_5 (η)
Ni	Ni_3Sn_2, Ni_3Sn_4, Ni_3Sn_7
Fe	$FeSn$, $FeSn_2$
Ag	Ag_3Sn

or any gold deposit, which is too thick (Section 3.6.6) are decidedly fatal: they are brittle and very weak. Any soldered joint in which they have formed is liable to crack.[11]

Reflowsoldering versus wavesoldering: the metallurgical consequences

There is an important difference between the metallurgical features of wavesoldered and reflow-soldered joints: with wavesoldering, most of the reaction products between solder and substrate, apart from those anchored in the intermetallic layer, are washed back into the large volume of the solder bath where they dissolve and very gradually add to its copper content (Section 3.2.3).

By contrast, every reflowsoldered joint forms a closed system, with a very small solder volume, in which all the products of the solder–substrate reaction stay trapped. As the solder solidifies, they precipitate out from solution and form crystallites which are dispersed through the joint. Their size depends on the speed at which the joint solidifies, and this varies widely between the different reflowsoldering methods. The relevant parameter here is the solidification interval, which can be defined as the time elapsed between the point when the joint temperature falls away from its maximum and the point when it passes through $183\,°C/361\,°F$.[13]

With laser soldering, the solidification interval is measured in milliseconds; with wavesoldering it is below 0.5 seconds, with infrared soldering it is about one to two seconds, and with vapourphase soldering it may be up to five seconds, depending on the design of the plant. The longer the solidification interval, the coarser will be the grain structure of the joint, and the larger will be the intermetallic crystals dispersed in it. This factor has a measurable effect on the long-term behaviour of a joint.[14] In the longer term, the microstructure of a soldered joint slowly changes after soldering: due to a process of slow diffusion in the solid state, the fine lamellar structure of the tin–lead eutectic becomes coarser. Also, because the energy level of the intermetallic phases ε and η is lower than that of both copper and solder, the phase change which led to their formation slowly continues in the solid state.[15] These phenomena, and their consequences, will be discussed in Section 3.3.3.

Factors which determine the microstructure of a joint

The thickness of the diffusion zone depends on several factors: first of all on the maximum temperature reached during soldering, and secondly on the length of time during which the substrate was exposed to the molten solder, i.e. the time which it spent above $183\,°C/361\,°F$. The latter will be termed 'confrontation period' from now on.

The higher the temperature reached during soldering, and the longer the confrontation period, the thicker will be the diffusion zone. Both the confrontation period and the maximum temperature reached during soldering vary widely between different soldering methods. So does the thickness of the diffusion zone (Table 3.6).

Table 3.6   *Relationship between soldering parameters and zone thickness*[12]

Soldering method	Confrontation period	Temperature °C	°F	Zone thickness µm
Wavesoldering	2–5 sec	250	480	0.3–0.8
Reflow:				
Laser	0.02–0.04 sec	250–350	480–900	≤ 0.1
Vapour phase	25–40 sec	215	419	0.7–1.5
Infrared	15–30 sec	250–300	480–570	0.5–1.5
Impulse	0.5–3 sec	250–300	480–570	0.1–1.0

What also matters is the speed at which a joint solidifies, i.e. its rate of cooling between its maximum temperature and the point at which its temperature passes through 183 °C/361 °F. The higher the speed of solidification, the finer is the grain structure of the solder in the joint.

3.3.3 Mechanical properties of soldered joints

The mechanical behaviour of the typical soldered joint is a consequence of its layered structure. The solder in the middle of the sandwich is comparatively yielding and ductile. This means that it can absorb stresses which may occur, for example, through a thermal expansion mismatch between a component and the substrate on which it is mounted. On the other hand, the diffusion zone on either side of the solder consists of crystals which are strong, hard and brittle. Their response to any deformation is to crack. From this it follows that a soldered joint is strong in shear and tension, but relatively easy to peel apart provided at least one of the joint members is flexible (as anyone knows who has managed to open a sardine tin by rolling back its soldered lid) (Figure 3.8).

The numerical values of the mechanical strength of solders and soldered joints under various manners of loading are given in Table 3.7. Some interesting facts emerge from these data, which result from early pioneering work on the properties of soldered joints.

Joints made with lead between copper members have a remarkably high tear strength. A strong flux or a strongly reducing atmosphere are needed to persuade lead to 'tin' copper, but the adhesion between the two metals is very strong once it has been achieved. There is no intermetallic layer at the interface, but there is a certain, very slight mutual solubility between the two metals.[16] The bonding between them is not exothermic as is the case with tin-containing solders. This means that in contrast to a soft-soldered joint, there is no long-term change in the microstructure of a Pb/Cu joint, which therefore has an excellent long-term stability (Section 3.3.4). Furthermore, the small amount of Cu dissolved in the lead which fills the joint adds to its strength. In this context, a nickel substrate behaves in a manner very similar to that of copper.

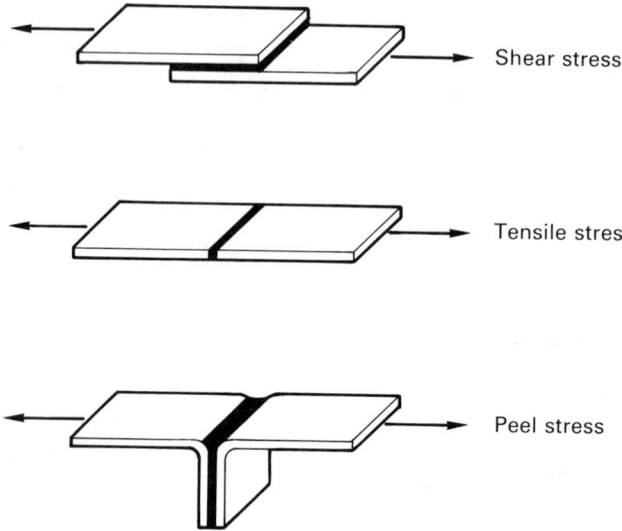

Shear stress

Tensile stress

Peel stress

Figure 3.8   *The mechanical behaviour of soldered joints under stress*

3.3.4   Soldering on substrates other than copper

Many SMDs confront the solder with substrates other than copper. Naturally, the molten solder interacts with them in quite a different way (Table 3.9). The high leach-out speeds of gold and silver mean that thin layers of them will disappear in molten solder almost instantaneously. Additions of palladium to silver (on chip-capacitors' metallized faces) or platinum to gold (on LCCC metallized surfaces) slow down this rapid leach-out without reducing the solderability too much, but considerably raising the cost. Nickel dissolves in molten solder by two orders of magnitude more slowly than silver. For that reason, it is often used as a barrier layer to protect the leachable noble metal from the solder. Because the solderability of nickel requires highly activated fluxes, it is often given a galvanic topcoat of copper and/or tin or tin/lead solder.

Two points need mentioning here. First, this topcoat of tin or solder will of course disappear immediately as the molten solder flows over it. The molten solder will then have to confront whatever metal surface it finds underneath. This surface must be absolutely clean, otherwise the solder will 'dewet' (Section 3.6.1). If it is not inherently solderable, like for example iron or stainless steel (Section 3.6.4), it must be copper plated before the topcoat of solder or tin is applied.

Secondly, plated deposits can be porous. Galvanically deposited solder is a conglomerate of discrete tin and lead particles, and is permeable to oxygen, water vapour, and sulphur compounds like sulphur dioxide or hydrogen disulfide. Once these penetrate through the solderplating, they can ruin the solderability of whatever metal surface is underneath. Hence reflowed galvanic, or hot-tinned

Table 3.7   *The mechanical properties of solder and soldered joints*

Solder	100% Pb	60 Sn/40 Pb	60 Sn/40 Pb + Ag	100% Sn
Bulk strength★				
tons/sq. in	0.9	4.0	4.5	1.1
N/sq. mm	140	620	700	170
Shear joint strength on copper★				
tons/sq. in	—	2.8	—	—
N/sq. mm	—	430	—	—
Max. sustained shear load on copper★★				
tons/sq. in	—	0.5	0.6	—
N/sq. mm	—	77	93	—
Tear strength on copper (Cadwick)★★				
N/mm width of joint	14	7.6	7.1	8

★Derived from data given by Nightingale, S. J. (1929) The Jointing of Metals, Part 1: Soft Solders and Soldered Joints. *Res. Report No. 3, Brit. Nonf. Met. Res. Assoc.*, Wantage, UK.
★★Derived from data given by McKeown, J. (1956) Properties of Soft Solders and Soldered Joints. *Res. Monograph No. 5, Brit. Nonf. Met. Res. Assoc.*, Wantage, UK.

Table 3.8   *Solderable terminals of different SMDs. (Tabular form of the data given in Section 2.3)*

Component	Nature of terminal
Melf resistors	Caps of Fe, Cu-plated (1 μm), galvanic surface coating of 90 Sn/10 Pb (1–3 μm)
TA and MKT condensers	End-caps of 89 Cu/9 Ni/2 Sn; hot-tinned with 60 Sn/40 Pb (2–8 μm)
Chip resistors and ceramic condensers	Thick-film Ag (80–90)/Pd (20–10), 30–100 μm thick, or Thick-film Ag, galv. coated with 1–5 μm Ni (barrier layer), top cover of galv. 3–10 μm Sn
LCCCs	Thick-film Au (62–87)/Pt (38–13)
SOTs, SOICs, PLCCs	Gullwing leads of 42 Ni/58 Fe sheet, often galv. coated Cu (1–5 μm), top-coat hot-tinned 60 Sn/40 Pb, or 98 Cu/2 Fe sheet, galv. or hot-tinned Sn (3–10 μm)
QFPs, TABs	Ditto, made from thinner-gauge sheet

Based on verbal communication, W. Richly, Siemens, Germany.

solder or tin coatings are best for long-term solderability. With both, the molten solder will find a ready-made intermetallic layer on the substrate over which it has to flow.

The problem of the solderability of the top surface of that intermetallic layer, which consists of the compound Cu_6Sn_5 (η) is dealt with under 'solderability', Section 3.6.

Table 3.9   *Solder interaction with substrates other than copper*[12]

Substrate	Intermet. phase adjacent to the solder	Speed of solder/substrate reaction	Speed of substrate leach-out at 250 °C/480 °F
Cu	Cu_6Sn_5	fast	0.15 m/sec
Au	$AuSn_4$	very fast	5.25 m/sec
Ag	Ag_3Sn	very fast	1.6 m/sec
Ni	$NiSn_2$	slow	0.01 m/sec

3.3.5   Long-term behaviour of soldered joints

The microstructure of a soldered joint changes slowly with time, in two respects:

1. The fine lamellar structure of the eutectic gets coarser, because there is a certain amount of energy tied up at any grain boundary; a soldered joint populated by a small number of coarse grains of lead and tin has a lower energy content than one consisting of many small tin and lead crystals. This energy difference is the driving force for the grain coarsening, which proceeds through the mechanism of solid diffusion.
2. The intermetallic crystals ε and η, which have formed at the solder/substrate interface, continue to grow. This growth, the result of a solid-state reaction, is also driven by an energy differential, because the tin/copper reaction is exothermic. This means that the tin–copper compounds have a lower energy content than the reaction partners by themselves.[17] Like all reactions, the speed of both grain coarsening and the thickening of the intermetallic layer increase rapidly at higher temperatures. Since at room temperature a soldered joint is already within 34% of its melting point in terms of absolute temperature (Section 3.2.1), it is not surprising that at, for example, 100 °C/212 °F the tin and lead atoms can move through the crystal lattice of the solder with about twice their room-temperature mobility.

Ageing and the consequent grain coarsening lower the mechanical strength and the ductility of a soldered joint.[15] The thickening of the brittle intermetallic layer does not weaken a joint, but it reduces its ability to absorb repeated deformation without cracking.

The mechanism of joint failure in service has been the subject of many investigations. A recent important contribution examines in detail the metallurgical features of the various types of joints between SMDs and their footprints, the mechanics of their failure, and the possibility of predicting their fatigue life under the conditions of practical operation. It is worth noting that the paper's conclusions end with the sentence 'Further work is needed.'[18]

3.3.6   Long-term reliability of soldered joints

The meaning of 'reliability'

The expression 'long-term reliability' of an electronic assembly, or of any soldered joint in it, can be given a precise meaning in this context: it represents its ability to function as expected, for an expected period of time, within an expected failure level. In a recent study, it has been stated: 'a solder joint in isolation is neither reliable nor unreliable: it becomes so only in the context of the electronic components that are connected via the solder joints to the printed wiring board. The characteristics of these three elements, together with the conditions of use, the design life, and the acceptable failure probability for the complete assembly determine the reliability of the surface mount attachment.[19]

An important factor in the reliability consideration is the anticipated conditions of service. Tables have been drawn up which list 'worst-case use environments' ranging from consumer electronics (temperature range 0 °C/32 °F–60 °C/140 °F, 12 hrs daily running, 365 operating cycles per year) to automotive under-hood electronics (temperature range −55 °C/−67 °F to 125 °C/257 °F, with up to 100 one-hour operating cycles per year).

These scenarios form the basis for accelerated test programs to assess the reliability of a given electronic assembly.[20]

The mechanics of joint failure

If a soldered joint on an SMD populated board fails in service, it is through fatigue damage (unless is was so badly soldered that it came apart at the first provocation). The stresses which produce fatigue failure can be caused by a mismatch between the thermal expansion of the board and the component, or by a temperature difference between the two at the beginning or after the end of a cycle of operation.

There are some mitigating factors: if the component is small, the lateral joint displacement is small. If the temperature cycle is relatively short, the elastic deformation of the solder in the joint can take care of the strain before the solder begins to creep. The same is true for low operating temperatures, where the elastic limit of solder is higher. If the component leads are compliant, they can absorb the strain.

Conversely, there can be aggravating circumstances: if, for some reason, the intermetallic zone fills a large part of the joint gap, the solder left in it is less capable of yielding to strain and the joint is likely to fail earlier. The stiff gullwing legs of SOTs or SOs are less compliant than the thin leads of multilead components. The fatigue aspects in the life of a soldered joint have been extensively investigated and described.[21 − 23]

Design for joint reliability

Given certain constraints, such as the operational environment and operational cycles, some factors are in the hands of the designer. For example, components with compliant leads should be a preferred choice, while melfs and chips with their stiff joints should be as short as possible to minimize the effect of the thermal expansion mismatch between them and the FR4 board. The leadless ceramic chipcarrier, soldered to an epoxy board, has suffered an unlamented extinction because of its large size. Attempts to create a suitable environment for the LCCC by developing substrates with matching coefficient of expansion, like vitreous-enamelled metal substrates or by giving it clip-on flexible legs, were not commercially successful.

For some reason, lead–rich solders with low tin content and small additions of silver such as the alloy known as LS4 in the last war (5% Sn, 1.5% Ag, bal. Pb, melting range 293 °C/559 °F–298 °C/568 °F), which has excellent fatigue properties and gives strong joints on copper, have not yet caught the eye of the electronics industry and its designers. The higher melting point need not deter them unduly, as far as its use for the impulse soldering of individual components is concerned. It would of course be an impediment with the wholesale soldering of complete boards by infrared or vapourphase.

3.4   The flux

3.4.1   *Tasks and action of the soldering flux*

Removing oxide skins

The success of any soldering operation depends critically on the condition and behaviour of two surfaces: that of the substrate and that of the molten solder. The flux has to deal with both of them (Section 3.1.1).

By the time the molten solder encounters the substrate, the flux should have completely removed all surface oxides from both the molten solder and the substrate, so that nothing prevents or interferes with the reaction between the two: this is the precondition for the formation of a soldered joint (Figure 3.9).

During the solder/substrate reaction, which takes place at a temperature above 183 °C/361 °F, the flux must prevent the formation of fresh oxide on both solder and substrate.

Soldering methods which exclude or at least drastically reduce the presence of oxygen from the scene of joint formation are a help with the second task: wavesoldering under nitrogen falls into that class. It can of course not remove any oxide which is already there before soldering starts, because nitrogen is purely protective and will not actively reduce oxides (Section 4.6).

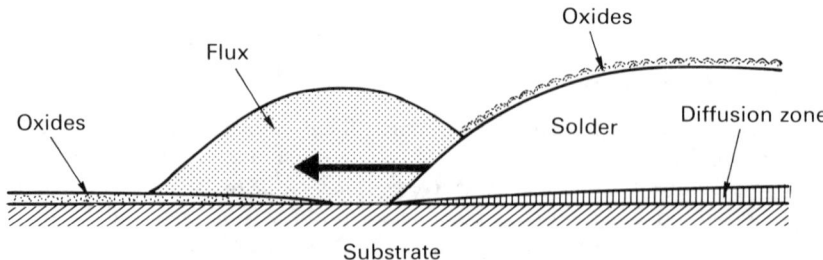

Figure 3.9   *The function of the flux in soldering*

With reflowsoldering under nitrogen or in a vapour phase, using a solderpaste, the situation is not so clear-cut: it could be argued that the flux portion of the solderpaste not only precleans both the substrate and the solderparticles in the paste, but when the solder melts, it is on hand to promote the union between the two, while at the same time excluding the atmospheric oxygen from the scene of action. It is claimed, however, and practice seems to confirm it, that the practically oxygen-free working vapour, and reflow under nitrogen, reduces the incidence of faulty joints and permits the use of less active solderpastes which reduce the need for cleaning after soldering in a large number of cases (Chapter 3.4.3).

The oxide skin on molten solder

In a normal atmosphere, a thin skin of oxide forms almost immediately on the surface of molten solder as soon as it is exposed to atmospheric oxygen. This skin acts like a tough, somewhat rigid envelope on the liquid solder and restricts its free movement. A simple but instructive experiment, which anyone can (and probably should) repeat on a benchtop, demonstrates this action.

Place a short piece – about 2 cm/1 in long – of solid solder wire (without a core of flux) on an untinnable surface, e.g. a small piece of stainless steel sheet or a ceramic wafer, and heat it on a hotplate or over a gas burner until the solder melts. You will find that as the solder wire melts, it retains its shape, looking very much like a small sausage in a somewhat crinkly skin. Now let one drop of soldering flux (e.g. an RMA or RA flux) fall on that sausage: the skin will disappear in an instant and the sausage turns into a shiny, perfectly spherical ball of liquid solder (Figure 3.10).

This experiment demonstrates two things:

1. Even a mild flux removes the oxide skin from molten solder by dissolving or dispersing it. It is mainly the tin oxide which contributes to the stiff nature of this skin. Lead oxide is more powdery.
2. The molten solder, as soon as it is free to move, forms a round sphere, this being the shape with the smallest surface area which a given volume of liquid can assume. The molten solder assumes that shape under the influence of its

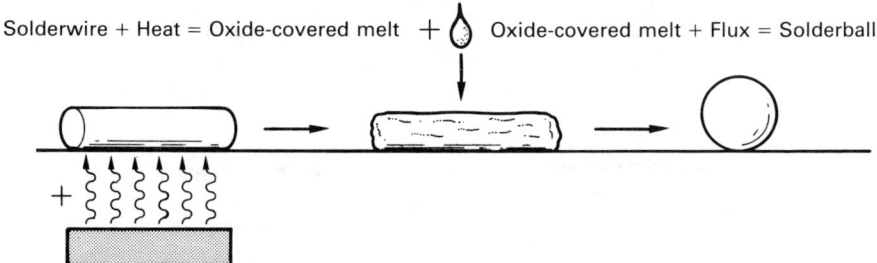

Figure 3.10 *The solderwire/flux experiment*

'surface tension', which it can follow without restraint as soon as the oxide skin has disappeared.

Surface tension is the result of the force with which every atom or molecule in a liquid attracts its neighbours. Any such particle which is well inside a given volume of liquid is attracted equally from all sides, so that the forces acting on it cancel out. The particles at the surface have neighbours only on the inwards-facing side, so that they are all being pulled inwards. Consequently, the liquid tries to make its surface as small as possible, and forms a sphere: in effect, the surface behaves like a tensioned skin which encloses the molten solder, hence the term 'surface tension' (Figure 3.11).

Surface tension, tending to reduce the surface of a given volume of molten solder to a minimum, is responsible for the suppression of icicles and bridges in wavesoldering (Section 4.4.3), and for the suppression of 'solderballing' and bridging in reflowsoldering with solderpastes. Hence the need for some flux to survive on the underside of the board during its passage over the solderwave, so as to be available to prevent bridges and icicles forming on the exit side (Section 5.2.3) (Figure 3.12).

Molten solderwire   Flux   Solderball

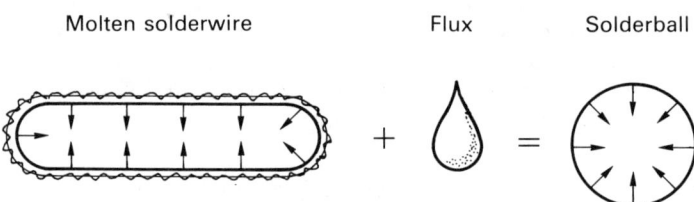

Figure 3.11   *Surface tension and the molten solder wire*

Not enough flux   Enough flux

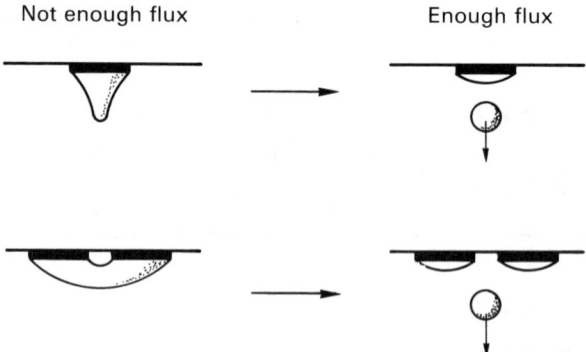

Figure 3.12   *Icicles, bridges, and surface tension*

3.4.2 Wetting and interfacial tension

It is sometimes asserted that the flux lowers the surface tension of molten solder, thus helping it to spread over a substrate or penetrate into a joint. This is not correct. All the flux does to the molten solder is to let it move freely, so that it can follow the forces which act upon it. These forces are not only its own surface tension, but also the so-called 'interfacial tension' or wetting force, which pulls the solder across the surface of the substrate.

The chemical affinity between the substrate and the tin in the solder is the driving force behind the formation of the intermetallic crystals. The energy set free, as the solder reacts with the copper (or nickel or silver) substrate, pulls it sideways into a flat coating or into a joint gap. (There is some evidence, though, that some fluxes initiate chemical reactions, which actively pull the molten solder across the substrate – Dr W. Rubin, verbal communication.) This wetting force overcomes the inwards pull of the surface tension of the solderglobule, forcing it to give up its spherical shape and to flow where it is needed. The wetting force is the same as the interfacial tension. What matters is that the interfacial tension is higher than the surface tension of the solder (Figure 3.13).

It is worth stating here once again that the flux has nothing to do with the strength or quality of the resulting bond. All it does is to clear the path for the solder to advance across the substrate, and to give it the mobility to do so.

3.4.3 Properties required in a flux

The essential function of a flux is of a chemical nature: it must remove a surface layer of a metallic oxide or tarnish, in most cases from copper, tin and lead, less frequently from nickel. It does this not only by dissolving, but also by lifting and dispersing oxide and tarnish from substrate and solder.

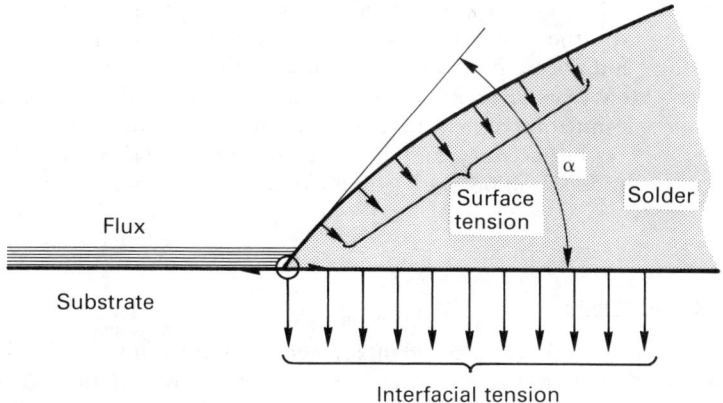

Figure 3.13 *The wetting process as a balance between surface tension and interfacial tension. α: Wetting angle*

The physical requirements which the flux must meet depend on the soldering method used:

1. With wavesoldering, the flux must be suitable for being applied to the entire surface of a circuit board by an automatic and reproducible method (Section 4.2). A liquid flux is best suited for this purpose. For reasons discussed in the relevant chapter, gaseous fluxes are not entirely satisfactory, at least not at the present state of the art.

    The liquid flux must wet the board evenly, and well enough to penetrate into narrow gaps and fill the vertical holes in throughplated double-sided boards. Given a liquid flux cover of up to 0.5 mm/20 mil thick, its volatile constituents should be capable of being driven off by heating to about 100 °C/210 °F within not much more than 30 sec. Sufficient of the thickened flux left after this must be able to physically survive the passage through a double wave, if one is employed, so as to stop the solder on the exit side from oxidizing and thus to prevent bridges, icicles, 'spider's web' or 'snails' trails' and solder balls.

2. At the soldering temperature, the flux must be a mobile liquid, and form a coherent cover for the front of the molten solder as it advances across the substrate, as shown in Figure 3.13.

3. In addition to this, the oxide-reducing capability of the flux must survive exposure to the soldering heat for as long as the soldering method demands: with a double wave in wavesoldering, this heat exposure may last up to five seconds at 250 °C/480 °F. In an infrared oven, the temperature may rise above 183 °C/361 °F for over 30 sec, with a peak of about 300 °C/390 °F.

4. Physically, the flux, having lost much of its solvent portion, must remain sufficiently mobile to make way for the molten solder, allowing it to flow across footprints and component leads, to rise up through every through-plated hole, and to penetrate into every joint gap. Therefore, it must not polymerize, become sticky, carbonize, or cake up during the temperature excursions of the soldering process. It must retain its mobility during soldering, and not become difficult or impossible to remove after soldering.

5. In the last few years, the removal of flux residues from soldered assemblies has become a dominating topic. There are a number of reasons why cleaning after soldering may be necessary, and these are fully discussed in Section 8.1.1. The need to consider cleaning has become a dominating factor when choosing a flux.

3.4.4 Rosin fluxes

Until the seventies, only rosin-based fluxes were considered suitable for electronic soldering. In their liquid form, rosin fluxes are, or were, solutions of rosin in alcohol, with solids contents of up to 25% or more. Their residue could safely be left on the finished board, where it formed a hard, protective coating. Indeed, up to 60 °C/140 °F, pure rosin has a bulk resistivity of the same order of, if not better than, FR4. Therefore, in the early days of wavesoldering, one liked to see the soldered boards coated with a hard, shiny and continuous varnish-like coating of solidified rosin. If considered necessary, this residue could easily be removed with an azeotropic mixture of CFC113 and an alcohol. Since the existing or impending banishment of the CFCs, this situation has changed drastically (Sections 8.3.5 and 8.3.6).

Natural rosin is a distillation product of the sap of pine trees. Chemically, it is a mixture of several organic acids, such as abietic and pimaric acid and their close relatives. Its composition depends very much on its place of origin and is also seasonally variable, so that buying good rosin is a matter of skill and judgement. Instead of a pH value, the acidity and hence the fluxing power of a rosin is defined as its 'acid value', which is measured in mg KOH per g, this being the amount of KOH required to neutralize a rosin solution in alcohol, per gram of dissolved rosin. An acid value of 160–180 would be considered adequate for a good rosin.

Synthesizing rosin in order to provide a standardized base material with reproducible properties has not been found practicable. A synthetic substance, pentaerythritol tetrabenzoate, a proprietary product, is a close match to rosin, however, and forms the basis of a series of synthetic resin fluxes.[24]

At room temperature, solid rosin is chemically inert, and as a flux residue it protects the soldered board against dirt and moisture. At temperatures exceeding 50 °C/120 °F, rosin becomes sticky and attracts dust and fluff, and is then no longer an attractive flux residue. In the molten state, above about 80 °C/180 °F, and as a solution in alcohol, rosin reacts with copper like a mild acid, and acts as a flux.

Because its fluxing action on untinned copper is not particularly efficient, so-called activators which enhance its efficiency are normally added to the rosin. These will be dealt with presently.

Problems with rosin fluxes

1. Rosin softens and becomes chemically active above about 70 °C/160 °F. This may happen if the assembly has to operate in a hostile environment: inside vehicles and aircraft parked in full tropical sun, and under the hood of motor vehicles anywhere, ambient temperatures can rise above 120 °C/250 °F (Section 3.3.4). In a warm and humid environment, rosin, left as residue on a board, can hydrolyze and lose its insulating character.
2. Rosin flux residue, being hard and highly insulating, obstructs the contact between ATE test probes and the test pads on the board, and it also tends to gum up the tips of the probes. This problem is more serious with wavesoldered boards, which are fully covered with flux residue, than with reflowsoldered boards, where the flux residue from the solder paste is confined to the footprints and their immediate neighbourhood.
3. The residue compromises the adhesion of protective conformal coatings.
4. The interaction between the rosin flux and the substrate during soldering produces complex metallic reaction products, which are not fully soluble in

many of the solvent systems employed in cleaning. They form the so-called 'white residue', which is the subject of a large volume of investigative work. A detailed account of it has been given by C. Lea.[25]

Activators

Because, as has been said already, rosin by itself is not a very efficient flux on bare copper, during the last war so-called activators began to be added to it, in amounts of generally up to 5 per cent weight of the rosin in the flux solution. The aim was to speed up its fluxing action and to keep up with the demands of armament production. Activators have been used ever since.[26]

Activators are polar, mostly high-molecular organic compounds, which may or may not contain a halogen atom. It is important that within the percentage added, the activator is fully soluble in the rosin and does not separate out when the rosin solidifies. This is intended to ensure that after soldering, the polar activator stays safely locked up in the solidified rosin residue, unable to make it conductive or form local centres of corrosive action. Test procedures which ensure that this requirement is met are laid down in most national flux standards (Section 3.4.6). The activator should of course also be soluble in the flux solvent, which is usually isopropyl alcohol. Finally, if the flux residue must be removed after soldering, the solvent used must be able to cope with the activator as well as with the rosin.

The role of halogen

Activators are fluxes in their own right, and by themselves they are more efficient than the rosin or its synthetic resin equivalent. Activators fall into two classes, halogen-containing and halogen-free.

Dibasic organic acids, such as adipic or succinic acid, have been found particularly useful as halogen-free activators. The latter vaporizes at 235 °C/455 °F, which makes it attractive as an additive in wavesoldering fluxes. At the normal soldering temperature of 250 °C/480 °F, it forms a cloud of gaseous flux, which surrounds the solderwave and reduces bridging on the exit side. Adipic acid, which melts at 153 °C/307 °F is a useful flux for wavesoldering in a controlled atmosphere (Section 4.6).

The choice of acids suitable for rosin activation is wide,[27] and often more than one are added together.

Most of the halogenated activators are either chlorides or bromides of nitrogen-containing compounds such as amines. The choice between Cl and Br depends largely on the idiosyncrasies of a flux manufacturer; often, bromides are considered less corrosive at a given fluxing efficiency than chlorides, but there is no concrete proof. There is a general belief in the industry that a halogen-activated rosin flux poses a higher risk of corrosion and of lowered surface insulation resistance (SIR) than a halogen-free one, and some soldering documentations specifically outlaw the use of halide-activated fluxes. This seems to be largely due to historical reasons.[28] Some halide-free modern activators are more corrosive by themselves than halide-containing ones.

3.4.5   Low-solids and no-clean fluxes

Wavesoldering applications

In common usage, fluxes with a solids content of 2% to 4% are called 'low-solids' fluxes. Recently, ultra–low-solids fluxes with about 0.5% solids content have appeared on the market: these are tailored for use in wavesoldering machines which operate in a nitrogen atmosphere with extremely low oxygen content (<50 ppm oxygen). Low-solids fluxes are preferably applied to the circuit boards by a sprayfluxer. Their behaviour in a foamfluxer (Section 4.2.1) is erratic, if not unsatisfactory.

The demarcation between low-solids fluxes and no-clean fluxes is blurred. Low-solids fluxes are the response to the ATE test-probe problem faced by wavesoldered boards. The recent advance of alternative test concepts has had no effect on the continuing popularity of low-solids fluxes.

Put crudely, low-solids fluxes are activated rosin fluxes with most of the rosin taken out but all of the activator left in. This results in a flux with a total solids content from 5 per cent down to less than 2. The proportion of rosin or its resin substitute in the solids portion of the flux can be well below one-half. Naturally, the only activators suitable for this stratagem are those which are by themselves not corrosive, and which form a thin, dry residue. The above-mentioned class of organic acids are one of the most popular candidates for this role. The main concern is that the flux residue left on the test pads should be easily penetrated by the ATE test probes, without raising their contact resistance or gumming them up too quickly.

A minimum of rosin in a low-solids flux for wavesoldering is necessary to give the flux some body and staying power: staying power to get some of it to the far side of the second solderwave, body in the form of a certain amount of liquid, in this case molten rosin, to keep the peelback of the wave in check and prevent it from flying apart in a shower of small solder droplets, which is one of the problems of wavesoldering in a controlled atmosphere (Section 4.6).

Naturally, it can no longer be expected that the rosin in the flux residue will retain the activator safely locked up in solid solution. Therefore, having become the main constituent of the flux and its residue, the activator must be strictly non-corrosive and non-conductive under the service conditions for which the soldered assembly is intended. Consequently, low-solids fluxes are normally halogen-free.

Low-solids fluxes require optimal solderability of both footprints and component leads. Most of them need a more intense preheat between the fluxer and the solderwave, either longer or, more commonly, to a higher temperature. Wavesoldering in a controlled atmosphere is an ideal method for getting the best results from low-solids fluxes. For wavesoldering, the description 'low-solids flux' can normally be equated with 'no–clean flux'. Nevertheless, it is advisable to carry out environmental testing of some soldered sample boards before deciding on a low-solids flux for a given application.

Use in solderpaste

Reflowsoldering with a solderpaste presents quite a different situation. ATE test probes are not the problem they are with wavesoldering, because care is in fact taken to ensure that the solderpaste does not invade a neighbouring test pad, so starving the joint of solder. Low-solids fluxes for the sake of just leaving less residue are uninteresting. On the other hand, formulating a paste which really needs no cleaning offers worthwhile rewards: cleaning of densely-packed SMD boards is difficult; boards have to be designed and components must be specified for cleanability (Section 8.1.2).

The task is not an easy one. The flux portion of a paste must have enough 'body' to provide the thixotropic properties which hold the solderpowder in suspension and make the paste printable without slump. So as not to cause solderballing, paste squeezed away from a footprint must remain moist and mobile until all the solder has been fetched back into the joint. Without cleaning, stray solderballs become vagrant and liable to cause damage with fine-pitch and ultrafine-pitch layouts.

Several 'no-clean' solderpastes are now on the market. Most of them require reflowing in a low–oxygen atmosphere. One of them demands reflowing in a special reactive atmosphere, which volatilizes most of the flux residue (Section 5.2.4). With all of them, there is very little apparent flux residue left after soldering, but what is important is that whatever there is sits on the pads or leads, and not in the spaces between them.

3.4.6   *Watersoluble fluxes*

If cleaning after soldering is obligatory, however non-corrosive the flux residue may be, it is sensible to choose a flux which can be removed as simply as possible (Section 8.7). The watersoluble fluxes fall in that class. Water, and the recently introduced polar solvents like the modified alcohols, offer the possibility of a greatly simplified and environmentally acceptable cleaning technology.

Any completely rosin- or resin-free flux would consist of polar substances only, and is consequently soluble in polar solvents like water or any of the alcohols. The choice of such fluxes is large, and again it depends largely on whether the flux is to be used for wavesoldering or for incorporation in a fluxpaste. For wavesoldering, any of the activators discussed previously, in alcoholic solution and with the addition of a longer-chain polar compound such as a glycol to give the flux 'body' and staying power in the wave (see Section 3.4.5), will be suitable. The presence of a halide is no requirement, but no disqualification either. What matters is that, after soldering, the board must not be wet or sticky.

The field of watersoluble wavesoldering fluxes is wide, and still largely unexplored. What is required here is that the industry overcomes some of its distrust of halogenated flux systems, which have been proved effective and extremely washable in other industries for many years.

The incorporation of watersoluble fluxes in solderpastes poses entirely different and less tractable problems. Several watersoluble solderpastes are commercially available, and their number is likely to grow. Here too, rosin was very useful in its dual role of providing a fluxing medium, and as a thickener to give 'body' to the paste. This role of the rosin has to be taken over by high-molecular watersoluble aliphatic compounds. Some of the watersoluble solderpastes which are at present on the market require a protective atmosphere for reflowsoldering.

3.4.7 Solvents used in fluxes

Generally, and until recently, isopropyl alcohol (isopropanol) was used almost universally as the solvent in every soldering flux. With some formulations, a portion of a higher-boiling alcohol was added; with others, water of up to 10% or more was added to improve their performance in a foamfluxer and reduce the 'climb-through' of the flux from the underside of the circuit board through the holes to the top, where it might interfere with the function of relays or contacts. Those latter fluxes would of course have to be free from rosin.

In 1992, isopropanol came under attack from two sides: first, isopropanol falls in the class of 'volatile organic compounds' (VOCs, Section 8.3.5), which are implicated in the creation of urban and industrial smog. In several American states (e.g. California), the use of VOCs is already severely restricted.

Secondly, though the soldering industry has been using flammable isopropanol-based fluxes for decades without any great qualms, observing of course all reasonable precautions, the recent emphasis on environmental aspects of industrial practices, together with stricter regulations concerning the storage of flammable liquids on manufacturing premises, has brought the use of any of the lower alcohols, known as 'Class I liquids', under renewed scrutiny as flux solvents in wavesoldering machines. As a result, waterbased wavesoldering fluxes have been introduced to the market.

Their ingredients must of course be fully watersoluble, and hence they cannot contain rosin. Secondly, since the heat of evaporation of water is over three times that of isopropanol, and its boiling point is 20 °C/36 °F higher, a waterbased flux must be predried with a more powerful heater and to a higher temperature than an alcoholic one (Section 4.3.1). To assist predrying, these fluxes are applied by sprayfluxers, as a very thin coating.

3.4.8 Flux standards

Problems of formulating flux standards

The main function of a standard specification of an industrial product is to provide a formal framework of requirements which the product has to meet, so that in principle it could form the basis of a contract between the seller and the buyer. With an engineering product, for example a screw, a standard specification will consist of set of physical dimensions and permissible tolerances. Compliance with the standard can be verified by a set of relatively simple measurements. With a chemical product like a solder, a standard specification will state the composition of the solder, with permissible tolerances, and limits for a number of impurities. Compliance can be verified by chemical analysis.

Soldering fluxes for use by the electronics industry are also chemical products, but with them the situation is far less simple, for several reasons. Some of them follow:

1. The technology of electronic packaging, including soldering, is far from static. Methods and requirements change constantly, often rapidly, and with them the tasks which soldering fluxes have to fulfil.
2. A number of flux formulations are patented and are thus the property of the manufacturer; others are closely guarded industrial secrets, being the result of much research and development and company-specific expertise.
3. Verification of the composition of electronic fluxes is largely beyond the scope of most users except the large corporations.

The standards institutions of the various industrial nations have chosen several ways out of these difficulties. In principle it is possible to frame two sets of requirement for a flux:

1. Set down what a flux must or may contain (for example rosin), and what it must not contain (for example a halide).
2. Set down what a flux must do (for example possess a given minimum fluxing activity), and what it must never do (for example cause corrosion or leakage currents).

Starting from these premises, different countries have chosen different ways of framing standard specifications for soldering fluxes. Some examples follow.

USA

Standard ANSI/IPC-SF-818 (February 1988) entitled 'General Requirements for Electronic Soldering Fluxes', is a comprehensive document which lists a scheme for classifying and characterizing these fluxes. IPC (the Institute for Interconnecting and Packaging Electronic Circuits) is an industry group based in the USA, but with international connections. IPC represents the American National Standards Institute (ANSI) at the International Standards Organization (ISO) and the International Institute of Welding (IIW) in all matters concerning the formulation of standards on an international basis. Alongside these industrial standards, the US military MIL and QQS specifications continue to play an important role.

Federal specification QQ-S-571 E (1986)

This specification divides fluxes into four classes:

R   Rosin flux
RMA   Mildly activated rosin flux
RA   Activated rosin or resin flux
AC   Non-rosin flux. Includes acids, organic chlorides, inorganic chlorides,
        etc.

The specification names the type of rosin to be used for the 'R' fluxes, but does not name the acids or salts contained in AC.

For the rosin-based fluxes, the resistivity of a water extract, prepared from the fluxes by a described method, is specified.

The specification does not pronounce on the suitability of a given flux for electronic soldering. However, it will shortly be published in a revised format as an American National Standard, and will not be based on composition, but on the functional requirements of the fluxes (private communication, Dr W. Rubin).

Military specification MIL-F-14256

This specification relates to liquid fluxes, which are arranged in two classes:

R   Rosin, non-activated
RMA   Rosin, mildly activated

These fluids are based on the requirements listed in QQ-S-571E.

For military applications, all fluxes which are activated need to have their residues removed by a suitable cleaning agent. Future issues of the specification will include watersoluble fluxes which are subdivided into those containing glycols and into glycol-free ones (private communication, Dr W. Rubin).

UK

British Standard Specification BS 5625: Purchasing Requirements and Methods of Test for Fluxes for Soft Soldering (1980)

This standard classifies fluxes according to what they contain and gives guidance on what type of product each should be used on:

Class 5a Rosin fluxes with halide-containing activators
Class 5b Rosin fluxes with halide-free activators
These fluxes are recommended for 'soldering and tinning' (sic) nonferrous components in the electrical and electronics industries. Suitable for use with high-speed mechanized soldering systems.
Class 6   Non-activated rosins
Recommended for soldering of copper in the electrical and electronics industries where lowest corrosion liability is essential.
Class 7   Other organic compounds

Recommended for 'general soldering operations where flux residues should have minimum corrosive action compatible with the application'.

This specification is currently (1993) under extensive revision (private communication, Dr W. Rubin).

Two comments may be worth making here:

1. The standard, issued in 1980, only lists rosin fluxes specifically for the soldering of electronic assemblies.
2. Like almost all other national and international standard specifications, it names only copper as the substrate in this context, as well as in the context of corrosion. Bare copper has now virtually disappeared as the principal solderable substrate in electronics (Section 3.6.6), though it continues to be useful as a convenient yardstick for assessing the corrosive action of a flux.

3.4.9 Testing soldering fluxes

The prospective user of a soldering flux wants to be assured of two things:

1. Will it work as I expect it to, in my particular manufacturing situation?
2. Will it be safe to use, and not cause corrosion or malfunctioning of my particular product?

Industrial (and military) standard specifications can give the user some guidance in these problems, but will not give a quantitative answer if he wants to know how effectively an individual flux will solder the components he is using to his circuit boards, and whether its residue will cause any damage. The wetting tests described in Section 3.6.8 will answer the first question, corrosion tests and the insulation tests described in Section 8.6.3 the second one.

The tests listed by the various standard specifications are designed to decide whether a given flux conforms to their requirements. Most of them are beyond the expertise and the facilities available to the average flux user; they appear to be designed mainly for use by the manufacturer or for arbitration purposes.

Soldering efficiency and the spreading test

If soldering efficiency is to be measured at all, a 'spreading test' must be used. In it, a pellet of 60% tin solder, mostly weighing 0.2 g, is placed on a coupon of copper sheet which has been cleaned by a prescribed procedure. A measured quantity of the flux under test is applied to it and the coupon is heated until the solder melts (up to 235 °C/455 °F in BS 5625). After a specified time, heating is discontinued and the solder is allowed to solidify. With BS 5625 the mean diameter of the roughly circular area covered by the solder is measured. With an activated flux, this should amount to $1.6 \, cm^2$, with plain rosin and other organic compounds $1.0 \, cm^2$ or another agreed value.

Other specifications do not indicate the area of spread, but simply demand a sharp, 'feathered' wetting angle.

Corrosive action

The test for corrosive action is again confined to observing what a flux will do to copper during soldering, or what the residue which is left on the copper will do in a moist atmosphere.

In BS 5625, flux residue, left on a copper coupon after having melted a small amount of 60% tin solder together with the flux under test, is stored in a humid atmosphere, at 40 °C/645 °F and 91% to 95% relative humidity, for three days. Corrosion is deemed to have occurred if the flux residue has changed colour, or if white spots have appeared in it.

With the American standard ANSI/IPC-SF-818, a drop of the flux to be tested is placed on a flat glass slide, on to which a thin film of copper, with a thickness of 0.05 μm/0.002 mil (500 angstrom) has been deposited by an evaporation technique, a so-called 'copper mirror'. Copper mirror slides are commercially available. The slide with the drop of flux on it is kept in a humidity chamber at 23 °C/73 °F and 50% relative humidity for 24 hours, and then examined. If the copper mirror has disappeared underneath the flux, it is deemed to have failed the test. A flux which passes the copper mirror test is an 'L-type' (low activity) flux, which group comprises all R-type fluxes, most RMA, and some R. If some of the copper mirror has gone, it is an 'M-type' (medium activity) flux, which may still be an RMA, but is mostly RA and sometimes a watersoluble or a synthetic activated flux. If the copper mirror has disappeared completely, the flux is an 'H-type' (high activity). Watersoluble and synthetic activated fluxes fall in that group. An important aspect of flux classification relates to the surface–insulation–resistance (SIR) properties of a flux.

Halide content

Determination by analysis

If a halide-free flux is specified, some standards give a detailed analytical procedure for quantitatively determining the halide content of the flux. If this exceeds 0.05% by weight of the rosin content of the flux, calculated as Cl, the flux does not conform to, for example, a BS 5625 halide-free flux. If it exceeds 0.5% calculated Cl on the solids content of the flux, it does not conform to an ANSI/IPC-SF-818 flux of Type L.

Silver–chromate test

This is a qualitative yes/no test, and does not indicate a specific halide percentage. Silver chromate ($AgCrO_4$) is a brick-red substance, which turns white or yellow in the presence of a halide. Silver-chromate impregnated testpaper is commercially available. If such a piece of paper turns white or yellow when a drop of the flux under test is placed on it, halide is deemed to be present, and the flux cannot be classed as R or RMA. There is a problem, though: certain acids and amines (which may well be free of halide) are also capable of causing the colour of silver-chromate paper to change. Because this test is relatively insensitive, a flux with up to 0.05% halide will still pass it as 'halide-free'.

Beilstein test

This test is also a qualitative test, and gives no indication of the actual quantity of halide present. However, it is more sensitive that the silver-chromate test, and unequivocal. On the other hand, it will also detect the non-ionic halogen in a halogenated solvent, if any should be present in a flux.

The Beilstein test detects the presence of halogen in an organic compound. It requires a small piece of fine copperwire gauze, which is heated in an oxidizing flame (e.g. the blue part of a bunsen-burner flame) until it ceases to turn the flame green. It is withdrawn, allowed to cool, and a small amount of the flux under test is placed on it. It is then put back into the flame. If the flame turns blue-green, the flux contains traces of halide. If not, it is deemed to be halide-free. The Beilstein effect depends on the formation of a volatile copper halide. (F. K. Beilstein, Russo-German chemist, 1838–1906.)

Solubility of flux residues

The average flux user needs guidance on how to assess the ease with which the residue of the flux he is using, or wants to use, responds to the cleaning method he is using or intends to use. So far, the international standard ISO 9455-11 : 1991 (E) is the only standard which is relevant to this problem.

This standard describes a method of heating a sample of the flux on a dish-shaped piece of brass sheet up to 300 °C/570 °F for a given time, placing the sample in a humidity chamber for 24 hours and then immersing it in the solvent which is to be used for cleaning. The presence of any residual flux left after cleaning is indicated by the ability of the cleaned test specimen to form an electrolytic cell.

With the growing importance of cleaning soldered assemblies, the further development of such tests is certainly desirable.

3.5 Soldering heat

Conventional soldered joints are made with molten solder. Hence, the soldering temperature must always be at least above the melting point of the solder, i.e. above 183 °C/361 °F. The immediate environment of the joint, and sometimes the whole assembly, must be brought up to the soldering temperature too. The exact temperature needed depends entirely on the soldering method used. It is rarely less than 215 °C/420 °F and is often much higher.

3.5.1 Heat requirements and heat flow

Heat is a form of energy, which is usually measured in one of the following ways. One calorie (1 cal) raises the temperature of one gram of water by 1 K (which is the same temperature difference as 1 °C, Section 5.4.2). One calorie equals 4.187 joule, or in units which are meaningful in the context of soldering, 4.18 watt.seconds (W.sec).

Table 3.10 indicates the amounts of heat required in some common soldering situations. In this context, it is useful to know the heat conductivity of the various materials involved, so as to be able to gauge the speed with which the heat input spreads within an assembly (Table 3.11).

The figures given in Tables 3.10 and 3.11 are worth studying. Table 3.10 shows that organic substances like FR4 have a much higher specific heat than metals. This has an important bearing on most soldering situations. The greater part of the soldering heat expended in making a joint is not used to heat the metallic joint partners, but to heat the FR4 epoxy board on which the copper laminate sits. Hence the need to preheat the boards before they pass through the solderwave (Section 4.3), but also the benefit of preheating the circuit board, at least locally, when soldering single multilead components (Section 5.7), or before carrying out repair work, i.e. desoldering and resoldering single components (Section 10.3).

The list of heat conductivities is equally illuminating. The heat conductivity of epoxy is two orders of magnitude lower than that of the ceramic substrate of a hybrid assembly. Hence the need for taking the thermal management of SMDs, which are mounted on an epoxy board, much more seriously than that of hybrid constructions, which were initially the beginnings of SMD technology.

Table 3.10 *Heat required to raise the temperature of a substance from 20 °C/68 °F to a soldering heat of 250 °C/482 °F*

1 g copper	88 watt. sec
1 g solder	102 watt. sec (including heat of melting)
1 g FR4	338 watt. sec
A soldered joint (volume 1 cub. mm)	0.7 watt. sec
A circuit board	27 kw sec
23.3 cm × 16 cm	
9.2 in × 6.3 in	
('Europa' format)	
1.2 mm/47 mil thick	

Table 3.11 *Some heat conductivities in watt/cm °C*

Copper	3.9
Aluminium	2.2
Brass	1.2
Steel	0.5
Solder	0.5
Ceramic (alumina)	0.25
FR4, rosin	0.002
Air	0.000 000 002

The figures also show how even the narrowest air gap prevents the flow of heat between two hot bodies. Hence the need to have a drop of molten solder on the tip of a soldering iron or thermode, or at least some flux on the joint to bridge that gap (Section 5.7).

3.5.2 Heating options

Equilibrium and non-equilibrium situations

The basic aim of every heating process is the transfer of heat from a heat source to the heat recipient, i.e. from a hot body to a colder one via a heat transfer medium. There are two basic heating situations: equilibrium and non-equilibrium systems.

In equilibrium situations, the temperature of the heat source is the same as the soldering temperature which must be reached. The time within which the joint reaches its soldering temperature depends on the efficiency of the thermal coupling between source and joint. The joint cannot be overheated, i.e. it cannot get too hot, but it can be 'overcooked', i.e. it can be heated for too long a time. The latter carries the risk of excessive growth of the brittle intermetallic compound, and thus an unsatisfactory joint structure and the risk of a shortened joint life-expectancy.

In non-equilibrium situations, the temperature of the heat source is higher, often very much so, than the soldering temperature itself. Whether the correct soldering temperature is reached or exceeded is a matter of timing the heat exposure. The higher the temperature of the heat source, the steeper is the temperature rise of the solder joint, and the more critical becomes the precise control of the duration of its heat exposure. Overheating may not only endanger the joint and its properties, but in severe cases it can damage the assembly itself (Figure 3.14).

Wavesoldering, vapourphase soldering, hot air or gas convection soldering, impulse soldering and handsoldering with a soldering iron present equilibrium heating conditions. Infrared soldering, laser soldering and flame soldering are non-equilibrium systems.

Heat sources

A thermostatically controlled electrical resistance heater is the most common primary heat source. This transmits its heat to the heat-transfer medium, whether it be the drop of solder on a soldering iron or the solderwave in a wavesoldering machine. The reader may be amused to learn, though, that the first few wavesoldering machines were gas heated.

Small, pointed butane- or propane-gas flames are used for soldering individual joints in awkward locations. Equipment using a very hot, needle-shaped hydrogen–oxygen flame is also commercially available. These flames, which represent extreme cases of non-equilibrium heating, may be hand-held, but more often are manipulated by programmed robots, and then of course equipped with controls which prevent overheating.

Laser beams present the ultimate in non-equilibrium heating. To speak of the 'temperature' of a laser source makes no real sense; what matters is the extreme

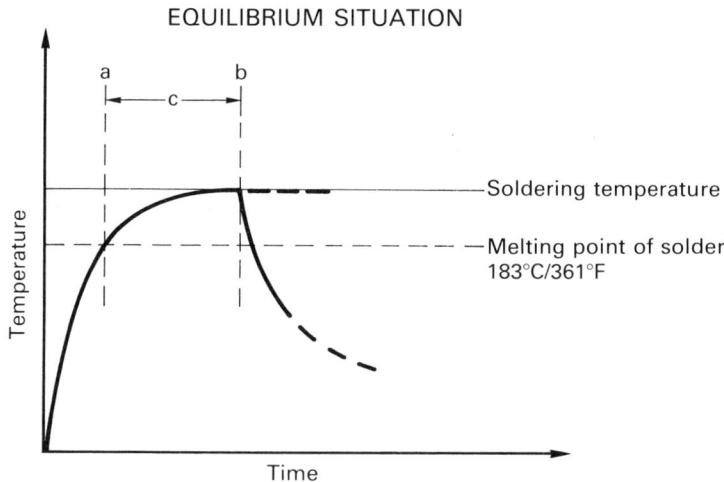

EQUILIBRIUM SITUATION

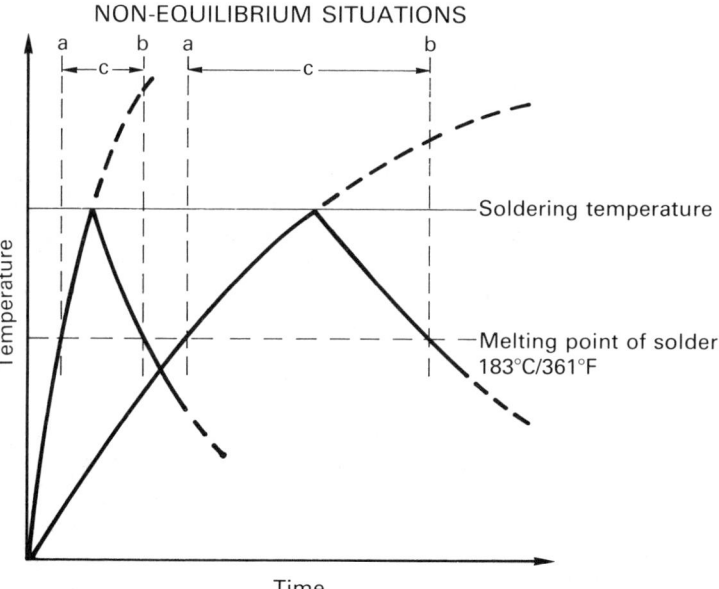

NON-EQUILIBRIUM SITUATIONS

Figure 3.14 *Equilibrium and non-equilibrium heating situations*

energy density of the spot of laser light, which impinges on the joint surface, and which may reach 10 kw per square millimetre (Section 5.6). A very precise energy dosage is of the essence, to avoid destruction of joint and burning a hole into the substrate. Exposure times are measured in milliseconds.

Heat transfer mechanisms

The soldering heat can be transmitted from the heat source to the joint by any one of three basic mechanisms: conduction, convection and radiation.

Conduction relies on a direct physical contact between a hot solid body or liquid and the surface of one of the joint members. The efficiency of heat transfer depends critically on the close fit between the heating and the heated surface. Any airgaps between them fatally affect the heat transfer. Molten solder is the best heat-transfer medium available: being a liquid, it conforms perfectly to whatever surface it has to heat. This is the virtue of the solderwave, as well as of the drop of molten solder on the tip of a soldering iron, which will come in very useful with repair soldering (Sections 10.2 and 10.3). Strictly speaking, convection comes into the heat-transfer mechanism of wavesoldering as well, because the solderwave consists of a body of moving solder. By contrast, dipsoldering in a stationary bath relies on heat transfer by conduction only, like a soldering iron.

3.6 Solderability

3.6.1 *Wetting and dewetting*

Wetting

The term 'wetting' describes the behaviour of a liquid towards a solid surface with which it comes into contact. In our case, we are naturally concerned with the way the molten solder behaves toward the substrate. Though everyone knows instinctively what is meant by 'wetting', it will be useful to examine in detail what is involved in wetting in the context of soldering, and how it can be quantified and measured.

Section 3.3.1 described the soldered joint as the result of a surface reaction between molten solder and a solid metallic substrate, and it was explained why an intimate contact between the two is a precondition for the joint to form. We must now amplify this by saying that wetting is the precondition for this intimate contact.

Wetting is not a 'yes or no' situation; there is a scale of wetting quality between total non–wetting and complete wetting. The yardstick for measuring the quality of wetting is the 'wetting angle', which is formed between the surface of the solid and that of the liquid along the line where they meet (Fig. 3.15).

A wetting angle of 180° is a sign of total non–wetting, while an angle towards zero denotes complete wetting. In the context of soldering, a wetting angle of less than 60–75° is normally, but arbitrarily, considered acceptable; anything up to 90° is doubtful, and beyond 90° definitely bad.[29] Whether and when 'bad' can or should be equated with 'non–acceptable' will be discussed in Section 9.3.

The wetting or contact angle between the molten solder and the substrate is the result of the opposing forces of the surface tension of the solder, which tries to pull it together into a globule (somewhat flattened by gravity), and the interfacial tension between the solder and the substrate, which tries to pull the solder across its surface, so that as much of the solder as possible can come in contact and react with it. The wetting angle can be interpreted in terms of the three surface energies

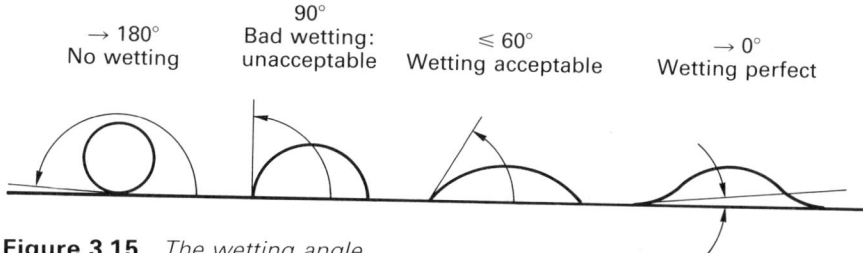

Figure 3.15 *The wetting angle*

involved: that of the molten solder, of the solid, and of the interface between the two. Klein Wassink[30] provides a detailed discussion of this aspect.

In practical terms, the significance of wetting can be stated very simply. Good wetting helps the solder to get to all the places where it ought to be; doubtful and bad wetting prevent the solder from entering a joint.

Dewetting

'Dewetting' is not the same as 'non-wetting'. As the term implies, in a dewetting situation the molten solder did get to where it ought to be, but it does not stay there. Instead, it pulls back and forms separate islands of solder, with areas of exposed intermetallic compound in between. This situation can occur in dip-tinning, e.g. in the hot-air levelling process for circuit boards (HAL), or in wavesoldering, but only rarely in reflowsoldering.

Dewetting is caused by local, untinnable spots of surface contamination, such as oxide particles, or surface dirt like traces of silicones or fingerprints. Non-metallic inclusions in galvanic coatings, for instance embedded colloids caused by unsuitable or badly controlled plating baths for copper, nickel or gold, can cause dewetting too.

Surfaces which dewet are at first completely covered with molten solder, which bridges the untinnable spots. Before it can solidify, its surface tension pulls the still liquid soldercoating apart, and away from the discontinuities (Figure 3.16).

3.6.2 Capillarity and its effects

A capillary is a very thin hole or a narrow gap (from *capillus*, Latin for 'hair'). If the surfaces of the hole or gap are wettable, interfacial tension quickly pulls the liquid

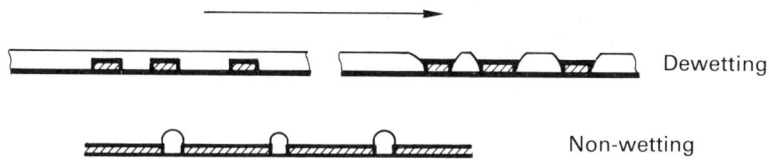

Figure 3.16 *Dewetting and non-wetting*

solder into it with considerable force, often against the force of gravity. On the other hand, if the inner walls of the gap are untinnable, the surface tension of the solder prevents it from entering it.

If one of the members forming the gap is movable, like the gullwing legs of an SMD during reflowsoldering, the interfacial tension pulls the walls of the gap towards one another, which means it pulls the flat end of the leg into the middle of its footprint. If, on the other hand, one or both are untinnable, the surface tension of the solder pushes them apart (Figure 3.17).

The consequences of capillarity for soldering are important:

1. If the joint surfaces are wettable, capillarity pulls the solder into the joint, against the force of gravity if necessary. If they wet badly or not at all, the solder cannot get into the joint, even if gravity would tend to pull it into the gap.

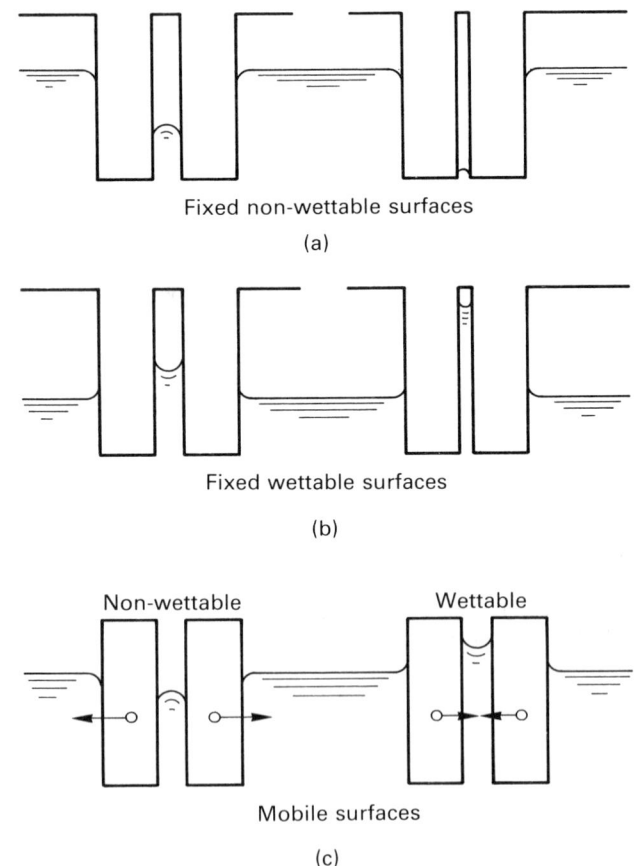

Fixed non-wettable surfaces

(a)

Fixed wettable surfaces

(b)

Non-wettable   Wettable

Mobile surfaces

(c)

Figure 3.17

2. If one of the joint members is mobile, as is the case in reflowsoldering, and if both are wettable, interfacial and surface tension pull the joint members together. If one of them is unwettable, they are pushed apart.

In reflowsoldering, capillary forces are the cause for the self-alignment of small SMDs, but also for 'tombstoning' and the floating of chips or melfs on badly designed layout patterns (Sections 6.4.2 and 11.2.2; also Figure 3.18).

3.6.3 Capillarity and joint configuration

Capillary joints and open joints

The way in which the solder flows into a wettable joint depends on the soldering method and on the shape of the joint itself. Basically, there are two types of joint: 'capillary joints' and 'open joints' (Figure 3.19). With a capillary joint, or lap joint, two flat and essentially parallel surfaces face one another, and the joint forms a two–dimensional gap. Tubular joints, like through-plated holes, are a special form of capillary joint, where the gap is cylindrical. With an open joint, or butt joint, one or both of the joint members are not flat, and they touch one another along a line or just in one spot.

With capillary joints, the escape route for the air and flux in the joint can get blocked if the molten solder closes all the edges around the joint gap before all the air and flux inside the gap have been pushed out by an orderly, frontal advance of

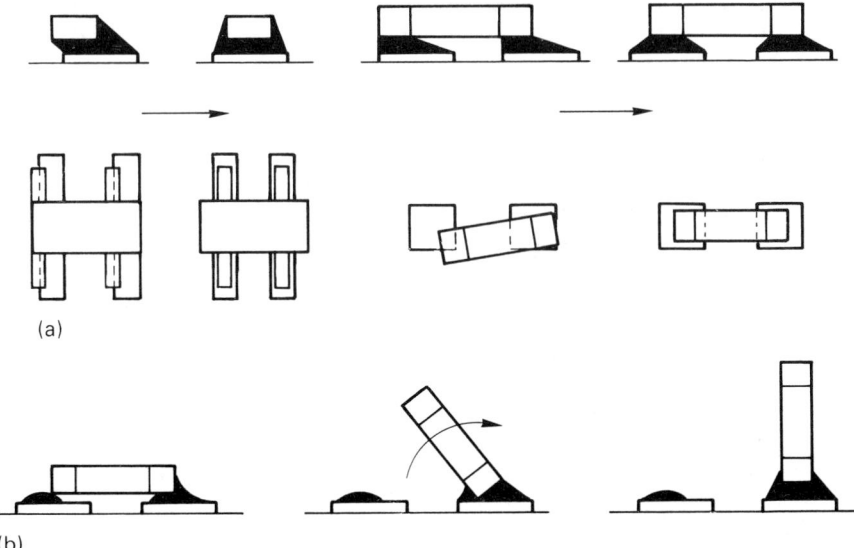

(a)

(b)

Figure 3.18 *The effects of capillarity in reflowsoldering. (a) Self-alignment of components; (b) tombstoning*

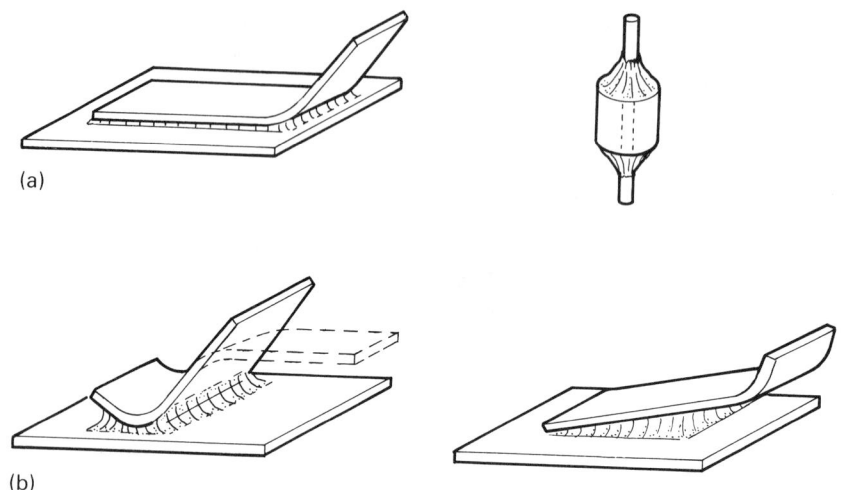

Figure 3.19 *(a) Capillary joints and (b) open joints*

molten solder into the gap from one (or at most two) sides only. With an open joint, there is no such problem: from whichever direction the solder enters an open joint, the escape routes for air and flux cannot be blocked.

With wavesoldering, all capillary joints, flat and tubular, fill from one side only. Both types will normally be sound, especially the latter, unless air or water vapour escape from the walls into the hole after the solder has entered it (blowholing), which is a matter dealt with in Sections 9.5.3 and 11.2.2.

The penetrating speed of molten solder into a flat capillary gap between two copper surfaces, 0.09 mm/3.5 mil apart, has been measured for various solders and at various temperatures by McKeown (see Reference 2). At 243 °C/470 °F, using a 63% tin solder and a concentrated zinc–ammonium chloride flux, McKeown measured a penetration speed of 3.5 sec over a gap length of 10 cm/4 in, which equals about 2 m/6 ft per minute, the average travelling speed of a circuit board across a solderwave. With a halide-free flux, the penetration speed is bound to be lower, which means that the solder advancing within the gap will be overtaken by the advancing wavefront outside. Well before the gap has been filled by capillary penetration, a good deal of air and flux will be trapped in it by solder which has closed all exits.

When reflowsoldering a capillary joint with solderpaste, solder and a partially volatile flux are already in the gap before soldering starts, and entrapment of gas and flux in the flat joint is almost impossible to avoid. The same is true for a reflowed capillary joint, where solid solder is preplaced on one of the joint members, because the molten solder tends to advance more quickly along the edges of a joint than in the middle (Figure 3.20).

Only with impulse soldering, where the joint members are pressed together during soldering, are joints less likely to be porous. Internal porosity in a capillary

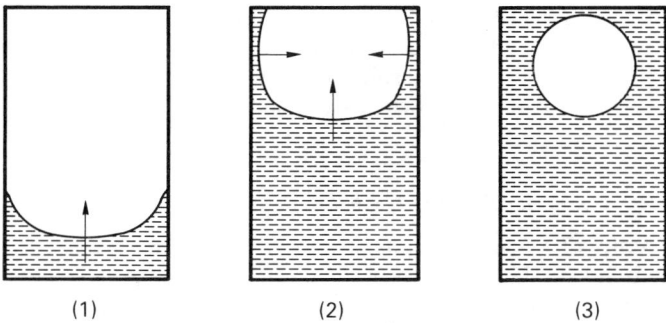

Figure 3.20 *How molten solder fills a capillary joint*

joint is almost undetectable by external inspection and only shows up under X-ray examination (Section 9.4.3). However, whether a porous capillary joint, even if it contains up to 50% voids by volume, is really inferior to a completely solid one is very debatable; this will be dealt with in Section 9.3.

Open joints, as which one can count not only the end-joints of melfs and chips, but also the joints of all SOs, because their horizontal legs are short and the joint gap is wedge-shaped, do not trap air. The few airbubbles found in such joints, if they are sectioned, are no cause for worry. On the contrary, they are more likely to arrest incipient internal cracks rather than starting them.

Open joints versus capillary joints

Open joints have several advantages over capillary joints in both wavesoldering and reflowsoldering. The solder gets into them more easily, and they are much more inspectable by optical means, unless they are partially or completely hidden under the component housing like the J-legs of a PLCC, or the pin-grid array under a large chip. The presence of solder in the joint and its wetting angle with the joint members are both readily verified. With capillary joints, the only external evidence for adequate wetting and penetration is the continuity and the wetting angle of its circumferential solder fillet.

Thus, a good case can be made for avoiding parallel capillary gaps and for designing wedge-shaped, open ones instead, unless the joints are to be soldered by impulse soldering under the mechanical pressure of a thermode. Here too, it may be worth while considering giving the faces of the thermodes a slant, thus making the impulse-soldered joints wedge-shaped, with any trapped flux and vapour pushed towards one end.

3.6.4 The importance of solderability

The ease with which molten solder wets a metallic surface is called 'solderability'. It has a decisive influence on the achievement of both soldering success and soldering quality, and thus on the fault rate of a soldering operation and its cost efficiency. It is therefore important to choose joint surfaces which are inherently solderable and to make sure that they remain in a solderable condition. To this end, one must be able to measure, or at least to assess, their solderability.

The inherent solderability of a metal

The solderability of a metal depends on two factors: first, on the nature and the chemical stability of the oxide layer on its surface; secondly, on the chemical affinity between the metal and the solder, in other words on the readiness of the solder to form a diffusion zone at its interface with the substrate.

The first factor determines whether a mild flux will do, or whether an active, highly polar and corrosive flux is needed to remove the oxide. The chemical affinity is based on the amount of energy set free by the reaction between the tin, more rarely by the lead in the solder, and the substrate. The inherent solderability of a number of common metals is shown in Table 3.12.

This listing assumes clean, though not oxide-free, surfaces. Solderability is governed by several factors, among them the following:

1. The ease with which surface oxides or sulfides are dissolved by a flux.
2. The surface energy of the metal surface (which means its readiness to react with whatever comes in contact with it), metals with low surface energies being more difficult to solder.
3. The metallurgical affinity between the metal to be soldered and the constituents of the solder. For example, lead is more compatible with nickel than tin, therefore lead-rich solders are better on nickel than pure tin or a eutectic tin–lead solder.

3.6.5 Oxide layers

The chemical stability of the various metal oxides differs widely. Gold and platinum do not form oxides under normal circumstances, so their chemical behaviour is irrelevant. Silver does not oxidize at room temperature, but ozone, an ingredient of urban smog, attacks and blackens it.[31] It readily reacts with sulfur which is always present in our normal industrial atmosphere. The familiar brown tarnish of silver sulfide which results from this is resistant to mild fluxes, which on the other hand deal readily with copper oxide and zinc oxide. Iron oxide is more difficult, and cast iron, because of the non–metallic graphite particles on its surface, is untinnable and unsolderable by the methods admissible in electronic soldering. Chromium and its alloys owe their resistance against tarnish to a transparent, but stable and tough, oxide layer which makes them almost unsolderable. The oxide layer on aluminium and its alloys is equally transparent, but can be dealt with by special, though corrosive, fluxes. On the other hand, surface oxides and sulfides of copper are readily removed, even by mild fluxes.

Table 3.12 *The solderability of common metals, listed in order of descending solderability*

A. Readily solderable with mild fluxes (R and RMA)
    Gold and its alloys
    Tin/lead solder
    Tin

B. Solderable with mild fluxes (RMA)
    Copper
    Silver
    Copper + 2% iron
    Silver/palladium (as thick-film on chips and melfs)
    Gold/platinum (as thick-film on LCCCs)

C. Solderable with activated fluxes (RA)
    Brass
    Nickel
    Cadmium

D. Solderable with active fluxes (OA etc.)
    Zinc
    Tin/bronze
    Nickel/copper alloys
    Nickel/iron alloys (Alloy 42)
    Nickel/iron/cobalt alloys (Kovar)
    Mild steel
    Alloy steels
    Beryllium bronze

E. Only solderable with special fluxes
    Aluminium bronze (fluxes based on phosphoric acid)
    Stainless steel (ditto)
    Aluminium and its alloys (fluxes containing zinc compounds)
    Cast iron (not solderable but tinnable with fluxes consisting of fused chlorides)

F. Unsolderable
    Chromium
    Silicon
    Titanium
    Manganese

3.6.6 Solderability-enhancing surface coatings

Uncoated soldering surfaces, like bare copper footprints or Alloy-42 gullwings, have become rare. Most surfaces intended to be soldered are coated with tin or a tin–lead alloy to improve and preserve their solderability. Silver or gold are also used sometimes, but their value as solderability preservers is doubtful. Silver is liable to tarnish, and the presence of gold in a soldered joint can lead to embrittlement if there is too much of it (see below).

Tin and tin–lead coatings

It is true to say that nothing is more solderable than solder itself. It is even more solderable than pure tin, because the lead in the solder makes it more resistant to atmospheric moisture and to corrosive environments. For that reason, a coating with a 50% or even a 40% tin solder is often preferred to one with 60% tin.

As soon as the molten solder encounters a tin or tin–lead coating, it melts and dissolves it. Provided the coating is thick enough, above about 25 μm/0.1 mil, the molten solder will lift off any surface contamination which might sit on the surface of the coating, and flow underneath it. This is called the 'lift-off effect' (Figure 3.21).

Galvanically deposited tin–lead coatings have their problems. They consist of discrete particles of tin and lead, and are therefore porous. All the damaging ingredients of the atmosphere, especially an industrial one, can and will slowly penetrate between the particles down to the base metal. The tin–lead deposit dissolves on contact with the molten solder, which is then confronted with the bare base metal, which was originally quite clean of course, otherwise the galvanic deposit would not have adhered to it. But after a period of unsuitable storage, contamination might have penetrated down to it, and its solderability will have suffered, if not disappeared.

The remedy is to fuse, i.e. to reflow, galvanic tin or tin–lead deposits in the presence of a flux cover, and turn them into a coherent layer of tin or tin–lead solder (Section 6.3). As a bonus, this creates a layer of intermetallic compound between the coating and the base metal.

If the substrate base is not readily solderable with an RMA flux (like iron or an iron–nickel alloy), a thin galvanic coating of nickel, possibly with a topcoat of copper, must be provided between the base metal and the tin or solder topcoat. Nickel dissolves only slowly in molten solder. While the topcoat disappears immediately in the molten solder once soldering starts, the nickel survives long enough to protect the possibly badly solderable base underneath. With a pre-fused coating, the solder will find a ready-made intermetallic layer already in place when soldering starts. The alternative to a fused galvanic tin–lead coating is one produced by hot tinning, by one of the processes of roller-tinning, immersion tinning or the HAL hot-air-levelling process (Section 6.3.1).

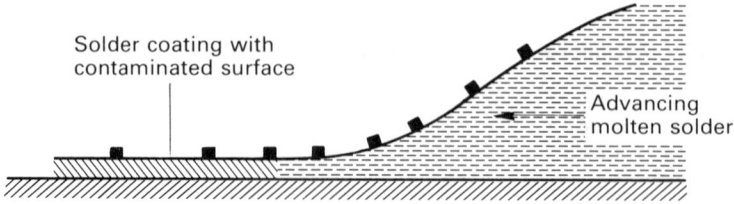

Solder coating with
contaminated surface

Advancing
molten solder

Figure 3.21   *The lift-off effect*

Silver coatings

Once popular, silver is now used less frequently, for several reasons. One is its low resistance to unsuitable storage conditions, as described above. Another is the danger of migration of silver between neighbouring conductors under the influence of an electrical DC potential between them. This leads to the formation of fibrous dendrites, which cause short circuits.[32]

Silver dissolves relatively quickly in molten solder. If it has to serve as a solderable coating on an otherwise unsolderable substrate, measures which are described below must be taken in order to prevent it from disappearing before soldering is completed.

Gold coatings

Gold as such is superbly solderable, with excellent storage properties. In spite of this, it is rarely used because it is associated with several problems, apart from its cost.

Galvanic gold deposits, mostly $> 1\ \mu m/0.04$ mil thick, can be almost unsolderable if they were produced by an unsuitable plating technique. Shiny gold deposits, such as are used in jewellery, contain colloid additions and fall into that class. Gold coatings thinner than $1\ \mu m/0.04$ mil are liable to be porous and soon become unsolderable.

Gold rapidly dissolves in molten solder and with $>3\%$ Au present in a joint, large, brittle flakes of the intermetallic compound $AuSn_4$ form in the joint gap when the solder solidifies. This situation can easily arise in a reflowsoldered joint.[11,33]

For this reason, gold-coated joint surfaces should only be considered if there is a compelling technical reason for them, or if the customer, often a military one, insists.

The rule, established in the seventies by the European Space Agency (ESO), that the thickness of any gold deposit on a solderable surface should bear a relation to the width of the joint gap seems to have been forgotten, since the outerlead bonding surfaces of TABs, which are meant to be soldered, are occasionally goldplated (see Reference 11).

3.6.7 Leaching effect of molten solder

The rate at which a substrate dissolves in the molten solder is called the leaching rate, and it differs from metal to metal. It depends on the rate at which substrate and solder react with one another, on the solubility of the reaction products in the solder, and of course on the soldering temperature. Table 3.13 lists the leaching rates of some substrates, in descending order. The figures show that silver disappears in the molten solder ten times and gold up to thirty times as fast as copper, and that the leaching rate rises quickly with the soldering temperature.

The leaching effect can have serious consequences, for instance if the thick-film solderable metallized surfaces on chips, melfs and the (now obsolescent) LCCCs

Table 3.13 *Leaching rates of some substrates*[33]

Substrate	Leaching rate in 60% Sn/40% Pb solder ($\mu m/sec$)	
	at 215°C/420°F	*at 250°C/480°F*
Au	1.7	5.25
Ag	0.75	1.6
Cu	0.075	0.15
Pd	0.025	0.075
Ni, Pt	≪0.01	0.01

disappear, making them unsolderable. There are two ways of reducing a high leaching rate:

1. An alloying addition. About one per cent of indium added to the solder slows down the solution of Au in the solder. The 1.65% to 2% of Ag added to electronic solders (Section 3.2.2) does the same for a silver substrate, but this effect is small.
2. Modifying the substrate. An addition of up to 35% platinum to the thick-film gold on LCCCs significantly slows down the leaching rate. Up to 20% palladium does the same for thick-film Ag on resistors and condensers. Cheaper and more effective, however, is a galvanically applied, leach-resistant nickel layer, topped with a tin or solder coating for optimal solderability.

The leach resistance of components can be tested by the immersion solderability test, described in the following section.

3.6.8 Measuring solderability

The solderability of a surface is based on the speed and reliability with which it is wetted by the molten solder using a given flux. Naturally, the milder the flux which one wants to or must use, the higher must be the solderability of all the surfaces involved. Solderability is one of the most important parameters in all soldering processes, especially the mechanized and automated ones which are used in mass soldering. The soldering success, and with it the reject rate and the economics of the whole operation, depends on a consistently good solderability of all footprints, lands, throughplated holes, and component wires, leads, and sintered surfaces. It is therefore essential to have a meaningful and reproducible method for measuring it.

In the age of handsoldering, it was mainly the flux vendors who measured solderability in order to assess the ability of their activated fluxes to cope with the often indifferent solderability of component wires and soldering lands. With the advent of the mechanized soldering of large numbers of joints, where soldering times must be as short as possible, and with no–clean soldering, where fluxes should be as mild as possible, the emphasis shifted to the soldering surfaces involved in the process: verifying and monitoring their optimal and consistent solderability became of paramount importance.

Uses of solderability measurement

Solderability measurement has several uses. The main one is the assessment of the solderability of a substrate, using a flux of known and standard efficiency. The other one is the assessment of the efficiency of a flux, using a substrate of standard and reproducible solderability. This latter requirement is not an easy one to fulfil and it has engaged the attention of flux manufacturers and drafting committees of standard specifications for quite some time (Section 3.4.7). Comparing the efficiency of an alternative flux with that of a flux with a proven performance, using components of known and consistent solderability for reference, is a simpler proposition, and is frequently practised in the industry.

Surfaces which need testing

Circuit boards

Given the present state of the art, the solderability of almost all commercially available circuit boards, that is their lands and footprints, can be assumed to be good, unless they are bare copper. In that case, a simple dipping test, which is described later, will verify solderability. With HAL pretinned boards and reflowed galvanic coatings, smooth and fully tinned footprints are a safe indicator of perfect solderability. Any defects in this respect can be assumed to have been spotted by the quality control of a competent board manufacturer.

Component leads and metallized surfaces

The solderability of components is not necessarily visually obvious. Only with Ag/Pd sintered thick-film faces on melfs and chips, a dark grey or brown tarnish means that sulfide has formed on them because of unsuitable storage and that they have become unsolderable. Should that have happened with a batch of loose, not belted, components, they can be saved by a short immersion in a photographic fixing bath (known as 'hypo'), or its equivalent, a 10% sodium thiosulfate solution in water, followed by a rinse first in de-ionized water, then in clean isopropanol, and drying off in air. Hypo is an efficient solvent for silver compounds, including the brown sulfide.

Wetting tests

Observing the solder meniscus

The contour of the surface of the molten solder along the line where it touches an immersed metallic body is called the 'meniscus'. The shape of the meniscus is an indicator of whether and how well the solder wets the metal (Figure 3.22). For a visual check of whether all is well with a doubtful leadwire or component, looking at the meniscus is a quick, simple test. The component lead or wire is dipped in an

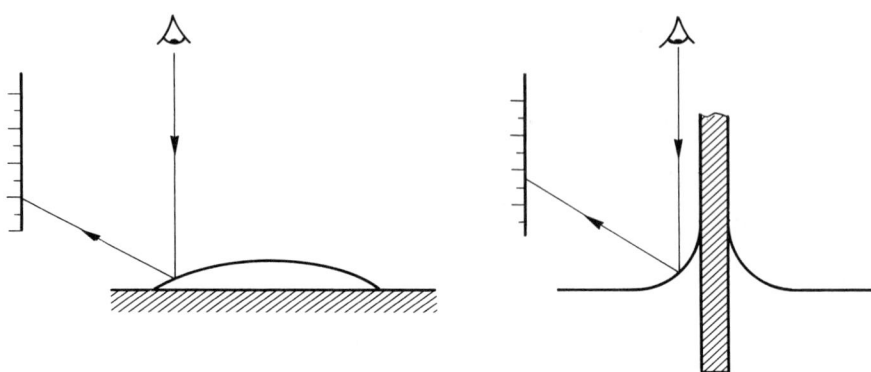

Figure 3.22 *The solder meniscus and its assessment*

MRA flux, conveniently the one which is used in production, allowed to dry for a short while, and dipped in a small bath of 63% Sn solder held at 250 °C/480 °F (it is useful to have such a bath handy in the quality control or production department; see below). The meniscus formed by the solder against the immersed body is observed visually. Opto-electronic equipment for measuring the deflection of a beam of light focused on the meniscus has been described.[34]

The wetting balance

The wetting balance has become the standard instrument for measuring solderability. By now, it has reached a high degree of technical perfection and is capable of measuring, recording and evaluating the wetting behaviour of almost any metallic surface involved in soldering. This includes the leads of all types of SMDs and the solderable endfaces of melfs and chips. Several makes of wetting balance are commercially available.

In principle, the test procedure is as follows. The lead, or surface to be tested, is fluxed with a standard RMA flux and briefly dried. The test specimen is then suspended from a sensitive balance or sensor, and immersed in an oxide-free surface of molten solder, mostly by raising a small, thermostatically controlled solder bath upwards against the suspended specimen (Figure 3.23), at a controlled speed. The sensor measures the vertical force acting on the specimen. A system of microprocessors plots this force during the test, and evaluates, prints and stores the result.

The graph of the force measured by the sensor is known as the wetting curve (Figure 3.24). It is convenient to disregard the weight of the specimen itself, as well as its buoyancy in the molten solder when fully immersed, which of course equals the weight of the solder it displaces on full immersion.

What is left is a curve which first dips downwards, showing a negative weight. This is the effect of the negative meniscus, which lasts until the specimen is warm enough to be wetted by the solder and for the flux to begin its work. The curve

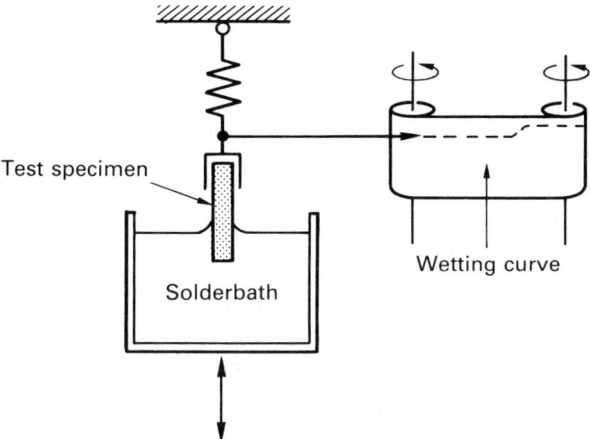

Figure 3.23 *The wetting balance*

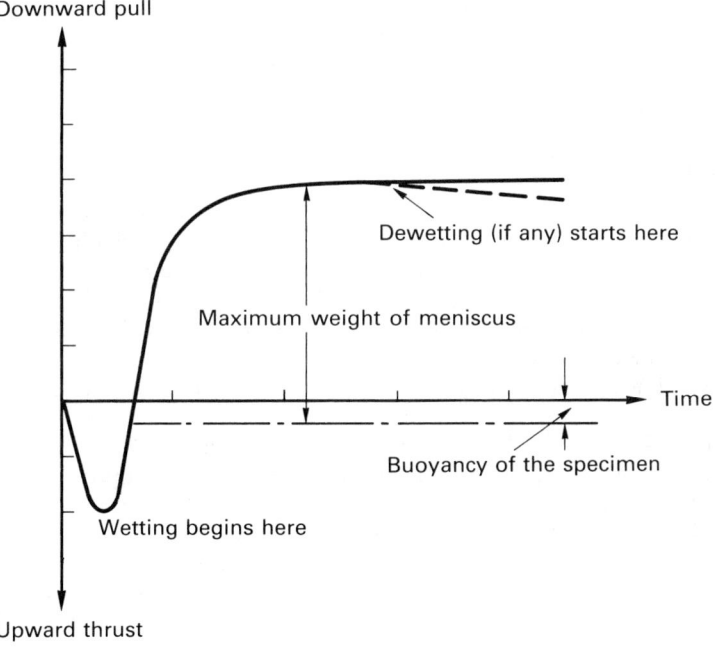

Figure 3.24 *The wetting curve*

then begins to climb at a rate which is governed by the efficiency with which the flux cleans the surface of the specimen, and thus the speed at which the solder meniscus climbs upwards. The meniscus stops climbing, and the wetting curve flattens out, as the final wetting angle is reached, or asymptotically approached. The force recorded at this point, reduced to a unit of length of the meniscus, is called the wetting force. Obviously, the blunter the final wetting angle, the lower the wetting force.

What the wetting balance really measures is not a mysterious force but simply the weight of the solder meniscus, which has climbed upwards on the specimen under the influence of the interfacial tension between solder and specimen, minus the upwards buoyancy of that part of the specimen which is immersed in the solder. The outer contour of the meniscus, and with it its volume and weight, are governed by the surface tension of the solder. The mathematical law which the contour of the meniscus follows has been calculated and described by Klein Wassink.[35] If the specimen tends to dewet, the meniscus will start to descend after a time, and the wetting curve begins to drop after it has reached its maximum.

Deciding on the best method for deriving a numerical value of solderability from a given wetting curve has been the subject of much discussion over the years. Account must be taken of the rate of rise of the wetting curve once wetting has set in, of its shape, of its maximum value and of the time taken to reach it. The computer of a wetting balance is programmed to deal with it all; a detailed account of the evaluation of wetting curves is outside the scope of this book.

The globule test

With chips, melfs, and SMDs like PLCCs, the small size and shape of the solderable surfaces makes the measurement of the meniscus force difficult. The surfaces concerned are small, often of complex geometry, and their buoyancy in molten solder is greater than the wetting force which must be measured. To cope with this situation, the 'globule test' (Figure 3.25) has been evolved as an alternative to the immersion method described above.

The specimen is fluxed, dried and suspended from the measuring head of the wetting balance. A heated anvil, normally held at a temperature of 235 °C/455 °F, which carries a small globule of 60% Sn solder, weighing 200 mg, replaces the solderbath. It is raised against the specimen from below until it touches the specimen. As soon as the globule begins to tin it, the surface tension of the bridge of molten solder, which forms between anvil and specimen, pulls it downwards. The solderability index is calculated from the time within which the resultant wetting curve reaches two-thirds of its maximum value (normally one-half to one second).

With chips and melfs, the solderability of both ends must be measured, because an asymmetrical solderability can be the cause of 'tombstoning' (Section 3.6.2). Normally, about ten specimens, taken from a batch or belt of SMDs, are tested in this manner. The solderability of single leads of PLCCs, QFPs, etc., can be measured in the same way.[36] A commercially available computerized wetting balance (Multicore) calls up the correct testing procedure and parameters once a

BEGINNING OF CONTACT

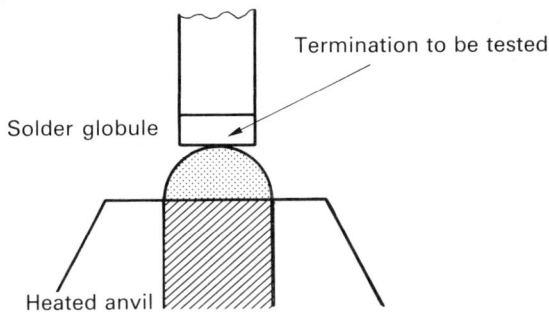

Termination to be tested

Solder globule

Heated anvil

WETTING COMPLETED

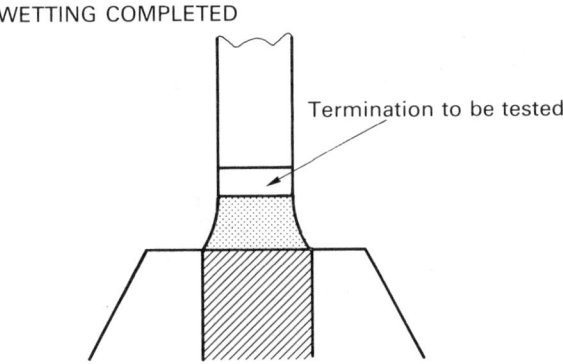

Termination to be tested

Figure 3.25 *The globule test*

test specimen has been identified and placed in the machine, and automatically produces and records its solderability value within about one minute.

The dipping test

A wetting balance is a sophisticated laboratory instrument, which represents a considerable investment. It is used mainly by vendors of components or fluxes, and by major users of either. The dipping test is a simple, but useful, alternative.

It is often called the 'dip and look' test. It does not provide a quantifiable numerical result, but it enables the user to make a reasonably objective, unambiguous judgement of the solderability of a component. It requires an electrically heated, thermostatically controlled solderbath of about 2 kg/4 lb capacity, filled with 60% Sn/40% Pb solder. This solderbath might well be the same as the one used in the solderballing test for solder paste (Section 5.2.3).

The best procedure for a dipping test is as follows. At least three components from a batch or belt are selected. The component to be tested is gripped with stainless steel tweezers, and dipped in the flux which is used in production or, in the case of reflowsoldering with solderpaste, in an RMA flux. Excess drops of flux are removed with a piece of filter paper, and the fluxed component is allowed to dry at room temperature.

Immediately before the test, the surface of the solder bath is cleaned of oxide by skimming it with a dry, clean stainless steel spatula. The fluxed test specimen is then dipped vertically in the bath, in the manner shown in Figure 3.26.[36] It is lowered into the bath steadily and slowly, at a speed of about 25 mm/1 in per second. It is kept immersed under the solder for about two seconds, and then withdrawn without jerking at about the same speed at which it had been lowered.

With a little practice, this procedure is easy to carry out. It is equally easy to mechanize the procedure with a simple motorized device. In some countries, simple dipping test equipment is commercially available.

The test parameters depend on the soldering method by which the components are to be used. They are tabulated in Table 3.14.

Having been dipped, every specimen is examined for signs of dewetting, under a magnification of about × 5. If more than 95% of the surfaces of every specimen are covered with a smooth continuous solder coating, the batch of SMDs from which the set of specimens has been taken can be assumed to be suitable for soldering. If more than five per cent of surface area has dewetted, the suitability of the batch of SMDs represented by the specimens is doubtful. It will be wise to repeat the test with a further three or more specimens. If the majority of those fail too, the batch should not be used. It is relatively easy to estimate visually whether a dewetted area represents above or below five per cent of the total, as Figure 3.27 shows.

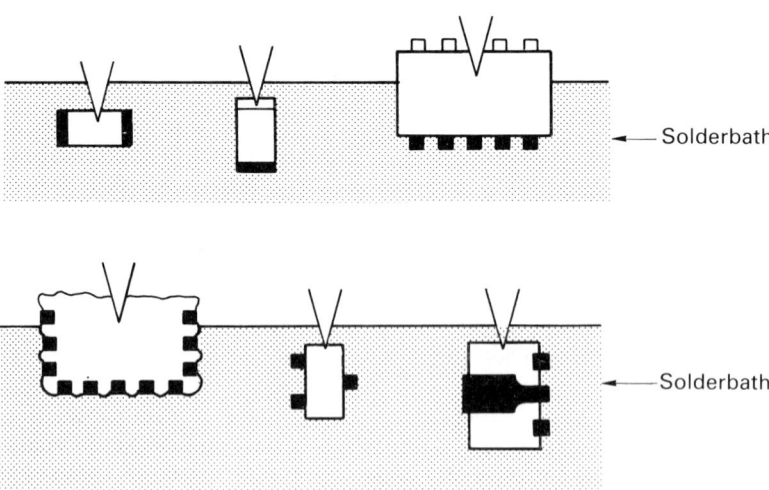

Figure 3.26 *The dipping test for SMDs*

Table 3.14   *Dipping test parameters*[36]

Purpose of test	Temperature of solderbath	Dwell time in solderbath
Solderability in vapourphase soldering	215 °C/ + / − 3 °C 420 °F + / − 5 °F	3 sec
Solderability in wavesoldering	250 °C + / − 5 °C 480 °F + / − 10 °F	2 sec
Tendency to dewet	260 °C + / − 5 °C 500 °F + / − 10 °F	5 sec
Leach resistance of chip metallization	260 °C + / − 5 °C 500 °F + / − 10 °F	30 sec

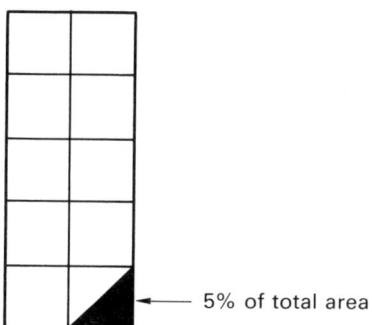

◄—— 5% of total area

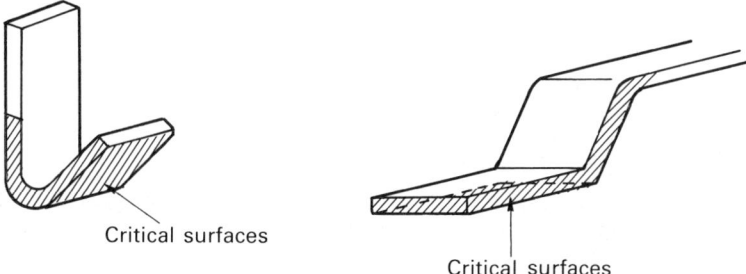

Critical surfaces

Critical surfaces

Figure 3.27   *Judging a dipping test result*

3.7   References

1.   Strauss, R. (1992) The Difference between Soldering Success and Soldering Quality; its Significance for Quality Control and Corrective Soldering. *Proc. 6th Intern. Conf. Interconnect. Technology in Electronics*, DVS Rep. 141, Duesseldorf, Germany (in German).

2. McKeown, J. (1948) The Properties of Soft Solders and Soldered Joints. *Brit. Nonf. Metals Res. Assoc.*, Research Monograph No. 5, Wantage, UK.

3. Raynor, G. V. (1947) The Pb–Sn Equilibrium Diagram. *Met. Abstr. (London)*, **19**, p. 150.

4. Earle, L. G. (1946) The Pb–Sn–Ag Equilibrium Diagram. *J. Inst. Met. (London)*, **72**, p. 403.

5. Smernos, S. and Strauss, R. (1984) Low Temperature Soldering, *Circuit World* (Ayr, Scotland), **10**(3), pp. 23–25.

6. Strauss, R. (1989) SMD Surface Mounted Devices, *Verlag Technische Texte*, Bonn, Germany, p. 47 (in German).

7. Thwaites, C. J. (1986) Some Metallurgical Aspects of SMD Technology, *Brazing & Soldering (UK)*, Spring 1986.

8. Bader, W. G. (1969) Dissolution of Au, Ag, Pd, Pt, Cu and Ni in a molten Tin–Lead Solder. *Welding J.*, **12**, 48, pp. 551–557.

9. Steen, H. A. H. and Becker, G. (1986) The Effect of Impurity Elements on the Soldering Properties of Eutectic and Near Eutectic Tin–Lead Solders. *Brazing & Soldering (UK)*, No. 11, pp. 4–11.

10. Campbell, A. N., Screaton, A. M. and Schaefer, T. P. (1955) *Canad. J. Chem.*, **33**, p. 511.

11. Schmitt-Thomas, K. G., Lang, H.-P. and Moedl, A. (1993) Metallurgical Examination of Thermally Stressed TAB Outer-lead Bonds. *Conference on 'Soldering, Science and Practice', Techn. Univ. Munich, March 1993*. DVS Rep. 153, Duesseldorf, Germany (in German).

12. Derived from data established by Knott, U. C. (1981) Dissertation, *Structure of Soft-soldered Joints*, Techn. Univ. Munich, Germany (in German).

13. Strauss, R. (1988) Wavesoldering v. Reflowsoldering – The metallurgical consequences of the Choice of Method. *Brazing & Soldering*, **14**, pp. 5–8 (in German).

14. Tanner, C. G. (1987) Reliability of Surface Mounted Component Soldered Joints produced by Vapour Phase, Infrared, and Wavesoldering Techniques. *BABS Intern. Conf., Nov. 87*, paper 29.

15. Ashby, F. A. and Jones, D. R. H. (1988) *Engineering Materials 2*. Pergamon Press, Oxford.

16. Hofmann, W. (1962) *Lead & Lead Alloys*. Berlin, Springer Verlag. (In German, English translation available through Brit. Nonf. Met. Res. Assoc., Wantage, UK.)

17. Ashby, M. F. and Jones, D. R. H. (1988) *Engineering Materials 2*. Pergamon Press, Oxford, UK.

18. de Kluizenaar, E. E. (1990) Reliability of Soldered Joints: A Description of the State of the Art. *Soldering & SMT (Ayr, Scotland)*, No. 4, pp. 27–38, No. 5, pp. 56–66, No. 6, pp. 18–27.

19. Engelmaier, W. (1993) Reliability of Surface Mount Solder Joints; Physics and Statistics of Failure. *Proc. Intern. Conf. Softsoldering, Munich*. DVS Report 153, Duesseldorf, Germany, pp. 149–160.

20. IPC (1992) Guidelines for Accelerated Reliability Testing of Surface Mount Solder Attachments. *IPC Document IPC-SM-785*.

21. Engelmaier, W. (1989) Performance Considerations, Thermal–Mechanical Effects. *Electron. Mats. Handbook, Vol. 1, ASM Intern., Materials Park, OH*, p. 740.
22. Wild, R. N. (1973) Some Fatigue Properties of Solders & Soldered Joints. *IBM Techn. Report 73Z000421.*
23. Solomon, H. D. and Sartell, J. A. (eds) (1986) *Electronic Packaging: Materials & Processes*, ASM.
24. Rubin, W. (1990) A No-Clean Review, *Proc. Conf. Electron. Manufact. & the Environment, Bournemouth, UK*, pp. 36–43.
25. Lea, C. (1992) After CFCs? *Electrochemical Publications*, Ayr, Scotland, pp. 94–98.
26. Zado, F. M. (1983) Increasing the Soldering Efficiency of Noncorrosive Rosin Fluxes. *Western Electric Eng.*, **27**(1), pp. 22–29.
27. Klein Wassink, R. J. (1989) *Soldering in Electronics, 2nd ed.*, Ch. 5.5.3. Electrochemical Publications, Ayr, Scotland.
28. Lea, C. (1992) After CFCs? *loc. cit.*, p. 301.
29. Manko, H. H. (1979) *Solders and Soldering*, McGraw-Hill, NY, p. 313.
30. Klein Wassink, R. J. (1989) *Soldering in Electronics, 2nd ed.*, Ch. 2. Electrochem. Publ., Ayr, Scotland.
31. Wise, E. M. (1948) *Metals Handbook*, ASTM, Cleveland, Ohio, p. 1111.
32. Klein Wassink, R. J. (1989) *loc. cit.*, pp. 217–218.
33. Strauss, R. (1988) Wavesoldering v. Reflow-soldering: Metallurgical Consequences of the Choice between them. *Proc. 4th Intern. Conference, DVS (German Welding Soc.), Report 110*, Duesseldorf, Germany, pp. 174–176 (in German).
34. Albrecht, H.-J. Scheel, T. and Freund, T. (1985) The Wetting Angle Measuring Unit. *Soldering & Brazing*, No. 8, pp. 8–15.
35. Klein Wassink, R. J. (1989) *loc. cit.*, pp. 46–49.
36. Lea, C. (1989) Solderability Measurement of SMDs, *Report DMA (D) 648*, Nat. Phys. Lab., Teddington, UK.

4 Wavesoldering

4.1 The wave concept

4.1.1 Wavesoldering before SMDs

Wavesoldering of printed circuit boards was conceived in 1956[1] at a time when circuit board technology was re-introduced from the USA into Europe, where it had originated in 1941[2] but had failed to gain a foothold after the war. SMDs had not yet appeared, and all solderjoints were of the push-through type, disposed on the flat, unobstructed underside of the board. Initially, they were all soldered by hand; joint-by-joint soldering was difficult and expensive to mechanize. The idea of pumping molten solder upwards through a nozzle, and passing the fluxed underside of a circuit board through the crest of the resulting wave, proved to be a practicable and comparatively simple method for soldering all the joints on a board in one operation. It opened the door to economical mass-soldering of printed circuit boards.

With all the joints disposed on a plane, unobstructed surface, the molten solder can readily reach and fill every joint and, with the smallest distance between joints at 2.5 mm/100 mil, bridging is easy to avoid. The early introduction of soldermasks made clean soldering and visual inspection still easier.

4.1.2 Wavesoldering after SMDs

The situation changed drastically with the introduction of SMDs, which had begun to be used in hybrid technology in the early and mid sixties, spreading to epoxy circuit boards from the early seventies onwards.[3] Initially, the designers of SMDs had only the reflowsoldering of hybrids in mind, and probably did not consider wavesoldering. SMDs carry their soldering surfaces on or very close to their bodies, sometimes in awkward corners and difficult for the molten solder to reach, often close together and difficult for the solder not to bridge. The two-dimensional open plain of the wired circuit board became a three-dimensional landscape with the soldered joints hidden in the valleys. The problems and the limitations of wavesoldering SMDs stem from this. Moreover, SMDs must be glued to the soldering side of the circuit board before they can be passed through the solderwave, which adds a further operation and an additional technology to be mastered.

The conclusion seems obvious: don't wavesolder unless you must. You must wavesolder if a board carries a significant population of inserted wired components as well as SMDs: inserted through-joints cannot be satisfactorily reflowsoldered by today's methods. Since the introduction of wavesoldering in a nitrogen atmosphere low in oxygen, which improves the surface tension of the molten solder and reduces the risk of bridging (Section 4.5), and with components becoming much flatter than they used to be (Section 2.1), and finally with the printing of solderpaste onto fine-pitch circuit boards having become an exacting task, the above advice can no longer be given without reservation. Nevertheless, it is still generally true for conventional wavesoldering machines.

If there are only a few push-through joints on a board, it may be cheaper and simpler first to reflowsolder all the SMDs on the board by one of the methods outlined in Chapter 5 and then insert and solder the few remaining wired components by hand, or with one of the now available soldering robots.

The difficulty can be overcome by adopting a design and manufacturing strategy in which the SMDs and the inserted components on a board are soldered in two separate operations. For example, a board can be so designed that the SMDs are first placed and reflowsoldered on the side of the board which subsequently receives the wired components. These are then inserted, and the board is reversed and wavesoldered on a simple solderwave, which need not be a double or special chipwave (see Section 2.2). During that second operation, the structure and the properties of the SMD joints, which ride over the solderwave on the top surface of the board, are in no way degraded.

Soldering demands that every joint be supplied with flux, solder and heat. In wavesoldering, these tasks are carried out in three distinct and consecutive stages: the fluxer applies an even, thin coat of flux solution to the joints on the underside of the board. The subsequent preheating stage raises the temperature of the board to between 80 °C/175 °F and 110 °C/230 °F and removes the bulk of the solvent from the flux coating. Finally, the solderwave supplies the solder in molten form, and at the same time the rest of the required soldering heat (Figure 4.1).

4.2 Applying the flux

The importance of applying a coating of flux to the underside of a circuit board in a reliable, consistent and reproducible manner, before drying off the flux solvent and passing the board over the solderwave, cannot be overstated. If the flux cover is uneven, patchy or variable, even the best of all possible fluxes and carefully optimized soldering parameters are of no use, and faulty joints are bound to result. Wherever the flux cover is too thin, there is a danger of incompletely filled joints, and bridges, icicles and 'spider's webs' of solder adhesions are likely to form. Local lack of flux leads to massive adhesions of solder in the affected areas.

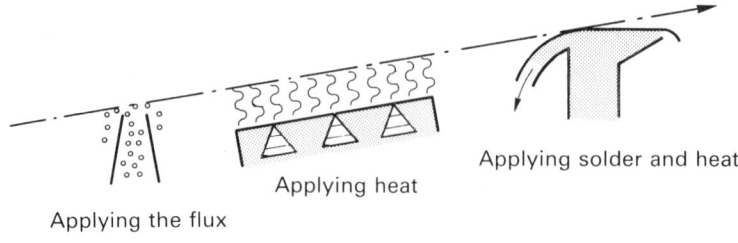

Applying solder and heat

Applying heat

Applying the flux

Figure 4.1   *The operational sequence of a wavesoldering machine*

For this reason, the technology of fluxing has reached a high level of sophistication and mechanization over the years. The fluxer has become an essential part of today's complex, costly, automated and computer–controlled wavesoldering line.

Attempts have been made in the past to coat circuit boards with a layer of solid flux in the form of a lacquer, before inserting or placing the components. So far, none have been successful because this method does not allow sufficient flux to reach the soldering surfaces of the components during the short time available when the board enters the solderwave. For this reason joint formation is erratic. On the other hand, fluxing need not necessarily always be directly integrated with a soldering machine. Several vendors offer fluxers as independent modules, which can be fitted into an existing soldering line.

With few exceptions, until recently isopropyl alcohol (isopropanol) was the universal solvent for all wavesoldering fluxes. Isopropanol has a flashpoint of $12\,°C/53.6\,°F$ and is therefore a flammable liquid of Class I, its flashpoint being below $38\,°C/200\,°F$ (see Section 8.3.1 under 'Flammability'). Concern is sometimes expressed about whether a fluxer, which contains and dispenses a flammable liquid, often in the form of an aerosol, constitutes a serious fire risk when in close proximity to the infrared heaters of the preheating stage. This is not the case, provided a few sensible constructional and operational measures are taken.

Nevertheless, towards the end of 1992 more stringent regulations concerning the storage of flammable liquids on manufacturing premises, and the classification of isopropanol as one of the volatile organic compounds (VOCs), which contribute to the formation of industrial and urban smog, and which have come under severe restriction in some states in the USA, have led to the development of alternative fluxes based on water (Section 3.4.7).

However, at the time of writing (1993), the vast majority of soldering fluxes continue to be based on isopropanol. As is explained in Section 8.3 it is not the flammable liquid itself, but its vapour, which burns. If the vapour concentration in air lies within the flammability limits, a flame or a glowing surface will ignite it. Provided the preheating stage is fitted with an air blower to expel the alcohol vapour given off by the drying flux, its concentration will remain below the lower flammability limit. In an ill–ventilated preheater, the flux vapour can, and occasionally does, ignite with a puff, without causing any harm.

With foamfluxed boards, a fire may start if the fluxer is ill-adjusted and not fitted with a brush (or, better, with an airknife) which removes clusters of foam hanging underneath the boards before they enter the heater. If such a foam cluster falls onto a heating element, especially if it is operated in the near infrared, high temperature mode, the foam is liable to ignite as it hits the heater.

Hence, with foamfluxers, especially when a high-solids flux with good foamstability is used, some means of preventing foam clusters from entering the preheater must be used. With an installation of this kind, especially with a not constantly supervised or processor-controlled installation, it is as well to fit a flameguard, which causes the preheating stage to be flooded with CO_2 and stops the air compressor of the foamfluxer (but not the conveyor). Needless to say, with sprayfluxed boards this contingency cannot arise.

In principle it is not necessary that soldering should follow immediately after fluxing and drying of the flux cover. There may be situations, especially with short or experimental runs of complex or costly boards, where fluxing and drying, maybe in air, is more satisfactory and safer than fluxing which is immediately followed by forced drying and soldering. This procedure allows inspection and maybe corrections or additions to a board before the irrevocable step of soldering it is undertaken. Gently predried boards can be stored safely in a dry environment for periods up to a few days without any deterioration in their soldering behaviour. Such prefluxed and dried boards require, if at all, only a gentle warming before entering the solderwave. It is of course important to test a given flux for its suitability for this procedure before embarking on it.

4.2.1 Types of fluxer

There are three basic types of fluxer to choose from: wavefluxers, foamfluxers and sprayfluxers.

Wavefluxers

Wavefluxers coat the underside of the circuit board with the help of a pumped wave of flux solution, analogous to a solderwave. This method has generally become obsolescent, and for SMD populated boards it is not recommended at all. Wavefluxers are liable to apply too much flux and in uncontrollable amounts, which needs an excessive intensity of preheating and leads to spitting in the solderwave and to problems with postcleaning.

Foamfluxers

With foamfluxers, the circuit board travels across the crest of a wave of foamed flux issuing from an elongated nozzle which straddles the path of the circuit board. The foam is created by forcing air through a cylindrical porous body (foaming stone) which is immersed in the flux, at the bottom end of a vertical elongated chimney which guides the foam upwards towards the foam nozzle (Figure 4.2).

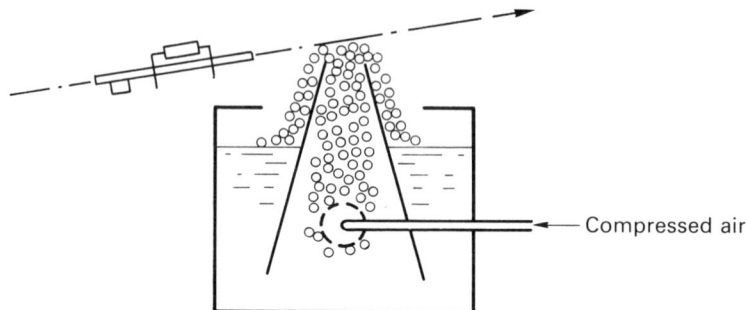

Compressed air

Figure 4.2   *Working principle of a foamfluxer*

Foamfluxers are constructed of either high-density polythene or stainless steel.

The working principle of the foamfluxer is simple and elegant; above all there are no moving parts. Nevertheless, its correct and reliable functioning depends on a few vital parameters, relating both to the flux and the design of the fluxer.

The height and stability of the head of foam depend above all on the foaming behaviour of the flux. Generally, rosin containing fluxes foam better than rosin-free ones, as do fluxes with high solids content compared with low-solids fluxes. For this reason, many low-solids fluxes contain small amounts of foam promoters (Section 3.4.1). The foam itself should preferably consist of fine bubbles rather than large ones, because the smaller the bubbles the larger will be the amount of flux carried in a given volume of foam. This quality of the foam depends not only on the size of the pores in the foamstone, but also on the foaming behaviour of the flux itself.

Apart from foamability, the stability of the foam is another critical property of a flux. The foam should not get coarser through the merging of bubbles as it rises towards the nozzle aperture. On the other hand, once the foam touches the board surface, the bubbles ought to burst quickly so as to form a thin liquid cover on the board.

An excessive stability of the foam, caused perhaps by an unsuitable or excessive foaming additive in the flux, is undesirable: clusters of bubbles hanging from the underside of the board are liable to drop down on to the heaters where they could ignite as the board travels through the preheating stage, as has been mentioned already. To prevent this from happening, foamfluxers should be fitted with a brush or, better, an airknife behind the foam nozzle to burst the bubbles and blow bubble clusters back into the fluxer.

Some design features of the fluxer are of help in getting difficult fluxes to produce a stable, sufficiently high crown of foam (Figure 4.3):

1.   The aperture of the foam nozzle should not be wider than 10–15 mm/0.4–0.6 in and the top edges of the foam chimney should be sharply inclined towards one another.

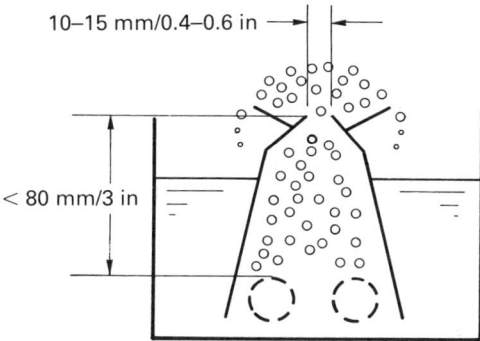

10–15 mm/0.4–0.6 in

< 80 mm/3 in

Figure 4.3 *Critical design features of a foamfluxer*

2. Two foam-supporting strips, about 10–15 mm/0.4–0.6 in wide and fitted to the outside of the foam chimney slightly below its upper edge will help to stabilize the foamwave and raise its height. Attaching a row of vertical bristles to either side of the foam nozzle with the aim of raising the foamwave is an obsolete practice which is not recommended: it is not very effective and it can interfere with loosely inserted wired components.

3. The pores in the foaming stones should be small (approximately 10–20 μm/0.5–1 mil) and of course as numerous as possible. The diameter of the foaming stone is normally 3–4 cm/0.25–0.3 in. With difficult fluxes, two smaller foamstones side by side are better than a single, larger one. Foaming stones are normally made from sintered ceramic or a polymer.

4. The distance between the foamstone and the top of the foaming chimney is a critical dimension. The shorter it is, the less is the danger of the bubbles bursting and the foam collapsing before it reaches the nozzle.

The amount of flux deposited by a foamfluxer depends primarily on the travelling speed of the board, more flux being deposited at lower speeds. However, it cannot be controlled directly by the operating parameters of the fluxer itself. A good head of fine foam will put down sufficient flux, even with low-solids fluxes. Flooding the board with too much flux is undesirable: the board should be fully wetted with flux, but not 'swimming' with it. Otherwise, the upwards tilt of the board on the inclined conveyor will cause the flux to flow towards the trailing edge of the board, which leads to uneven drying and consequently to uneven soldering behaviour along the length of the board.

The absence of moving parts makes the operation of a foamfluxer comparatively simple, but a few points require attention.

In principle, the compressed air required for operating a foamfluxer may be drawn from the works' compressed air supply if one is available, via a reducing valve. Unless that air is strictly free from moisture, oil and dust, efficient filters and a desiccator must be provided. Even traces of oil will cause the foam to collapse. Moisture will be absorbed in the hygroscopic flux solvent. Though small amounts of water in the flux do not affect its soldering behaviour and may even make it foam better, they increase its density and thus distort the relationship between the density and the solids content of the flux. This in turn complicates the monitoring of flux quality (Section 4.2.2). With low-solids fluxes in particular, misinterpretation of the flux density can lead to lowering of the solids content below the danger limit and trigger a steep rise in soldering defects. Pickup of moisture should therefore be avoided. Dust will gradually clog the pores of the foamstone and in consequence reduce the head of foam. This in turn can cause local gaps in the flux cover, with dire consequences for the soldering result.

Usually, therefore, the required compressed air is supplied to a foamfluxer from an external dedicated small compressor, or through an integral airpump. Either must be fitted with both dust filters and a desiccator, the latter specifically when operating in a humid environment.

A foamstone, once it is saturated with flux, must not be allowed to dry unless the absorbed flux has been thoroughly flushed out with a suitable solvent, such as hot, slightly acidified water in the case of rosin-free fluxes, or in the case of rosin-containing fluxes with one of the solvents mentioned in Section 8.3. As a rule it is good policy to keep the foamstone always immersed in the flux. That means that the fluxer is not emptied during normal working. Should this become necessary, for cleaning or servicing, when changing the flux or during longer breaks in production, the foamstone is kept immersed in isopropyl alcohol, in a container with a well-fitting lid. Once a flux-soaked foamstone has been allowed to dry, it may be difficult to clear the pores. Boiling in water to which a dishwashing detergent or sodium metaphosphate has been added can be tried as a last resort, though foamstones made from certain polymers may not survive this treatment.

Contact with a hot surface causes an immediate collapse of the head of foam, which will then take some time to re-establish itself. This circumstance can arise if a soldering carriage or a template which is still hot from its previous passage through the solderwave is sent through the conveyor with a fresh board.

The pressure and rate of flow of the foaming air depend very much on the dimensions and the porosity of the foamstone, and it would be unsafe to give precise data. However, a pressure indicator is an essential requirement with every foamfluxer. The pressure of the foaming air is the principal operating parameter of a foamfluxer, and for this reason it must be monitored and controlled. With processor-controlled soldering lines in particular, loss of pressure in the foaming air should trip an alarm signal and stop the feed of further boards to the fluxer. Obviously, the conveyor must not stop while boards are still travelling through the preheater and across the solder wave.

Some foamfluxers are fitted with sensors which turn on the foaming air when a board approaches and switch it off after it has passed, but this facility must be used with caution: some low-solids fluxes, and also some fluxers, may need quite some time to build up a sufficient head of foam after the air is turned on.

Sprayfluxers

Since the introduction of SMDs, sprayfluxing has gained in popularity, because flux droplets can be propelled into the narrow gaps between closely-spaced SMD components, whereas foam bubbles may burst before they reach the joints at the base of the component housings. With sprayfluxing, the foaming behaviour of the flux is irrelevant, an important point in view of the present tendency towards ever-lower solids contents and increasing foaming problems. With modern sprayfluxers, the amount of flux deposited per unit of board area can be controlled.

Finally, and this is its main advantage, a sprayfluxer always delivers virgin flux to the board, since there is no run-back of flux into the flux reservoir, as is the case with foamfluxers. Thus, there is no need for the constant monitoring and adjustment of the flux density, either manually or by an automatic flux monitor, nor the regular check of the water content of the flux, which is so critical with low-solids fluxes. Because the flux is not constantly aerated, as is the case with foamfluxers, the flux does not pick up water from the blowing air. Water contamination not only falsifies the density/solids–content equation, but can also seriously reduce the solubility of the flux residue, making cleaning more difficult. The flux in a sprayfluxer can be used to the last drop. There is no need to discard exhausted or contaminated flux, which would have to be disposed of safely and professionally. Users of sprayfluxers report this as an important financial saving.

With sprayfluxing, the user is confronted with a choice between several systems, each of which have their specific virtues. Several vendors of wavesoldering machines offer more than one type of sprayfluxer, which may be supplied integral with the machine or as separate units for retrofitting to an existing soldering line.

A few constructional and operational features are common to all sprayfluxers: almost all of them operate in an enclosure, which is fitted with an exhaust. Without this precaution, an aerosol of flux droplets is liable to spread in the neighbourhood of the fluxer and the soldering line, with obvious and undesirable consequences. Also, measures must be taken to avoid overspray beyond the width and length of the boards, as they travel across the spray zone. This involves sensor-operated controls for starting and cutting off the spray action to coincide with the arrival of the leading edge of a board and the departure of its trailing edge, and provision for suiting the width of spraycover to the width of the board.

Common to all sprayfluxers is the need for an efficient exhaust system, which gathers overspray and the aerosol of small flux droplets which unavoidably forms wherever flux is atomized. This airborne flux must not be allowed to settle on the top surface of boards or on the soldering equipment at large, where it could obviously cause a great deal of trouble. By now, most makers of sprayfluxers provide their machines with exhaust systems of sufficient size and suction power to prevent the escape of airborne flux from the unit. It is important that such exhaust systems are fitted with efficient filters which retain the flux droplets, and that these filters are easy to reach and to service. In particular, when rosin-based fluxes are used, regular servicing and cleaning of the whole exhaust system is of the essence: suction fans and conduits dripping with rosin constitute a serious and well-known fire hazard.

Compressed-air sprayfluxers

With this type of fluxer, the flux solution is fed to the spraynozzles by a metering pump, so that the rate of flux delivery can be controlled (Figure 4.4). The ability to control the rate of flux delivery and the thickness of the flux cover on the board has become very important with the growing use of fluxes with low solids contents and the existence of water-based fluxes (Sections 3.4.5 and 3.4.6).

Obviously, at a given rate of flux delivery, the amount of flux deposited per unit of board area is inversely proportional to the travelling speed of the board. The spraynozzles are arranged in a row which spans the width of the solderwave. The width of the spraypattern can be adjusted to match the width of the boards passing through a line by cutting off the air and flux supply from the appropriate nozzles at either end of the row. Before the advent of low-solids fluxes, a problem with compressed-air sprayfluxers arose from the need for time-consuming cleaning of nozzles, which tended to clog during rest periods because of the drying of resinous fluxes. Since the introduction of low-solids fluxes, compressed-air sprayfluxers have become more popular again.

Recently, a sprayfluxer which operates with a single, movable nozzle has become available (Blundell). The nozzle, which is mounted on a slide, traverses the width of the board forward and backwards, so creating an overlapping zig-zag pattern on the travelling board which ensures full coverage. The width of the scan by the nozzle, and the onset and cut-off of the spray action, are controlled by sensors. Flux is fed to the nozzle by a metering pump at a pressure sufficient to ensure atomization without the use of compressed air.

Rotating-drum sprayfluxer

This is a simple and very effective method of projecting a spray of fine flux droplets

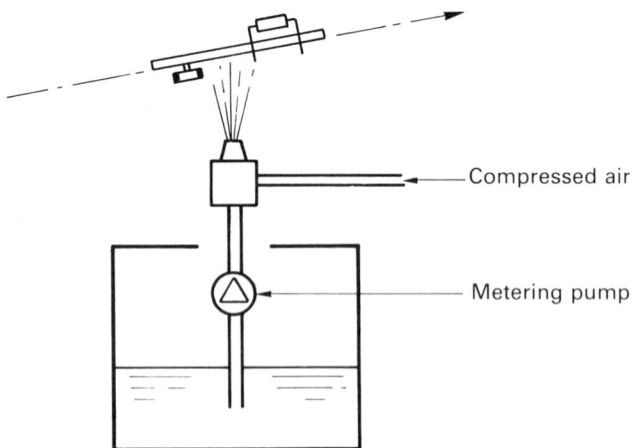

Figure 4.4   *Working principle of a compressed-air sprayfluxer*

against the underside of a circuit board. A cylinder, formed from a woven close-mesh stainless steel sieve, rotates in a shallow bath of flux, with its lower portion immersed in the flux to a depth of 1–2 cm/0.5–1 in (Figure 4.5).

The diameter of the cylinder may be 5–10 cm/2–4 in, and its length corresponds to the width of the solderwave. An airknife, located inside the cylinder near its apex, directs a jet of compressed air upwards against the inside of the cylinder, either through a line of closely-set small holes or a longitudinal narrow slit. This airstream blows the flux from the wiremesh, on which it has been carried out of the fluxbath, against the underside of the circuit board which travels a short distance above the cylinder.

The rate of flux delivery depends on the speed of rotation of the cylinder, which is driven by a variable-speed motor. The onset and the cutoff times of the spray are controlled by sensors so as to coincide with the arrival and departure of a board above the line of spray and to avoid overspray. The width of the sprayed area can be set by blanking off the ends of the airknife by, for example, movable sleeves, so as to match the width of the boards travelling across. The setting of this parameter can also be mechanized and controlled by sensors.

The cylinder need not be removed during rest periods and overnight. The flux on it will of course dry, but rotating the cylinder for about 15 minutes before starting work again will clear it. For longer breaks in production, the cylinder is removed from the fluxer and cleared of flux with an appropriate thinner, which as a rule is supplied by the flux vendor.

Rotating–brush sprinklers

Figure 4.6 explains the working principle: a rotating cylindrical brush, carrying fairly stiff nylon bristles, and of a length corresponding to the width of the solderwave, is arranged at right angles to the travel of the circuit board conveyor. The lower portion of the brush dips in a container of flux. The sense of rotation is contrary to the direction of travel of the board conveyor. Somewhat before the bristles reach the apex of their rotation, they pass the straight edge of a blade, which

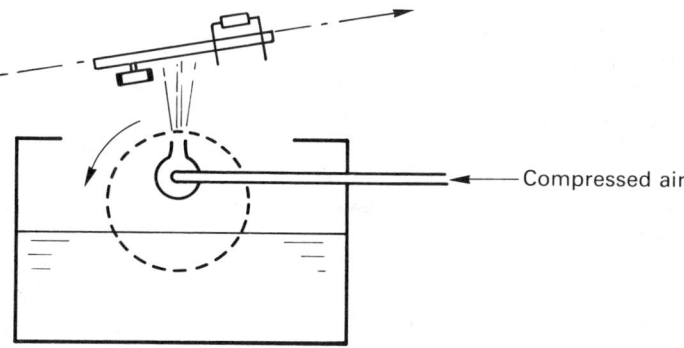

Compressed air

Figure 4.5 *Working principle of a rotating-drum sprayfluxer*

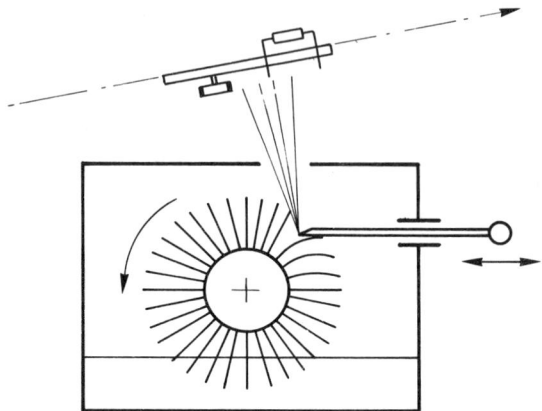

Figure 4.6   *Working principle of a sprinkling fluxer*

can be pushed into the path of the bristles so as to bend them backwards. Having passed the blade, the bristles spring forward and fling the flux they have picked up from the reservoir upwards against the underside of the circuit board which passes overhead.

A sensor-actuated mechanism pushes the blade against the brush when a board arrives above the aperture of the sprinkler and retracts it as soon as the board has passed. The width of the spray is governed by the length of the blade, which is adjustable to match the width of the boards to be fluxed. The amount of flux delivered is governed by the controllable speed of rotation of the brush, while the depth of immersion of the bristles in the flux determines the size of the flux droplets to some extent. It is customary to keep the brush rotating during short breaks in production. During longer breaks, the brush is removed and stored in a container filled with thinners, and provided with a well-fitting lid. Should the bristles harden by being left to dry in air, a brief period of rotation in the fluxer will soften them again.

Sprayfluxers, which propel the flux droplets in a straight path and at some speed against a circuit board, have occasionally met with some objections. Because of their straight line of flight, some droplets may reach the upper surface of the circuit board through apertures such as unoccupied through-holes, vias, or milled slots in boards which are to be broken into separate units after soldering.

Stray flux on the upper side of a board is undesirable. It can cause problems with relays, trimmers, or any other component which is sensitive to physical contamination. Directing the flight of the drops against the board at an angle reduces the problem, but does not entirely eliminate it.

Another difficulty arose with the introduction of wavesoldering in an oxygen-free atmosphere (Section 4.4). Blowing atomizing air into the oxygen-free machine interior runs contrary to the concept, and atomizing with compressed nitrogen is costly.

Ultrasonic spray fluxers

The development of ultrasonically driven fluxing systems was motivated by these problems. With ultrasonic atomization, a metered supply of flux is fed to the vibrating surface of an ultrasonic generator. The vibrational energy is transmitted to the film of flux which forms on that surface and breaks it up into an aerosol of very fine droplets, which form a cloud of aerosol above the generator (Figure 4.7).

With some ultrasonic fluxers, a gentle stream of nitrogen (or air with a conventional wavesoldering machine) wafts that aerosol against the underside of the board as it traverses the sprayzone. With others, the atomizing surface of the ultrasound generator is so shaped as to gather the aerosol cloud and to propel it towards the circuit board.

The fluxing head of some ultrasound systems traverses the width of the board in a zig-zag pattern, as has already been described; with others the shape of the aerosol cloud is given a fanlike shape, so that one or two atomizing heads suffice to straddle the width of a board. Sensor-actuated control of the width and duration of flux application are common to all ultrasound fluxers.

4.2.2 Monitoring and controlling flux quality

The solids content of a flux, given its type and formulation, is its most telling and decisive parameter. With the exception of one reservation which will be discussed presently, there is a direct relationship between the density of a flux and its solids content. Every flux has a characteristic density/solids–content curve, which ought to be given in the datasheet supplied by the vendor.

As a rule, these curves are correct for a temperature of 20 °C/68 °F and, strictly speaking, the flux sample should be warmed or cooled to that temperature before its density is measured. Vendors can save their customers a good deal of time and trouble if they provide flux-density/solids–content curves for a range of test temperatures (Figure 4.8).

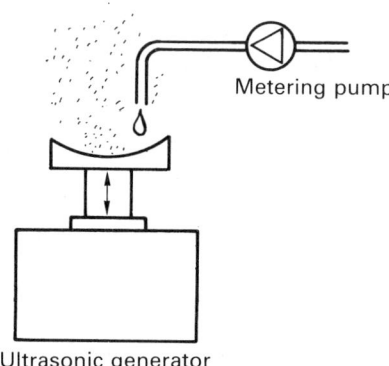

Metering pump

Ultrasonic generator

Figure 4.7 *Working principle of ultrasonic atomization*

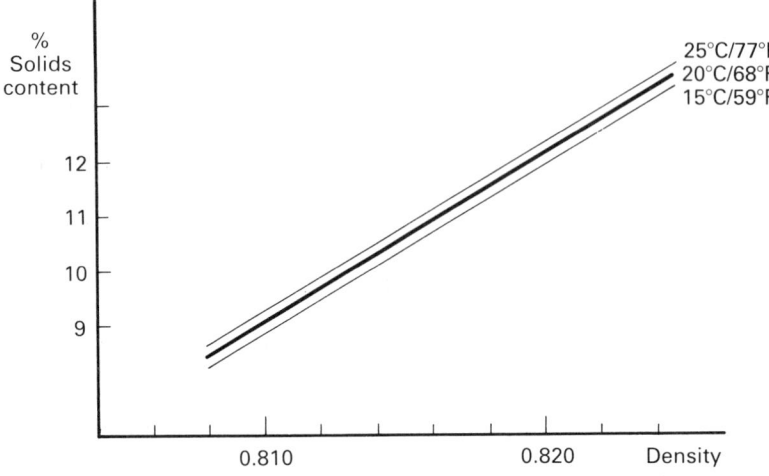

Figure 4.8 *Flux-density/Solids-content curve*

Whether and how often the flux density needs checking depends on the type of fluxer used. Wavefluxers and foamfluxers, where excess flux runs back from the circuit board into the flux reservoir, demand a regular check of the quality and purity of the flux. With these systems flux is constantly exposed to the ambient air, if not actively aerated. Solvent may evaporate, flux constituents may oxidize, moisture may be absorbed, impurities in the form of solids or contamination may be washed off the board surface back into the fluxer.

Flux density is usually checked with a floating aerometer of a suitably chosen range, often supplied by the flux vendor, together with a convenient measuring cylinder. It is important that the scale on the shaft of the instrument should be sufficiently open so that the flux density can be read accurately to the third digit after the decimal point. With the recirculating fluxers described above, the flux density should be checked every day before sending the first boards through the soldering line and, depending on circumstances and the workload, a second time after the lunchbreak.

Thickening of the flux through loss of solvent does not in most cases affect soldering quality, but it increases the amount of flux residue left on a board, which in turn affects appearance and testability on an adaptor bed, and makes higher demands on any subsequent cleaning procedure. Thickening is compensated by adding an appropriate amount of solvent or thinners, usually supplied by the flux vendor. This addition of thinners is often taken care of by an automatic flux-density controller, which can be retrofitted to a machine if necessary. That such density controllers must be temperature compensated goes without saying, because a drop of 1 °C/1.8 °F in temperature raises the density of a flux by approximately 0.0008 g/ml.

In this context, the distortion of the density/solids-content relationship through water picked up by the flux has important consequences. One per cent of water added to a flux based on isopropyl alcohol raises its density by 0.003 g/ml.[4] Adding thinners to a low-solids flux under the mistaken assumption that it has thickened, when in reality it has become heavier through water pickup, can have fatal results: the concentration of active ingredients in these fluxes is delicately balanced at the minimum which will ensure satisfactory soldering. Lowering it by adding thinners is very likely to lead to a rapid rise in soldering defects and bridging. This is exactly what an automatic flux-density control apparatus will do, if it is misled by water contaminated flux into the assumption that the solids content of the flux is too high.

With low-solids fluxes, which are being used on an ever-increasing scale (Section 3.4.5), density is no longer a reliable indicator of their solids content. With such fluxes, even a slight drop below the correct solids concentration is fatal. For all these reasons, flux control systems have been developed, and are increasingly being used, which estimate the solids content of the flux by chemical means, such as by monitoring its pH value or some other chemical parameter which is a measure of the percentage of its active ingredient. Many of these instruments are specific for a given make of flux, and their operating parameters must be adjusted to fit the exact type of flux in the foamfluxer or wavefluxer.

It is worth noting at this point that, apart from its effect on the flux density, the presence of water in a flux has no deleterious effect on its performance. On a warm humid day, the water content of a flux has been known to rise up to 10% with a foamfluxer, without any ill effect on the soldering quality.

Sometimes it is useful to know the water content of the flux if only to correct the result of an aerometer reading. Several flux vendors supply simple titrating kits complete with reagents which make it easy, even for an operator untrained in chemical analysis, to determine the water content of a flux with sufficient accuracy.

It needs no stressing that none of these complications arise with sprayfluxers, which always deliver virgin flux to the circuit board. Nevertheless, this does not relieve the user from the obligation to check a new canister of flux for the identity and density of its contents, before charging it into the fluxer. An error here can ruin a day's production.

Rosin-based fluxes, more so than rosin-free types, are subject to a certain degradation through constant exposure to air, mainly by oxidation, which reduces their fluxing power. The rate at which this happens depends largely on the type of rosin used. A visible effect of this degradation is a progressive darkening of the flux, which gradually changes from the pale yellow of the fresh solution to a dark brown. It may be useful to keep a sample of fresh flux in a small, well-sealed bottle which is stored away from daylight as a reference specimen to check and judge the darkening.

With wavefluxers and foamfluxers, only a fraction of the flux circulating through the fluxer remains on the board. The bulk returns and is constantly recirculated. This means that the underside of all boards passing through the soldering line is constantly being washed by the flux, which thus removes and accumulates all the contaminants such as dust, drilling and cutting swarf, grease or oil, and possibly small pieces of copperwire, etc. which adhere to the board underside. It is therefore advisable to empty a wavefluxer or foamfluxer after about 1500–2000 sq. m/15 000–20 000 sq. ft of board area have passed through it. Some companies operate on a time basis and, depending on the workload and on the nature of the product, they replace the filling of such fluxers after a certain number of weeks of continuous operation. However, even with discontinuous operation, after about two months' use in a wavefluxer or foamfluxer, a wetting or spreading test should be carried out with the flux (Section 3.6.1) to check its performance.

After emptying a fluxer, its interior must be cleaned and accumulated solids removed. The tank of some fluxers is fitted with a removable tray for this purpose. Discarded flux must of course not be dumped, but must be handed to a qualified and registered disposal specialist. Some flux vendors are prepared to take back discarded flux when delivering fresh supplies.

General operating hints

Most modern fluxers are designed in such a way that solvent losses through evaporation are reduced to a minimum. Nevertheless, it is advisable to cover the fluxer with a well-fitting lid during stops in production. Many makers provide such lids as a matter of routine. During longer rest periods and holidays, it is best to empty the contents of the fluxer into a closed container, while cleaning the interior of the fluxer at the same time. Some fluxers are fitted with an integral reservoir, into which the contents of the fluxbath can be drained during such intervals.

4.3 Preheating the board

4.3.1 Heat requirements

A freshly fluxed board cannot be wavesoldered successfully unless its underside has been heated to a temperature above about 80–100 °C/170–210 °F before it enters the solderwave. Many reasons have been put forward for this undisputed fact of life at one time or another, but by now there is general agreement that there are mainly physical, but no chemical, reasons for the need to preheat: the solderwave must raise the temperature of the board together with the joints to the full soldering heat of 250 °C/480 °F within at most a second. It can only do this if it is relieved of the task of boiling off the solvent contained in the flux, and of supplying some of the heat needed for raising the temperature of the board itself, which may be a heavy multilayer laminate, from room temperature to soldering temperature.

Preheating cushions the thermal shock, which would hit the board and the components on it, if it had to confront the solderwave straight from cold. Instead, as the board travels through the preheating stage, its temperature rises at the relatively gentle rate of about 2 °C/4 °F per second to approximately 80–100 °C/175–210 °F. This is particularly important with components which are sensitive to thermal shock, such as ceramic multilayer condensers.

Tables 4.1 and 4.2 give the order of magnitude of the amounts of heat involved in the preheating and the soldering stages of wavesoldering.

The data summarized in these tables underline the importance and quantify the function of the preheating stage: they show that the heat needed to boil off the flux solvent represents a considerable portion of the total heat demand, and that preheating reduces the heat demanded from the solderwave during its few seconds of contact with the board by almost one-third. Without an efficient preheating stage, conveyor speeds of up to 4 m/12 ft per minute would not be possible, nor could the molten solder be persuaded to rise through the plated holes in a multilayer board to form a meniscus on its top surface during the short time available for it.

Should an excess amount of flux solvent be left on a board through insufficient preheat, a vapour blanket is liable to form between the board and the solderwave. This not only slows down the heat transfer between the molten solder and the board, but it can also cause the solder to spit and thus provide one of the causes of small globules adhering to the underside of a board (for others, see Section 4.6.1).

Finally, the mobility of an insufficiently predried flux coating render it more liable to be washed completely off the board by the solder, so that there is not enough left on the exit side of the wave. Some presence of flux is, however, needed there to ensure the mobility and high surface tension of the solder in the region of the 'peelback' which prevents bridging and solder adhesions (see Sections 3.4.1 and 4.3.2). This aspect is particularly important with many low-solids fluxes, especially the rosin-free ones, which demand a sharper preheat than conventional high-solid rosin-based fluxes. By contrast, with the latter, too fierce a preheat is liable to bake the flux cover into a hard, partially polymerized lacquer which makes postcleaning more difficult, if not impossible (see Section 8.1.2).

Table 4.1 *Thermal properties of substances involved in wavesoldering*

	Air	*Water*	*Iso-propyl alcohol*	*FR4*	*Copper*	*Solder 60% Sn*
Specific heat						
cal/g/°C	0.24	1.0	0.57	0.35	0.092	0.042
W. sec/g/°C	1.00	4.19	2.39	1.47	0.385	0.186
Heat of fusion						
W. sec/g	—	—	—	333	205	46
Heat of boiling						
W. sec/g	—	2254	681	—	—	—
Thermal conductivity						
W/cm/sec × 10^{-3}	0.0058	5.4	1.7	0.3	390	52

Table 4.2 *Thermal audit of wavesoldering*

The data below are calculated for a standard 'Europa board' (160 mm/6.3 in wide × 233 mm/9.2 in long, surface area 373 sq. cm/58.0 sq. in). The board is assumed to be 1.2 mm/57 mil thick FR4 and to have been given a 0.1 mm/4 mil thick coating of rosin flux in isopropyl alcohol as solvent.

Volume of the board laminate	44.7 ml
Weight of the board laminate	80 g
Volume of the flux cover	3.7 ml
Weight of the flux cover	3.2 g

Thermal input during preheating the board from 20 °C/68 °F to 100 °C/212 °F:

Heating the laminate	9.4 kW sec
Heating the fluxcover to its boiling temperature	0.6 kW sec
Evaporating the flux solvent	2.2 kW sec*
Total	12.2 kW sec = 27% of total

Thermal input from the solderwave (solder temperature 250 °C/482 °F):

Heating the board from 100 °C/212 °F to 250 °C/482 °F	17.5 kW sec = 73% of total
Total heat demand	29.7 kW sec

The heat demand from the circuit tracks, the leadwires and the SMDs has been neglected in this calculation because of their comparatively low specific heat.

*The heat of evaporation of water is 3.3 times that of isopropyl alcohol. Should a flux have absorbed 10% of water, e.g. in a foamfluxer, 2.7 kW sec would be needed to dry the flux cover instead of 2.2 kW sec, a negligible difference in the context of the total heat requirement.

4.3.2 Heat emitters and their characteristics

In practice, preheating is effected by passing the fluxed boards over a bank of infrared heaters, at a distance of approximately 5 cm/2 in. These heaters are backed by a heat-reflecting metal panel, which ought to be easy to withdraw for the periodical removal of flux drippings.

It has become customary to direct a gentle stream of warm air through the space between the heaters and the boards travelling above them (Figure 4.9). There are several reasons for this. By removing the solvent-laden air from this space, drying is accelerated. Most importantly, this venting prevents the build-up of potentially explosive solvent/air mixtures. Naturally, with a wavesoldering machine operating in a nitrogen atmosphere (Section 4.6) this precaution is neither possible nor necessary.

With most preheaters, a reflector, made from polished aluminium or stainless steel, is fitted above the board conveyor. This not only conserves thermal energy, but reduces the temperature difference between the underside and the top side of the board. It reduces warping and helps the solder to rise to the top of through-plated holes. With very heavy or multilayer boards, especially if they carry massive internal copper layers, a top reflector is essential. In some cases, a few

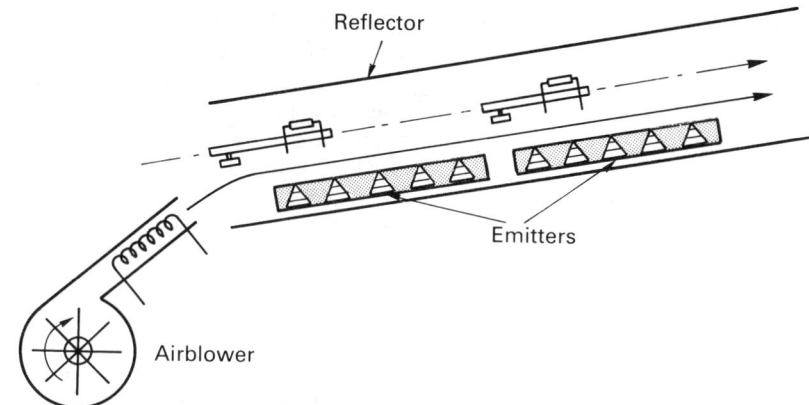

Figure 4.9 *The preheating section*

infrared heaters mounted above the board conveyor may be necessary to provide the necessary topheat to get the solder to rise to the surface of the board. This measure is preferable and kinder to the board and its components than raising the solder temperature or slowing down the conveyor.

The heating elements of the majority of wavesoldering machines are internally heated metallic or ceramic infrared emitters, fitted with a heat-reflecting backing. Most machines carry heat-sensors, which permit thermostatic control and the display of their temperature on the control panel of the machine.

As a rule, the heaters operate in the temperature range between 300 °C/570 °F and 500 °C/770 °F, and thus in the middle and far infrared range of the spectrum. At these wavelengths, the radiated energy is readily absorbed by both the flux and the epoxy laminate, which ensures an efficient heat transfer. The thermal energy given off by the surface of an emitter rises from 0.6 W per sq. cm/3.75 W per sq. in at an emitter temperature of 300 °C/570 °F to 2.0 W per sq. cm/12.5 W per sq. in at 500 °C/930 °F. The details of the physical laws which govern infrared heating are covered more fully in Section 5.4.2.

With some makes of machine, tubular resistance heaters are installed, either in a zig-zag pattern which straddles the maximum board width, or in straight lines at right angles or parallel to the direction of board travel. In the latter case, the width of the irradiated area may be adjusted to match the width of the boards being soldered.

Whatever the arrangement of the heaters, it is important that all parts of a board receive the same dose of thermal energy, because uneven preheating is a dangerous source of soldering faults. Most modern soldering lines give a warning if a heater in the preheating section should fail; some prevent further boards from entering the line in case of a heater failure. The boards still in the line must of course continue to travel forward, if they are not to be fried to a crisp or get stuck over the solderwave.

Internally heated infrared emitters have necessarily a high thermal mass, and their response to changes in the heating current is correspondingly slow. Depending on its design, an element of this type may require up to 15 minutes to reach its full operating temperature after being switched on. For this reason, some processor-controlled soldering lines are fitted with high-temperature tubular quartz heaters in their preheating section. These heaters consist of a spiral of tungsten wire located inside an evacuated quartz tube. Usually, these tubes are arranged in groups, at right angles to the direction of travel of the boards (Section 5.4.4).

They operate at temperatures between 800 °C/1500 °F and 1100 °C/2000 °F, and their emitted thermal radiation is in the near-infrared part of the spectrum. They reach their full operating temperature within less than a second after being switched on, and they respond very quickly to changes in the operating current. Depending on the type of heater and its temperature, the energy emitted lies in the range 15–50 W per cm/38–125 W per in length of tube. Because of this high energy density and their fast response to changes in heating current, quartz–tube heaters operate in short bursts, which must be accurately and reliably controlled so as not to bake the flux into a hard coating, or even burn the boards and damage expensive SMDs.

4.3.3 Temperature control

When conventional rosin-based fluxes used to have solids contents of upwards of 10%, it was customary to aim at a heating regime which raised the underside temperature of the boards to between 80 °C/180 °F and 90 °C/195 °F. For modern, low-solids fluxes, which may contain only small amounts of rosin, or none at all, flux vendors recommend higher underside temperatures, of up to 110 °C/230 °F. The aim is to consolidate the thin flux coating sufficiently to ensure that enough of it survives underneath the board after it has passed through one or two solderwave crests. Otherwise, bridging or the appearance of a thin 'spider's web' of solder, adhering to the surface of the soldermask, can become a real danger.

A given setting of the heating power in a preheating line is of course only valid for the conveyor speed at which it was established: slowing down the conveyor means that the boards get too hot; speeding it up leaves them too cold. Balancing and optimizing such operational parameters is fully dealt with in Section 4.7.

With the old rosin fluxes, an exit temperature of 80 °C/180 °F could be ascertained conveniently by lightly touching the underside of the emerging board with a fingertip: if the flux coating felt like honey, its temperature was just right; if it felt mobile, it was not hot enough; if the finger stuck to the board, it was too hot. Checking an underside temperature of 110 °C/230 °F in this manner could be painful, and more sophisticated methods are needed.

Temperature indicators in the form of self-adhesive labels are very convenient and are often available from the vendors of fluxes or soldering accessories. They record the exit temperature through an irreversible and distinct colour change, from white to brown or black, or from one colour to another. Sets of labels with a convenient range of colour-change temperatures are on the market. A board of the same size and thickness as the production boards, but without components, is used for a trial run through fluxer and heater. The temperature indicators are stuck to various strategic locations on the board. Having been read after the run, they are removed, and one board can be used many times over.

On many processor-controlled wavesoldering lines the temperature of the board underside is scanned and monitored by remote sensing, which is linked to the machine control. As has already been said, low-temperature heaters respond only slowly to current adjustments, and this must be considered in the control software. High-temperature quartz heaters are more suitable for this technique.

Compact, self-contained temperature logging equipment has been available for some time. These systems employ a temperature sensing and recording unit, which is housed in a heat-insulated casing. It can ride along with a sample board through the length of the wavesoldering machine, sampling and storing the output of a number of thermocouples, normally six. These are glued to strategic positions on the circuit board with a thermally conductive adhesive. The logged data can be transferred from the logging unit to a PC and stored, displayed or printed out, thus providing a complete temperature/time profile of not only the preheating stage, but of the whole soldering process. A number of such logging systems are commercially available, mostly from the makers of soldering machines or from flux vendors. Obviously, the same equipment can be used for establishing and recording the temperature profile of any type of reflowsoldering installation as well (Sections 5.3, 5.4 and 5.5).

4.4 The solderwave

4.4.1 Construction of the soldering unit

Solderwaves are produced by forcing molten solder upwards through a vertical conduit which ends in the so-called wavenozzle. Figure 4.10 shows the general principle of a widely used type of wavesoldering machine. Originally, the wavenozzle had the form of a narrow slot, arranged at right angles to the travelling

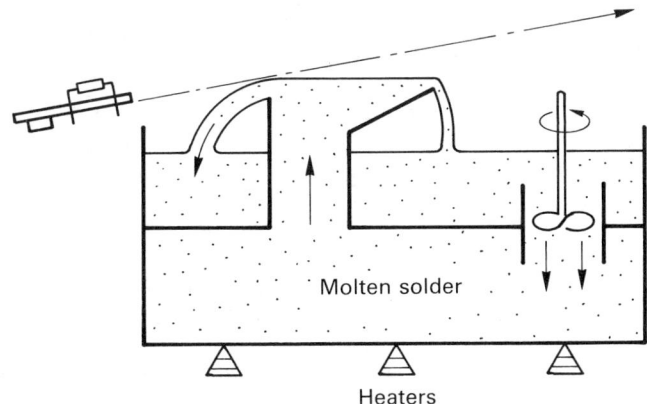

Figure 4.10 *Working principle of a wavesoldering machine*

direction of the board, with the emerging solder forming a hump of molten metal and falling in a symmetrical wave over both sides back into the main container. The symmetrical wave was soon replaced by the asymmetrical wave shown in the drawing, which gives tidier joints, reduces bridging and permits higher soldering speeds.

With most types of wavemachines, an axial impeller pump, driven by a variable speed motor, propels the solder downward into a pressure chamber, from which it flows through a vertical conduit upwards towards the wavenozzle. This arrangement keeps the movement of the solder towards and over the weirs at both sides of the nozzle as free from turbulence as possible. Before the advent of SMDs, this waveform, with the board skimming on an upwards inclined path over the crest of the overflow, was the best way to achieve clean, bridge-free soldering at conveyor speeds of up to and over 2 m/6 ft per minute. Waveforms and the way in which they had to be adapted to the demands of SMD soldering are discussed in Section 4.4.3.

The capacity of the solder tank may vary from 20–40 kg/40–80 lb with small benchtop machines to 200 kg/400 lb and over for high-capacity soldering lines for large circuit boards. In most cases the solder is heated by external heaters clamped to the sides of the solder tank. Many wavesoldering machines are constructed from stainless steel sheet. Mild steel, suitably protected against the action of the molten solder, is equally satisfactory.

The solderwave itself must fulfil two tasks. The first one is to get the already preheated board hot enough to permit the solder to fill every joint completely. The second one is to enable the solder to reach and to completely fill every joint on the board, and afterwards to let it drain away from all the places where it must not remain.

4.4.2 Thermal role of the solderwave

The reservoir of molten solder which supplies the solderwave is normally maintained at a temperature of 250 ± 5 °C (480 ± 10 °F). By general consent, this has been accepted worldwide as the most convenient temperature which meets almost all normal requirements of wavesoldering. Particularly temperature-sensitive components or substrates may need a lower soldering temperature, and this in turn demands special solders with lower melting points than the normal tin/lead solders. Since the efficiency of a flux is temperature dependent, special fluxes are required for low-temperature wavesoldering. A detailed discussion of these techniques, which lie outside the scope of this book, has been published elsewhere.[5]

Temperatures above 250 °C/480 °F are hardly ever used. Particularly heavy demands of soldering heat can almost always be met by slowing down the conveyor or intensifying the preheat.

Molten solder is by far the best heat transfer medium in the soldering business: it provides conductive transfer of heat by metal-to-metal contact, with perfect conformity between the heating and the heated surface. Copper may have a higher heat capacity and thermal conductivity than solder (Table 4.1), but molten solder adapts to the shape of every surface it encounters. Both the ancient soldering iron and some of the latest techniques of desoldering and resoldering SMDs make use of this convenient fact, by always working with a soldering bit well covered with molten solder.

Heating by metallic contact represents an equilibrium situation where the temperature of the heat source is the target temperature for the heat sink, with the temperature difference between the two quickly dropping to zero. This means that the duration of heating is not the critical factor it is with non-equilibrium systems such as infrared or laser soldering.

The mechanism of heat transfer between the solderwave and the circuit board relies on a mixture of conduction and convection. The measured rate of heat transfer between the crest of a normal non-turbulent solderwave and the copper laminate bonded to an FR4 board has been determined at approx. 2 W/sec. mm^2 (Pascoe, G. and Strauss, R., 1958 unpublished). With a turbulent wave the rate of heat transfer is somewhat higher. On this basis, the quantity of heat available to a normal solderpad and its inserted wire during its two seconds' contact with the solderwave adds up to about 15 W sec. The actual heat required to raise the temperature of the copper lining of the hole together with the footprints at both ends and the inserted wire amount to no more than 2 W sec. In this calculation, the thermal needs of the FR4, which surrounds the hole, and of the soldermask which covers the board and the conductor tracks can be neglected: though the specific heat of FR4 is four times higher than that of copper, its thermal conductivity is lower by three orders of magnitude.

These figures show that, thermally, the solderwave can cope with all likely heat demands. With SMDs, these are in any case lower than with wired components. The mass of the metallized surfaces on passive components, and of the legs of SOs, PLCCs or QFPs is measured in mg, whereas the surfaces available for contacting the solderwave are in the sq. mm order of magnitude. Thus, provided full contact between the molten solder and all solderable surfaces can be achieved, implying that the solder wets all these surfaces, the heatflow available is more than adequate to get them all hot enough within the time available: in short, the thermodynamics of wavesoldering SMDs are no problem.

The difficulties of wavesoldering SMD-populated boards are of a different nature: they arise firstly from the problem of physically getting the solder to every single joint, and secondly from ensuring that it does not stay behind in the wrong places after the board has emerged from the wave. In other words, it should not leave joints empty or form bridges, solder prills or 'spider's webs' adhering to the board.

4.4.3 Interaction between molten solder and the circuit board

Entry of the circuit board into the wave and the 'shadow effect'

Whether or not the molten solder reaches every single joint on the board is decided at its point of entry into the wave. SMD-populated boards are three-dimensional (see Section 4.1), with many SMD joints located in recesses at the sides of the components where the surface tension of the molten solder prevents it from reaching them. The air and flux vapours which are trapped between component and solder increase its reluctance to enter such corners. This phenomenon has been named 'shadow effect' (Figure 4.11).

For instance, the outer ends of the component leads of SOs, e.g. SOT 23 or SOT 143, do not extend beyond an aspect angle of 60° below the top edge of the component body (Section 2.1). In a laminar, non-turbulent wave, the surface tension of the molten solder gives it a circular contour, which cannot reach the entry to the joint. The problem is even worse with PLCCs, where the ends are bent inwards and tucked away under the body of the component. The problem is made worse by the air and the flux vapours which are trapped between the SMD and the advancing solder, without means of escape. The exact way in which the surface tension acts in the case of SMDs and prevents the solder from getting where it is needed has been quantitatively treated by Klein Wassink.[6]

With component leadwires inserted in through-plated holes these problems did not exist: the projecting wire end, once wetted by the solder, pulls it into the hole,

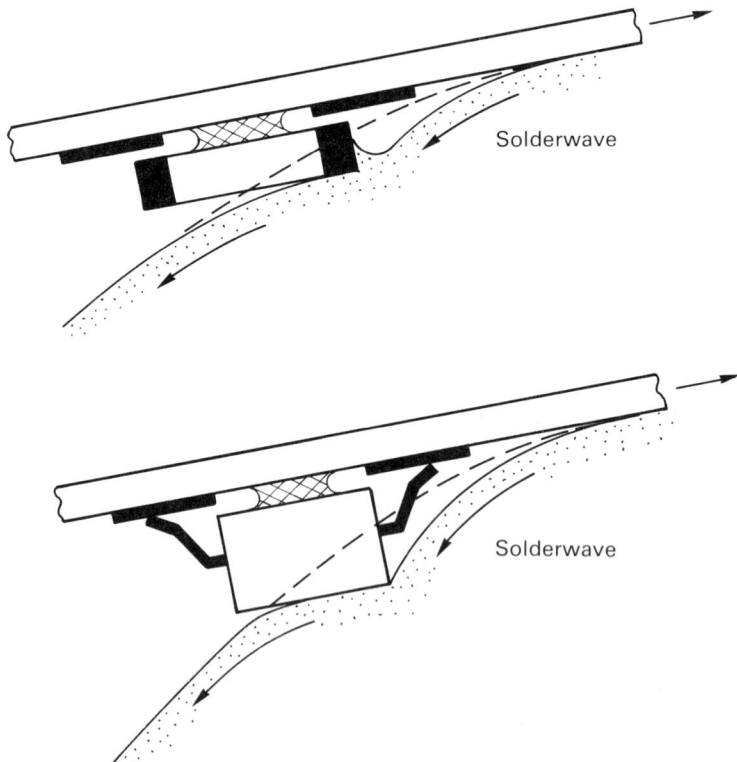

Figure 4.11   *The shadow effect*

and the flux floating on the advancing meniscus helps it to rise to the top of the board (Figure 4.12).

In order to propel the solder into shadow areas of SMDs, it must be given a mechanical impulse of sufficient momentum and in the right direction. A number of waves of specific configurations and flow patterns, known as chipwaves, have been developed over the years to achieve this (Section 4.3.3). Adapting the layout of the board to the idiosyncrasies of wavesoldering also helps, as will be dealt with in Section 6.4.1.

Exit of the circuit board from the wave and the peelback

As would be expected, the circumstances on the exit side of the solderwave determine whether any solder stays behind on a board in the wrong places, mainly in the form of bridges and, if so, where this is likely to happen.

Along the line where the board parts from the wave, both surface tension and the cohesive forces within the body of the molten solder play a role. The solder which has wetted the joints and the metallic surfaces of uncoated tracks and the backs of component legs is carried along with the travelling board until its weight overcomes the cohesive forces within the melt and the surface tension which hold it together. This volume of solder which forms in the nip between the back of the wave and the departing board is called 'peelback' (Figure 4.13).

Soon after the introduction of wavesoldering it was found that peelback, and its tendency to cause bridging and excessive solder pickup, can be minimized by letting the board travel slightly uphill. Over the years, a rising angle of $7 \pm 1°$ has become established worldwide as a useful standard, and it is neither necessary nor advisable to depart from it except under special circumstances. The $7°$ angle is equally suitable for SMDs.

If bridging becomes a problem with close-pitch SMDs which are near the limit of wavesolderability, e.g. <0.75 mm/30 mil, raising the angle of $8°$ or a little above and slowing down the conveyor at the same time might help. With some computer-controlled soldering lines, the optimal conveyor angle, once experimentally determined, can be stored in the control processor and recalled for a given type of board, e.g. by a barcode carried by the board. All these measures aim to

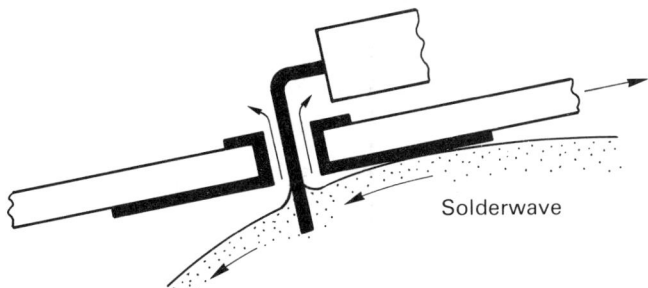

Figure 4.12 *Wired joint entering the solderwave*

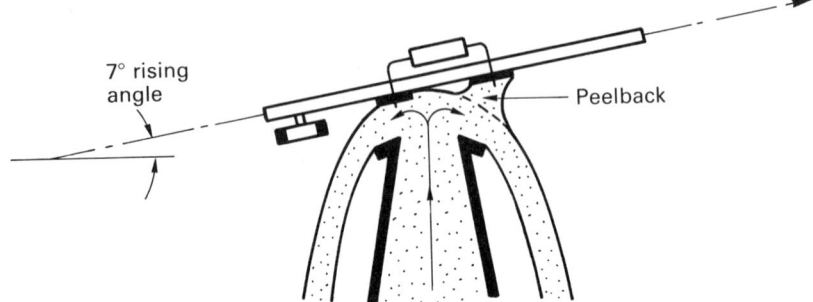

Figure 4.13   *Peelback*

make it as easy as possible for the solder to find its way back to the wave.

The flowpattern of the solder in the wave is an important factor in clean and efficient joint formation. Before SMDs appeared, the so-called 'asymmetrical' wave (Figure 4.14) was the most widely used type of wave. On entering the wave, the board encounters the solder as it falls in a smooth, laminar, non-turbulent stream over a weir, moving in the opposite direction to the travelling board.

Where the board leaves the wave a few centimetres further on, the solder forms a flat horizontal surface, which flows, again without any turbulence, in the same direction and ideally at the same speed as the board. This is achieved by placing an adjustable weir at the exit side of the wave. In this way, every joint on the board is lifted vertically out of a level pool of solder, the surface of which moves in the same direction and at the same speed as the board. This arrangement minimizes the pick-up of solder by the board.

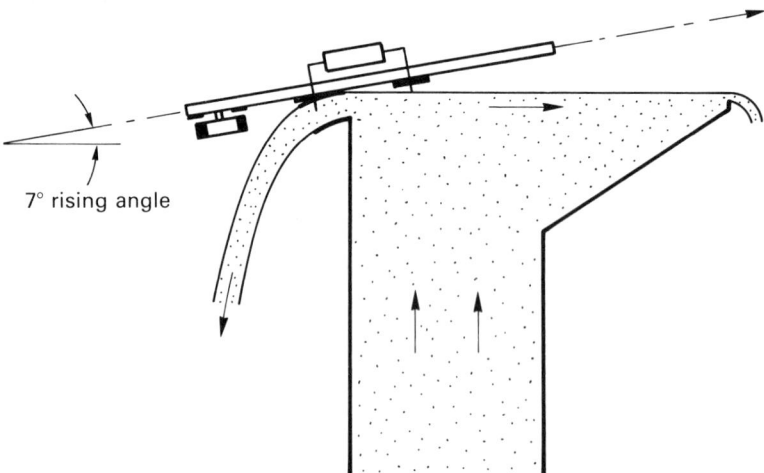

Figure 4.14   *The movements of board and solder in an asymmetrical solderwave*

The horizontal flow of the solder surface on the exit side has a further purpose: in the absence of flux, a static surface of molten solder is soon covered by a thin but tough skin of oxide, which can cause bridges and icicles. Though such flux as has survived the passage of the board on its underside can deal with most of the oxide skin, the slow constant movement of the solder surface prevents the oxide from forming a stationary skin.

With wired, inserted components, where the distance between leads is normally 2.54 mm/0.1 in, bridging is no problem with a well-adjusted asymmetrical wave, even at conveyor speeds of 2 m/6 ft or more per minute. With SMDs and closely spaced solderpads, these considerations are irrelevant and the cohesive force of the molten solder in the peelback becomes the dominant factor. When the gaps between neighbouring pads are narrower than the pads themselves, the peelback will span a number of pads before the weight of the molten solder carried along with the board overcomes its internal cohesive force and lets most of the solder fall back into the wave (Figure 4.15).

Several effects come into play under these circumstances. With a row of pads aligned parallel to the direction of travel, the peelback jumps from pad to pad until the last two are reached. There is no further pad to jump to and the peelback is reluctant to drain from the last two pads in the row, so a bridge is likely to form between them. This bridging can be prevented by lengthening the last pad in the

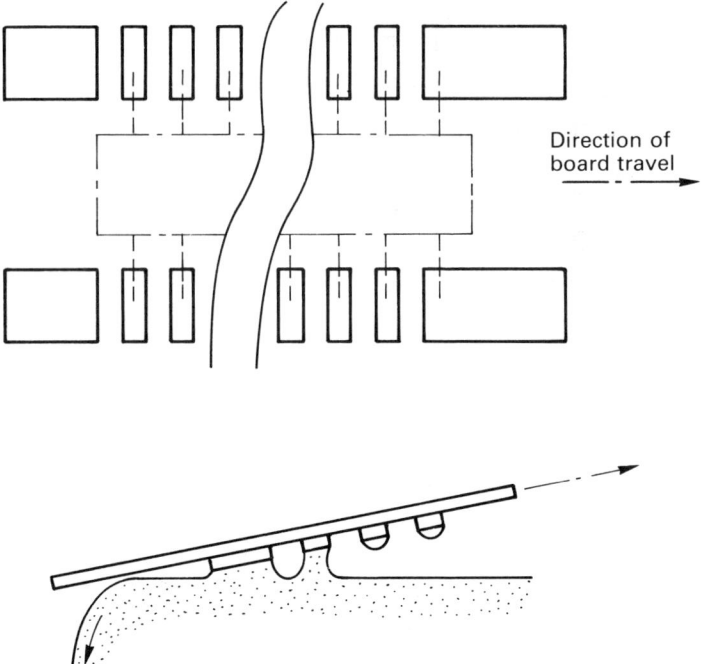

Direction of board travel

Figure 4.15 *Peelback and solderthieves*

row so that the suspended solder drains towards the trailing end of the pad, or by adding a blind or dummy pad, called a solderthief, to the end of the row. The thief serves no connecting purpose and renders the bridge, should it form, harmless.

If a row of pads lies at a right angle to the direction of travel, the peelback reaches all pads in the row at the same time, and has nowhere to jump to. Therefore, bridging is difficult to avoid between parallel pads which leave the solderwave simultaneously (Figure 4.16).

This idiosyncrasy of the solderwave is the reason for the layout rules for SMDs, which are fully dealt with in Section 6.4.1. Placing a multilead component diagonally to the direction of travel of the board is a widely and successfully practised compromise. It is, however, wasteful of space, a serious consideration with the steadily-rising cost of board real estate. For this reason, one machine maker offers a soldering machine with the solderwave placed at a 45° angle to the direction of travel (Figure 4.17).

As the peelback releases its grip on a land or pad, the suspended thread of solder may, as it parts, form one or more small globular droplets of solder. Normally, these drops fall back into the exit pool of the wave, but when soldering is carried out in an oxygen-free atmosphere and with a low-solids flux, they can end up sticking to the solder resist on the underside of the board. This problem is dealt with more fully in Section 4.5.

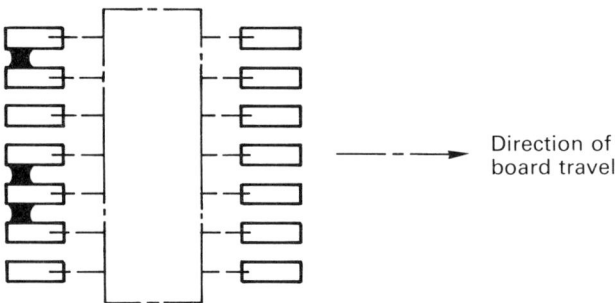

Figure 4.16 *Bridging in a row of footprints aligned at right angles to the direction of travel*

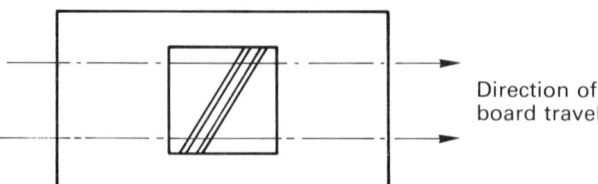

Figure 4.17 *Slanting solderwave*

4.4.4 Chipwaves

Double waves

As the preceding chapter explains, the way in which a board enters the wave decides whether the solder gets to all the places where it is needed. The manner in which the board leaves the wave on the exit side determines whether any solder remains in places where it is not wanted. The concept of the double chipwave is the logical embodiment of this truth: the first or primary wave makes sure that the solder finds its way to every joint on the board; the second or secondary wave, which follows closely after the first one, allows the solder to drain away from the board without leaving any bridges or other unwanted accumulations behind.

The primary wave achieves its purpose by providing the solder with a good deal of kinetic energy at the point where it meets the board. Having passed through this wave, the board is somewhat untidy, with some bridging and unsightly joints. These blemishes are tidied up in the second wave, which as a rule is a standard asymmetrical wave as has been described in the preceding section. Its smooth exit conditions iron out the imperfections left by the primary wave. Both waves follow one another as closely as the construction of the soldering unit will allow.

Two types of primary wave have by now become established. One is a 'symmetrical' wave, with an intentionally turbulent wavecrest (Figure 4.18). Its aim is to provide a zone of high kinetic energy with a vigorous, multidirectional flow of solder underneath the board, which propels it into every recess and at the same time flushes out trapped air and solvent vapours. This turbulence is often created by narrowing the nozzle exit so as to accelerate the solder as it is propelled upwards against the board. Sometimes a rotating or pulsating mechanical element is located close to the nozzle exit.

The alternative is the so-called jet or hollow wave. Here, the solder is projected obliquely upwards against the underside of the board. The path of the circuit board intersects the parabolic trajectory of molten solder somewhat below its crest. The momentum with which the solder hits the board and the SMDs which sit on it propels it into the recesses where the joints are located. There are two possible kinds of jetwave: with the unidirectional wave, the solderjet moves in the same direction as the board; with the counterflow wave, it moves in the opposite direction. It has

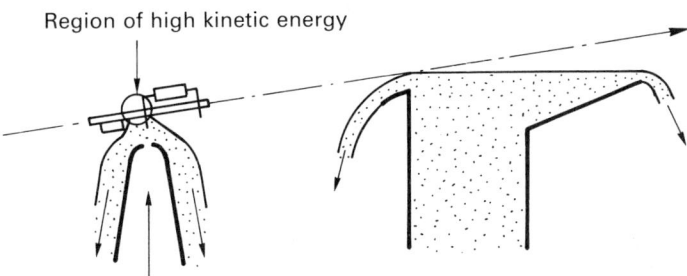

Region of high kinetic energy

Figure 4.18   *Turbulent primary wave*

become customary to use the unidirectional wave with all double-wave machines (Figure 4.19) and the counterflow wave on single-jet machines.

With the unidirectional wave, where the leading edge of the board intercepts the jet as it travels forward and moves in the same direction as the board, there is no danger of the solder flooding the top surface of the board at the point of entry. It can be advisable, however, to clip a low baffle to the trailing edge of the board to prevent the solder from washing over it as it leaves the wave.

Many double-wave machines have two pumps, one for each wave, but both drawing on a common solder reservoir. A jetwave requires a considerable hydrostatic pressure of solder to give the trajectory its required height, which must be maintained with reasonable accuracy. With some makes of chip-soldering machines the primary wave can be switched from turbulent to jet operation without having to fit a new nozzle.

Single counterflow jetwaves

A unidirectional wave has an irregular, uneven peelback zone. This leaves the joints well filled, but with an ill-defined amount of solder, which makes the secondary asymmetrical wave necessary. A counterflow wave has one great advantage: there is no peelback, because on the exit side the solder is constantly driven back towards the joints, at a speed which is higher than the speed of forward travel of the board. Since there is no peelback, there is no danger of bridging or excess solder left under the board, and consequently no need for a secondary wave (Figure 4.20).

However, provision must be made to prevent the solder from flooding the top surface of the board as it cuts into the wave. This can be done by fitting a deflector blade to the leading edge of the carriage which carries the board through the soldering line. The maximum conveyor speeds which can be achieved with a single contraflow jetwave are somewhat lower than the speeds of double-wave machines.

With one well-known make of single-jet contraflow machine (Kirsten, Switzerland), the solder is propelled not by a mechanical impeller pump, but by a

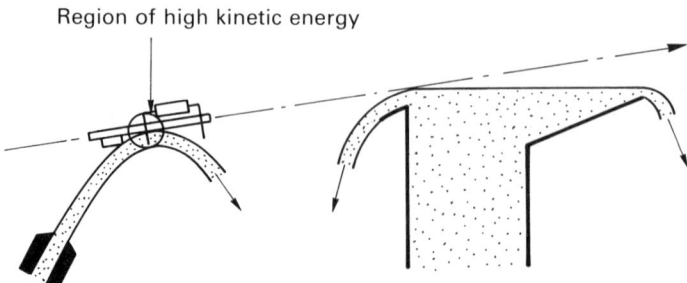

Region of high kinetic energy

Figure 4.19 *Unidirectional jetwave as primary wave*

Region of high kinetic energy

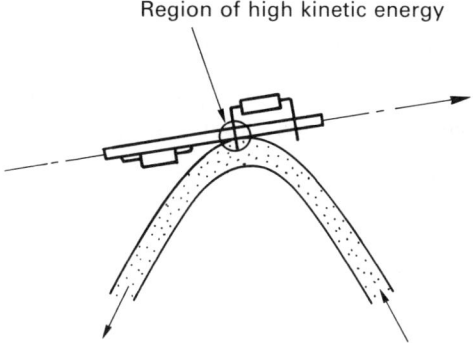

Figure 4.20   *Single-wave soldering with a counterflow jetwave*

so-called Faraday pump, which has no moving parts: a strong magnetic field is applied at right angles to the flow of the solder, while an electric current is flowing through the molten solder in its conduit at right angles to both the solderflow and to the magnetic field. The resultant electrodynamic force propels the solder through the conduit and into the jet nozzle (Figure 4.21).

Combination waves

Combination waves manage to combine the functions of both the primary and secondary wave into a single asymmetrical one. This is achieved by creating and confining a zone of mechanically-generated multidirectional kinetic energy

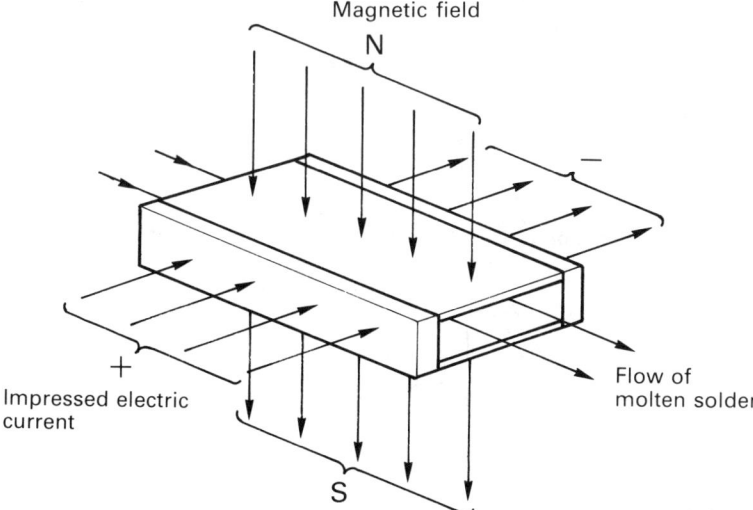

Figure 4.21   *Electrodynamic solder pump (Kirsten)*

within the normally smooth, laminar overflow at the point where the board enters the wave. The exit side of the wave follows the normal asymmetrical pattern with a smoothly moving surface.

One example is the 'Omega Wave'© (Electrovert). Here, the energy field in the entry zone is produced by a narrow vertical blade which is placed close below the surface of the overflow and extends over the full width of the wave (Figure

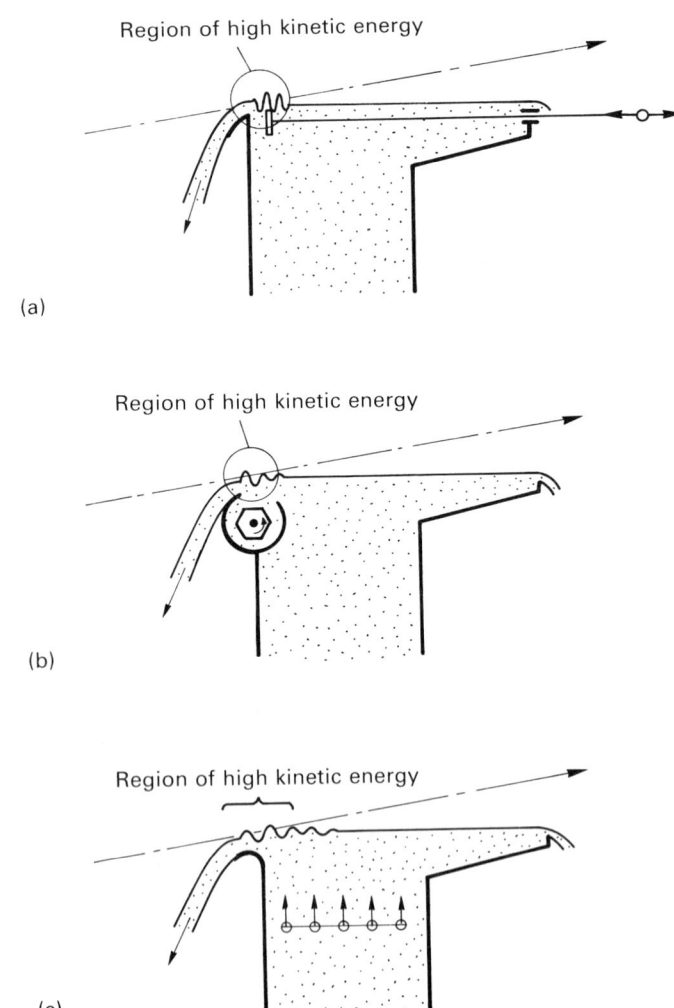

(a)

(b)

(c)

Figure 4.22 *Three examples of combination waves. (a) Omega Wave ©; (b) Smart Wave ©; (c) CMS Wave ©*

4.22(a)). This blade oscillates horizontally at 50 Hz/60 Hz at a controllable amplitude, driven by an electromagnetic vibrator. The oscillations create standing waves in the solder overflow with an adjustable vertical vector along the line where the board enters the wave.

Another example is the 'Smart Wave'© (Soltec). Here the solder is made to well up in a pulsating movement along the line of overflow where the board enters the wave. This movement is produced by a horizontal hexagonal bar which extends across the width of the wave and which rotates at an adjustable speed within a cylindrical baffle, just below the surface of the overflow. A slot in the baffle allows the solder to well upwards against the board. The speed of rotation of the hexagon determines the degree of turbulence in the wave overflow (Figure 4.22(b)).

A third example of the combination-wave concept employs no moving elements at all: with the CMS© Wave (Blundell) the solderpump is driven by an electronically-controlled stop–start motor, so that the whole solderstream itself pulsates (Figure 4.22(c)). Both the frequency and the amplitude of the pulses can be varied. Practice shows that the salient parameter is the pulsing rate, which can be suited to the type of SMDs being soldered. As with the other two, the wave itself is of the classical asymmetrical configuration, and with all three types of wave the board lifts off on the exit side from a smoothly moving stream of solder.

4.4.5 *Formation and control of dross*

In a normal atmosphere, molten solder quickly acquires a tough surface film of mixed tin and lead oxides. As soon as the solder is moved or disturbed, the oxide skin ruptures and tangles with the solder underneath. The resultant mix of oxides and clean solder is called dross.

Since wavesoldering involves moving solder around and letting it fall back into a bath of molten solder from a certain height, the formation of dross is unavoidable unless measures, which will be discussed later, are taken to protect it from the atmosphere.

Physically, dross is a grey, heavy metallic sludge which floats on top of the solderbath and sets into solid heavy lumps when it cools below 183 °C/361 °F. Chemically, dross consists of between two to five per cent metal oxide, the rest being clean solder. Normally, dross also contains a certain amount of flux residue, which results from the flux washed off the boards as they pass through the solderwave. With rosin-based fluxes of high solids content, flux residues can form a considerable portion of the dross. With low-solids and/or rosin-free fluxes, the flux content of dross falls to insignificant proportions.

On pump-driven units, a small accumulation of black powdery dross is liable to form where the pumpshaft enters the solderbath. This black powder, which consists largely of lower tin and lead oxides, may spontaneously begin to glow and turn into a grey or yellow powdery mixture of higher oxides. This glowing need not alarm a machine operator; it does not mean that the solder is beginning to burn up and turn completely into dross.

The amount of dross which forms on a wave machine depends on several factors:

1. Even minute amounts of zinc, aluminium and cadmium, the so-called 'skinformers', can lead to a serious increase in drossing. Copper or iron, if present above their safe limits, can also cause excessive drossing. The safe limits of all these impurities are fully dealt with in Section 3.2.3. If at any time, the formation of dross on a wave machine begins to rise above the usual level without an obvious reason, an immediate check analysis of the solderbath is strongly recommended.

   As a general rule, the lower the level of impurities in a given solder, the less is its tendency to form dross in a wave machine. Special grades of high-purity solder for wavesoldering are available from the main solder vendors. Some makes of solder are claimed to have special 'low-dross' characteristics, due to specific alloying and melting procedures.

   A miracle solder which does not dross at all in normal atmosphere is, and will probably remain, an unfulfilled dream. Small amounts of added phosphorus (about 0.002%) will reduce drossing to a minimum for a time, but the additive soon oxidizes and disappears. To maintain the effect, regular additions of phosphorus in the shape of phosphor–tin pellets, which are available from some solder vendors, have to be made. This method is rarely practised in industry, because there is a suspicion, though unconfirmed, that phosphorus may encourage dewetting of solder on copper.

2. Constructional features of the wavenozzle assembly have a pronounced effect on the drossing characteristics of a given soldering machine. If the solder which flows over the weirs of an asymmetrical nozzle or through the apex of a jetwave is allowed fall back into the solder reservoir from its full height, a great deal of turbulence will arise at its point of impact. A large amount of dross is bound to form, and much of it will be dragged below the surface of the solder, together with some air. In a short time, a thick layer of drossy sludge would form on the bath surface, making soldering impossible. For this reason, the free falling solder is intercepted on most soldering machines by suitable baffles and catchment devices, which guide it back to the level of the solderbath with minimum turbulence.

3. Covering all or most of the exposed surfaces of the solderbath with a layer of oil, molten wax, or some other suitable organic substance which does not smoke, smell, oxidize, decompose, polymerize or otherwise misbehave at 250 °C/480 °F is an old-established method of dross control which is still practised. A number of suitable cover oils or waxes are available from most flux vendors. Some of them are water soluble to facilitate their subsequent removal from the boards, should they have contaminated them.

   For many years, one machine maker (Hollis) provided his wave units with a facility to inject a measured amount of cover oil directly into the pump conduit, where it was dispersed in the solder and emerged in the form of small oil droplets in the solderwave. This method was very effective in suppressing the formation of dross, but it did leave a slight deposit of oil on the soldered board, which as a rule made cleaning obligatory. Since the advent of wavesoldering in controlled atmospheres (Section 4.5), the use of a cover oil on or in a solderwave is practised less frequently, if at all.

4. With almost all wave machines, baffles or bulkheads are provided at the level of the solder surface to prevent dross, which forms where the solder drops back into the bath, from getting near the intake of the solderpump. Any dross particles sucked into the solderpump are liable to cause trouble: they stick to the sides of the solder conduit and to the pump impeller, and disturb the smooth flow of the solder. This reduces the efficiency of the pump, causing the solderwave to drop, flutter or form a ragged crest. Above all, grey or black dross particles appear in the soldered joints, where they constitute a serious soldering defect.

Allowing the level of the solderbath to drop below the value recommended by the maker will increase the danger of dross being sucked into the pump circuit. Even with machines fitted with an automatic solderfeed, and with warning circuits to guard against a dangerous drop in solder level, it is worth while checking the solder level visually once a day.

With many wave machines, a circular baffle surrounds the rotating pumpshaft to stop the dross from reaching the pump inlet. Machine operators sometimes cover the space between that baffle and the shaft with a special high-melting wax to prevent the formation of the incandescent powdery dross which has been mentioned above.

The amount of dross which arises in the course of a day's normal running of a wavemachine varies widely. On smaller machines, the daily take-off of dross can be kept below 1 kg/2 lb. Even on large installations, it should not rise above a few kg or lb per day. An oil cover will of course reduce the amount of dross considerably.

A great deal depends on the way in which the machine is operated. If the wave is allowed to run 'dry', that is without fluxed boards passing over it, for long periods, more dross will form. This is a good reason for fitting wave machines with sensors, which turn on the wave only when a board approaches, and turn it off again after it has passed through the wave. Skimming the dross off the bath surface very frequently, so as to make it look clean and tidy (for example in order to impress passing management or visitors), is counterproductive: fresh dross will quickly form again. If the dross cover is left undisturbed, it protects the bath underneath from rapid further oxidation. As a rule, skimming the bath surface twice a day, e.g. before the midday break and before switching it off at the end of a shift, is sufficient. If within that period the layer of dross becomes so thick as to impede operation, it is time to check the impurity levels of the solder and its temperature.

Questions concerning the removal, handling and disposal of dross are dealt with in Section 4.8.

4.5 Wavesoldering in an oxygen-free atmosphere

4.5.1 Origins and development

Normal non-industrial air contains approximately 78 vol % nitrogen and 21 vol % oxygen, the remaining less than one per cent being taken up by the gases argon, carbon dioxide, hydrogen, neon, helium, krypton and xenon, listed in descending order. The habit of almost all metals of acquiring a skin of oxide in normal air, especially when they are molten, is the bane of the life of every practitioner of soldering. Without a flux to free the surfaces of the joint partners and of the molten solder itself from the oxide which covers them, soldering would be impossible. In response to this problem, a large number of fluxes have been developed over the years which cope admirably well with the needs of every conceivable soldering situation.

However, during recent years, flux has become an increasingly unwelcome ingredient of the electronic soldering process. The physical presence of flux residues on a soldered circuit board is an impediment to ATE testing with needle probes where it can be a source of serious errors. It also interferes with subsequent coating or encapsulating of the soldered board. The chemical nature of flux residues, being potential electrolytes, makes them unacceptable on boards populated with components of high impedance and close-pitch leads. Hence the growing need to remove the flux residue from soldered boards by a cleaning process.

Removing the flux residues with solvents based on chlorofluorocarbons (CFCs) was an efficient and relatively simple way of dealing with them, until in the late eighties their serious environmental effects were recognized. Consequently, their use is now being progressively and rapidly phased out (Sections 8.3.5 and 8.3.6). Several alternative cleaning methods have been and are still being developed (Sections 8.3.7 and 8.4), but nobody likes to clean a soldered board unless there is no alternative to it. Cleaning is beset with problems, and its costs continue to rise.

Since oxygen is the culprit in this situation, it seemed logical to consider soldering in its absence, without any flux or at least with so little that cleaning is no longer necessary. Soldering in a vacuum is not practicable, but leaving out the oxygen from normal air and soldering in nitrogen, either pure or with other gases added, has been proposed and practised for some time.

Attempts to conduct the wavesoldering process in an oxygen-free environment go back several decades. Except for a few specialized soldering tasks, these attempts did not lead to practical and generally applicable methods of wavesoldering until recently. Oxygen-free soldering entirely without flux assumes that all joint surfaces are completely or almost free from oxide before they enter the oxygen-free soldering environment, or that all of them are sufficiently heavily pretinned so that the molten solder can flow underneath an existing oxide layer (Section 3.4.1). Reducing any existing tin-oxide or lead-oxide layers by adding between 5% and 25% vol hydrogen to the nitrogen soldering atmosphere requires temperatures above $350\,°C/660\,°F$,[7] which are not practicable with normal electronic circuitry. The method is practised with certain hybrid circuits which carry ceramic components only, using a high-melting lead-rich solder and working in an atmosphere consisting of 90% nitrogen and 10% hydrogen (forming gas).[8]

In the recent past, various plasma treatments of assembled circuit boards in specially formulated atmospheres immediately prior to wavesoldering have been described as methods of fluxless soldering.[9,10] Equipment based on [10] has recently (1993) become commercially available. The term 'plasma' describes a volume of gas (which can be normal atmosphere), which is in a state of intense ionization. Plasma treatments have been used for some time for cleaning purposes in electronic manufacture, e.g. for the 'desmearing' of drilled holes in circuit boards prior to plating them.

4.5.2 Wavesoldering in nitrogen

Machine designs

The first practical and generally acceptable wavesoldering line which operated in a nitrogen atmosphere was introduced to the market in 1988 (SEHO, Germany, Figure 4.23). It found immediate and wide acceptance. Since then, a number of other leading wavemachine makers have followed suit.

The atmosphere in all these machines is practically pure nitrogen, with an oxygen level, often automatically monitored, ideally below 10 ppm. With the SEHO machine, as with several other makes, the complete soldering line, comprising the fluxer, the preheating zone and the solderwave unit, operates in a nitrogen-filled tunnel. On machines where the boards enter and leave the tunnel through horizontal slots at both ends, the nitrogen unavoidably leaks away through them and needs constant replenishment. Consequently, all nitrogen wave machines are provided with one or more strategically placed nitrogen inlets, which admit metered amounts of the gas. The amount of nitrogen loss can and is being minimized with many designs through careful dimensioning of the ports and by controlling the pressure of the nitrogen feed. Reported or claimed nitrogen consumption is normally at about 20 cub. m/550 cub. ft per hour (SEHO).

One make of machine (Soltec, Holland) keeps the nitrogen loss to a minimum by placing airlocks at both ends of the soldering line. When passing through a lock, the board enters through its open front bulkhead, which then closes, whereupon the lock is evacuated and subsequently flushed with nitrogen. The back bulkhead then opens and admits the board to the nitrogen-filled soldering tunnel. The same procedure is repeated in reverse order as the soldered board passes through the exit airlock. The full evacuation of an airlock may take up to thirty seconds. In order not to let this slow down the operation of the machine unduly, the conveyor system of this machine is divided into separate zones, so that a board which has

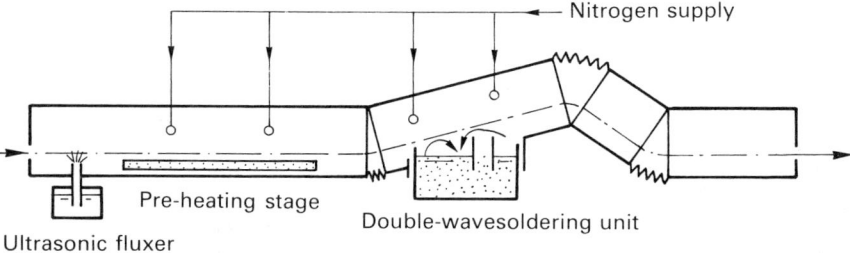

Figure 4.23 *Wavesoldering under nitrogen (SEHO, Germany)*

passed through the entry lock can proceed through fluxer, heating stage and solderwave while the next board enters the lock and goes through the entry procedure and the preceding one passes through the exit lock. The trade-off for the engineering effort and consequent investment cost lies in the reduced consumption of nitrogen.

Some makers offer conversion kits for retrofitting existing soldering lines to working with nitrogen. These conversions may take the form of installing a nitrogen tunnel to cover the whole length of the soldering line, or they provide a hood which covers the waveunit only (Electrovert). The simplicity and consequent lower investment requirements of the latter are offset by a higher nitrogen consumption, which is quoted at 75–100 cub. m/2000–2700 cub. ft per hour. Providing for the storage and the supply of nitrogen represents a significant portion of the capital investment involved in setting up a nitrogen soldering line, as does the cost of nitrogen consumed in its running costs.

The advantages to be set against these costs are considerable. Several makers of nitrogen-filled wavesoldering machines claim that the additional investment necessary for the nitrogen facility can be written off in a comparatively short time through the saving gained by the elimination of dross and the consequent loss of solder.

There is general agreement that dross formed on a wave running in nitrogen is only a small fraction of the dross created by the same solderwave in air. For one thing, this drastically reduces the solder consumption and thus the running costs of the soldering line. For another, weekly servicing can replace the twice-daily routine of dross removal and machine cleaning. The virtual absence of drossing means that oil cover is no longer needed when working in nitrogen; in fact if it were used, its vapours would severely contaminate the whole system. Many users report that they operate their nitrogen machines with a drastically pruned maintenance schedule and a consequent reduction in down time.

The most valuable trade-off with nitrogen wavesoldering is the virtual elimination of bridging. This has extended wavesolderability to components with a pitch approaching 0.5 mm/20 mil and thus has shifted the balance of the wave-versus-reflow decision to some extent back towards the solderwave (Sections 5.1.1 and 5.9).

The solderwaves

There is evidence that the nitrogen atmosphere increases the surface tension of the molten solder, as witness its reduced tendency to form solder bridges. This means that more kinetic energy is needed to push the solder into the recesses of SMD populated boards than would be needed for soldering in normal air. In consequence, the double wave on some nitrogen machines comprises a primary turbulent wave followed by a composite wave with a vibrating wavecrest (Section 4.4.3). This arrangement is claimed to help the solder to rise through the full length of throughplated holes and form a good meniscus at the top end.

Fluxes and fluxers

As was pointed out above, nitrogen does prevent the molten solder from oxidizing, but it cannot free the surfaces of component leads and solderpads from their existing and unavoidable cover of oxide. Only the above-mentioned plasma precleaning treatments may be able to do that and thus make truly fluxless soldering possible. Meanwhile, flux is still needed, but given the usually good solderability of both components and circuit boards, fluxes for soldering under nitrogen can be weaker and thinner than fluxes for working in air.

Initially it was hoped that formic acid, an organic acid with a very simple structure and a boiling point of 100.5 °C/212.9 °F would be ideal. It could even be injected into the soldering tunnel together with the nitrogen, and would form a fluxing vapour in the hot atmosphere. Provided the boards are hotter than 100 °C/212 °F as they leave the tunnel, there would be no flux residue at all. Unfortunately these hopes were not fulfilled: the formic acid vapour did not remove the oxide from all component leads, while it increased the surface tension of the molten solder to such an extent that its flow into the joints was seriously affected. Moreover, formic acid has a very sharp, penetrating odour, which can be unpleasant if too much of it escapes through the entry and exit ports. Finally, its health aspects have come under suspicion. For these reasons, formic acid has fallen into disfavour, even as an adjunct to more conventional fluxes.

For some time, adipic acid in a week alcoholic solution was a favourite alternative to formic acid. Adipic acid is a slightly more complex organic acid, which forms white crystals with a melting point of 153 °C/307 °F. It is a popular activator of rosin fluxes and also a frequently used ingredient of many low-solids fluxes. Its fluxing action is adequate, but in due course more complex fluxes have been found which give better soldering results, while being less demanding as far as perfect solderability of all component leads was concerned. At the time of writing, a number of fluxes with very low solids content are commercially available for wavesoldering in nitrogen. Some are completely free of rosin, while others contain very small rosin additions which are claimed to improve the flow behaviour of the solder while reducing the danger of bridging with fine-pitch components.

The need to operate the fluxer in a controlled atmosphere without disturbing it narrows the choice of the fluxing system. Foamfluxers and compressed-gas atomizers would need nitrogen as their driving gas, in quantities which would affect the economics of the machine. In consequence, most nitrogen wavesoldering machines use one of the ultrasonic atomizing systems which have been described in Section 4.2.1. Some makers offer the user a choice between different systems of sprayfluxers.

The virtues and vices of wavesoldering under nitrogen

One of the problems of wavesoldering in a nitrogen atmosphere is the appearance of small globules of solder, which adhere to the underside of the board. Why they appear and how they can be avoided has been and continues to be the subject of many technical papers and public discussions, but no firm conclusions or recommendations have emerged to date. It seems agreed that they originate at the point where the solder in the peelback on the exit from the second wave breaks up. As the last thread of molten solder parts, some of it tends to form one or more separate small drops (Figure 4.24).

Accepting that the surface tension of the molten solder is decidedly higher in nitrogen than in air, and that the absence of any oxide skin makes it more mobile, the kinetic energy released by the breakup of the peelback has a more pronounced effect under nitrogen than it would have in air. Moreover, any flux residue present at that moment would be in the form of fine dust rather than of a coherent skin, and therefore would not impede the mobility of the solder either. The latter argument is supported by the observation that the more rosin a flux contains, the less solderballing there is. This of course runs counter to the whole reason of nitrogen wavesoldering, which aims at a minimum of flux with a minimum of rosin, if any at all.

An explanation why these small drops should decide to fly upwards towards the board has to assume an electrostatic charge, a not unreasonable assumption in a situation which is essentially an atomization process. Finally, why do they stick to the board once they hit it? It is an observed fact that solderballs are only found on the solder resist, be it lacquer or film, but hardly ever on a bare board surface.

Van der Waals forces furnish a plausible explanation for the following reasons. It is a known fact that, in the absence of flux, molten solder wets the underside of a board and sticks firmly to it after it has set solid. Equally, it is known that molten solder wets clean glass and firmly adheres to it after solidification. The same forces might very well be assumed to act in the case of a droplet of molten solder which hits the smooth solder resist, flattens a little on impact before it solidifies and is held fast on its small but equally smooth interface.

Unlike the smooth, unstructured surface of glass or solder resist, both of which are basically undercooled melts, the surface of the board laminate is structured, consisting of a large number of microscopic peaks. The high surface tension of a droplet of solder which hits the laminate prevents it from conforming to the irregular surface of the laminate. A large number of point contacts replaces the smooth interface, and the van der Waals forces cannot act.

A few countermeasures against solderballing can be derived from this hypothesis of its causes. Raising the rosin content of the flux has proved to be helpful, but this expedient runs counter to the whole purpose of soldering under nitrogen. The abrupt tearing of the thread of solder in the peelback can be softened by replacing all straight edges of solderpads which run at right angles to the direction of board travel by pointed peaks. This measure require pre-knowledge of

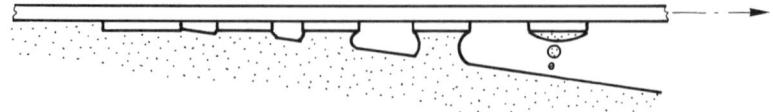

Figure 4.24 *Breakup of the peelback*

the board's direction of travel, and it occupies valuable board real estate (Figure 4.25).

Another measure which suggests itself would be a drastic reduction of the area covered by the solder resist. If the solder resist were to cover only the conductor tracks and any other metallic surface which must not be tinned by the solder, solderballing would be much reduced. Above all, it would be confined to areas where solderballs do not pose a risk of short circuits or reduced insulation values, as is for example the case between closely spaced solderpads. Finally, accepting it as true that solderballs originate in the peelback, and considering that a counterflow jetwave has no peelback, why are there no nitrogen wavesoldering machines fitted with a counterflow jetwave, at least as a secondary wave?

How dangerous or undesirable solderballs really are is still a matter of debate. At all events, a motorized brushing unit for removing them mechanically is on the market (Ascom, Switzerland). This unit consists of sets of rotary brushes set both parallel and at right angles to the direction of board travel, and is fitted with an exhaust to collect the loosened solderballs.

The virtues of wavesoldering under nitrogen are proven and can be precisely defined: until the dream of completely fluxless wavesoldering which would make cleaning fully superfluous is fulfilled in a commercially viable manner (as might now be the case), soldering with much less flux and much thinner fluxes than had previously been thought possible has become a reality with soldering under nitrogen.

That it makes cleaning fully and always unnecessary cannot be stated as an absolute truth. Whether cleaning can be omitted must be established by corrosion and SIR tests from case to case. Experience shows that the need for cleaning is very much reduced for the normal range of electronic products. Boards soldered under nitrogen are certainly ATE-testable and will accept conformal coatings or lacquers without any difficulty. It must be added at this point that the continuing progress in flux technology has in many cases given a similar reduction in flux residues, with the same benefits resulting from it, without the use of a nitrogen atmosphere.

Nevertheless, all users are agreed that with soldering under nitrogen, the incidence of faulty or skipped joints has been drastically reduced. The consequent

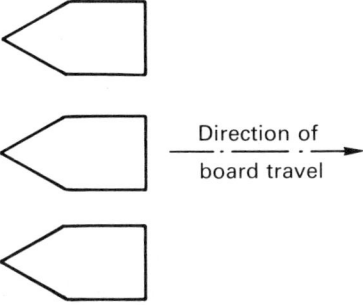

Figure 4.25 *Shape of solderpads to minimize solderballing*

saving in rework, together with the reduction of dross losses and the savings in maintenance costs and down-time, are reported to offset the cost of the nitrogen and to enable the writing-off of the machine and installation investment within a reasonable time, depending of course on the usage of the equipment. As far as skipped joints are concerned, users as well as machine vendors agree on several points. The solderability of all joint surfaces must be impeccable, especially with leadwires, throughplated holes and vias. Also, the double-wave system must be finely tuned to suit the demands of the three dimensional board geometry for each particular board pattern to obtain the optimal results which the nitrogen concept as such is capable of achieving.

Finally, as has been said already, users report that with soldering under nitrogen, the pitch of the component leads, which allows wavesoldering without the danger of bridging, has been extended downwards. When wavesoldering in normal atmosphere, bridging is difficult to avoid with a footprint pitch below 0.65 mm/26 mil. In a nitrogen atmosphere, bridge-free soldering is no problem with that pitch, and good results have been reported with 0.5 mm/20 mil pitch.

Wavesoldering under a cover of hot glycerine flux

An example of wavesoldering in a non-atmospheric oxygen-free environment, if not an atmosphere, merits mentioning here. In the 1960s, IBM conceived a method of wavesoldering under a cover of a high-boiling flux based on glycerine (boiling point 290 °C/554 °F). The flux is a 10 per cent solution of EDTA (ethylene diamine tetra-acetic acid) in glycerine, a reasonably efficient, halide-free, water-washable flux. The solder chosen is the tin–bismuth eutectic (43% Sn, 57% Bi) which has a sharp melting point at 138 °C/281 °F, and forms strong joints on copper. An alloy with a lower melting point than that of the tin–lead eutectic has been chosen so as to be able to run the wave well under the boiling point of glycerine, which starts to decompose when it boils. The operation is in fact run at temperatures well below 250 °C/480 °F. It is used for the wavesoldering of a type of board which carries no SMDs.

IBM have developed a dedicated type of wavesoldering machine in which the wave, complete with its metal pump and the board conveyor system, is fully immersed in the hot flux. The company have made the information on which the system is based generally available.

4.6 Board conveyor systems

4.6.1 Functional requirements

Boards must travel through a wavesoldering unit at a known, controllable and steady speed, without jolts or periodic flutter. While they traverse the solder-wave(s), their line of travel must be accurately defined and reproducible from board to board. Unless these conditions are met, the quality of soldering is bound to be erratic, however efficient and well controlled the fluxer, the preheating unit and the solderwaves may be. Furthermore, jolts or a rough ride while the joints solidify after their passage through the wave will spoil their appearance and may cause incipient cracks within the joints themselves.

The conveyor is an essential part of a wavesoldering unit, and it must be a carefully designed, sturdy piece of precision engineering. On most machines, boards travel on one single continuous conveyor from one end of the machine to the other. An exception is the nitrogen system with entry and exit airlocks. Here, the conveyors which take the boards through the entry and exit lock are separate from the interior conveyor, to avoid bunching or hold-ups along the machine.

Because the depth of immersion of every part of the board surface in the wavecrest critically affects the soldering result, geometrical precision matters most where the boards travel over the crest or crests of the wave. Here, the position of every board in the vertical axis must be defined, in reference to both its longitudinal edges, to within ± 0.3 mm/12 mil. Assuming the wavecrest to be straight and horizontal, any sideways tilt of the board relative to the wavecrest must also be held within these limits. They should not be exceeded, because some unsteadiness of the wave, and warping and bowing of the board itself, must also be accommodated.

The travelling speed of the boards need not be numerically defined or known to within less than a few per cent. What matters is that it should not drift upwards or downwards by more than those amounts without automatic correction or at least a warning. The conveyor speed must be reliably monitored and displayed. The preferably tachometrically controlled conveyor drive must be free from slip, and should not be affected by variations in the mains voltage.

4.6.2 Board-handling systems

Individual board carriages or pallets

Board carriages which grip a single board, or at most two boards of equal width in tandem, between two opposed, sprung longitudinal clips have been around since the beginning of wavesoldering (Figure 4.26). Their evolutionary development since then has been confined to refining their design and making them lighter. The clips are made from an unsolderable metal, such as stainless steel, sometimes

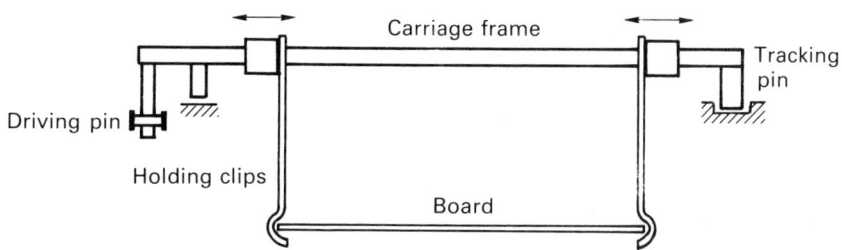

Figure 4.26   *Board carriage*

chromium-plated steel, rarely sheet-titanium. They grip the boards between V-shaped jaws.

The adjustable distance between the clips is set manually, with the correct lateral pressure against the sides of the board being judged by the operator: if it is too loose, the vertical position of the board in the V-jaws is ill defined; if it is too tight, the board, which softens and expands in the preheater and above all during its passage over the wave, is liable to bow, causing an uneven depth of immersion across its width. With particularly wide or heavily laden boards, this effect can be mitigated by clipping temporary stiffening edges to the leading, and sometimes also to the trailing, edge of each board.

Every board is manually loaded into its carriage by snapping it into the jaws of the sprung clips. Unloading is manual too. Carriages which provide for quick-action loading and unloading are available with some high-performance equipment.

This handling of individual boards on unloading after soldering has one very important point in its favour: every single board is seen by an operator before it leaves the soldering machine for storage or the next manufacturing step. This means that a malfunction of the soldering machine, which does not automatically set alarm bells, if any, ringing, such as a collapsed foamwave, a blocked sprayfluxer or an uneven solderwave is spotted at once, and not after a number of scrap boards arrive at the next workstation, which may be distant from the soldering machine in both space and time.

The carriages may either glide along the conveyor on rails, being pulled on either side by an endless link chain, or else their sides rest on endless moving fabric belts. The latter arrangement avoids mechanical damage to carriage and drive should one or more carriages jam. A conveyor must of course have a fail-safe provision which prevents a board coming to rest in the preheating section or over the solderwave without immediately switching either of them off.

Because of the close tolerance with which the board must travel over the wave, board carriages are pieces of precision engineering, which must sit securely on the guide rails without rocking. Should one be dropped accidentally, it must on no account be put back on the line before being closely checked for its correct alignment, preferably by the works' toolmaking department. Since the holding clips pass constantly through the fluxer and the wave, board carriages must be cleaned regularly to prevent buildup of flux residues which, especially with rosin fluxes, can become almost insoluble after a number of passages through the solderwave. Passing the carriages through the board-cleaning installation, if one is available, at regular intervals, will suffice. Alternatively, cleaning equipment for board carriages is on the market.

Finger conveyors

Board carriages represent a considerable investment, especially on large soldering lines, and their manual loading and unloading interrupts the smooth flow of an automatic production line. For this reason finger conveyors (Figure 4.27) are

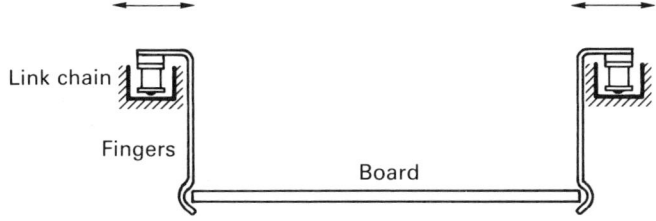

Figure 4.27 *Finger conveyor*

becoming increasingly popular. By now they have evolved into sophisticated and flexible systems, which fit in easily with integrated automatic production lines. The boards are carried between sprung individual finger clips, which are mounted on two endless link chains running along either side of the conveyor. The distance between the chains can be varied, manually or mechanically, to suit the width of the boards passing through the machine. With computer-controlled machines, the conveyor width can be called up together with all the other preset operating parameters, to suit a given batch of boards.

The fingers themselves are almost always made from titanium, which is untinnable with any type of flux. A good deal of development has gone into the shape of the fingers themselves to ensure that the position of the underside of the board in relation to the chassis of the machine is accurately defined, without undue pressure against the sides of the board. One maker (Sensbey) has designed the fingers as chucks, which release their lateral hold as the board passes over the wave while still defining their vertical position. This allows the board to expand in the soldering heat without warping. Chain conveyors are, or should be, equipped with a mechanical cleaning device for each chain, normally a rotary brush, to prevent buildup of flux residue on the fingers, which would interfere with their function and accuracy.

Soldering templates

Soldering mixed batches or short runs of circuit boards may need frequent resetting of the holding clips in every board carriage, or changing the distance between the conveyor chains. Sets of templates or frames with a milled or sawed aperture which corresponds to the outline of a given type of board are frequently used to cope with this situation (Figure 4.28). These frames are conveniently made from thick FR4 (2–4 mm/100–200 mil thick), with holding clips and lugs, as illustrated, and with fingerholes to facilitate the loading and unloading of the boards. They are also very useful for dealing with boards with irregular, non–rectangular contours. The fit of a board in its template aperture should of course not be too tight, but must allow for the expansion of the board during preheating and soldering. A set of templates made from good-quality FR4 will last for a year or more, even with constant use, if not maltreated.

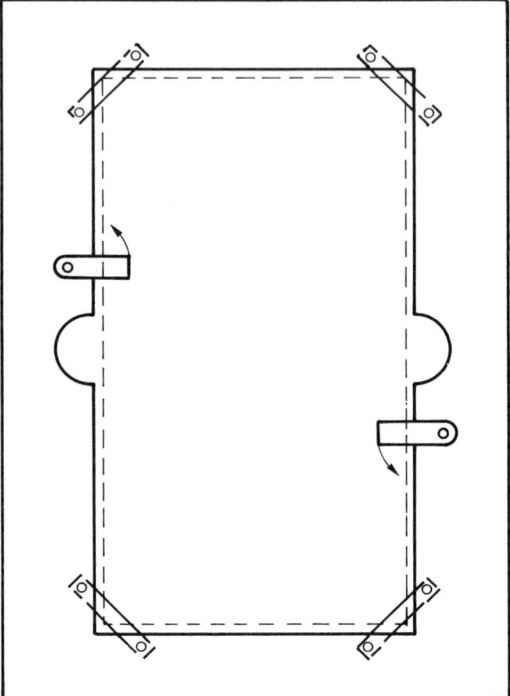

Figure 4.28   *Soldering template*

Board-handling robots

Several vendors offer handling robots which grip the edges of a board, and which can be programmed to move over the crest of a solderwave along a non-linear path and with a non-constant speed. The idea is to provide the optimum wavesoldering parameters for every section of a board with complex and non-uniform soldering requirements on different parts of its surface. Such handling robots require a considerable investment in both hardware and programming, and the niche they may have found is a narrow one.

4.7   Wavesoldering practice

4.7.1   Operating parameters and their role

Wavesoldering has more variable parameters than any other soldering method. Furthermore, most of these parameters are interdependent in their effect on the soldering result. Their list makes formidable reading:

- Angle of the conveyor towards the horizontal
- Conveyor speed
- Type of flux
- Density, water content and freshness of the flux
- Amount of flux deposited per unit of board area (provided the machine allows for the variation of this parameter)
- Intensity of preheat
- Composition of the solder
- Impurity level of the solder
- Solder temperature
- Height and stability of the solderwave, as governed by the speed and the running behaviour of the solderpump
- Profile and flowpattern of the solderwave, as governed by the setting of the wavenozzle weirs
- Depth of immersion of the board in the crest of the solderwave

Several of these parameters are best decided on once and for all, and not changed thereafter:

1. The standard conveyor angle of 7° towards the horizontal is the result of many years of wavesoldering experience, and it should not be altered except for some special reasons, which will be discussed later. The angle is best left alone, and several machine makers recognize this fact by selling their machines with a conveyor rigidly fixed at 7° towards the horizontal.

2. Choosing the right type of flux for a given production line, soldering task or type of board involves a great deal of effort, time, and very often expense. The type, layout and solderability of the board, the standard and constancy of component solderability, the cleaning, testing and postsoldering treatments and circumstances, and the final use of the soldered assembly all determine the choice of flux. Once this choice has been made, with flux vendors, production managers, purchasing and quality control all having cooperated in the task (or not, as the case may be), the flux specification should not be changed except under dire necessity. Even with a quite small company, changing the flux involves much time, effort and cost in checks on corrosion, performance and quality. With a large organization, changing a flux specification may take up to a year to finalize and can cost very large sums of money indeed.

3. The choice of solder is simpler than the choice of flux. Over the years, a solder alloy of eutectic composition, containing 63% of tin and little if any antimony, has been found to be universally suitable for almost every soldering task (see Sections 3.3.1 and 3.3.2). Unless special circumstances or a specific customer demand a solder with a different melting point or maybe a certain percentage of silver, the user will be well advised never to change the specification of his solder.

4. The standard wavesoldering temperature of 250 °C/480 °F plus or minus a few degrees is, like the conveyor angle, the result of over four decades of practical

wavesoldering experience. Without a compelling need, it is advisable not to depart from it.

4.7.2   Choosing and monitoring operating parameters

Condition of the flux

Given that the choice of flux is settled, the contents of the fluxer should at all times match the density and/or the acid value which is specified in the vendor's data sheet. Section 4.2.2 discusses in detail how this requirement can be met, by automatic equipment if required. It is worth restating at this point that the success of wavesoldering depends critically on the consistent quality of the flux, and that this constancy is assured more easily with sprayfluxers than with foamfluxers.

Amount of flux per unit of board area

This parameter also affects the soldering success, though to a lesser degree than the density and the activity of the flux. Too much flux means more solvent in the flux cover and, unless the preheater is adjusted accordingly, a risk of boiling and solder-prill formation as the board passes through the solderwave. If boards have to be cleaned after soldering, too much flux reduces the cleaning efficiency. Too little flux, uneven fluxcover or, worse, unfluxed patches inevitably cause soldering faults, such as bridges, icicles, solder adhering to the board and open joints, especially with low-solids fluxes. With these, the margin of error is much narrower than with high-solids fluxes.

The thickness of the flux cover can be controlled to some extent with the various types of sprayfluxer, but foamfluxers permit very little, if any, control over this parameter. At the time of writing (1993), there is no equipment on the market for automatically monitoring the thickness of the flux cover. A frequent visual check of the overall appearance of the soldered boards is the best method of ensuring the stability of this important factor. Automatic video surveillance of the output of a soldering line should be capable of giving warning of a malfunction of the fluxing unit.

Intensity of preheating

Insufficient preheat leaves too much solvent in the fluxcover, which is therefore more liable to be washed off in the solderwave, leading to bridging or open joints. This factor is particularly critical with double waves, where a substantial portion of the flux cover must survive the passage through the first, turbulent wave. Moreover, if the board is too cool, the solder may not rise through all plated holes and form the required solder meniscus on the upper board surface.

Too sharp a preheat can cause trouble with rosin-based fluxes: overbaking such a flux will cause the rosin to polymerize. This reduces its mobility, so that it may obstruct the solder in tinning all solderpads or in rising to the top surface of the board. It will certainly make cleaning less efficient.

By contrast, fluxes with a low solids content and very little rosin, and the so-called 'no-clean' fluxes (Sections 3.5 and 8.1) which are mostly rosin-free, require more intense preheating to ensure that the flux coating is not washed off in the double solderwave. With these fluxes, most vendors suggest that the underside of the board should have a temperature of 120 °C/250 °F on emerging from the preheating stage.

The methods of controlling the intensity of preheating are dealt with in Section 4.2.3.

Parameters of the solderbath and the wave

The level of molten solder in the machine should at all times be kept strictly at the height recommended by the maker. Many machines are fitted with an automatic solder feeder, which maintains the correct solder level. Failing an automatic level control, the solder level must be regularly checked at intervals depending on the usage of the machine, and if necessary topped up. Unless fitted by the maker, it is advisable to install a simple solder-level sensor, which gives an audible or visible warning as soon as the solder level drops below the maker's danger mark.

If the solder level drops too low, dross and flux–residues which float on the solderbath can be sucked into the inlet of the solderpump. Once in the solder stream, they tend to deposit on the solder conduits and the pump impeller. These deposits interfere with the steady running of the solderwave, as will be discussed below. Particles of dross and flux which reach the wave nozzle emerge in the wave as small, but conspicuous, black spots, which pop up in the wavecrest and finish up on the surface of the solderjoints. Such dross or flux inclusions do not necessarily threaten the function or reliability of the affected joints, but they are a legitimate cause of rejection by quality control or by the customer.

The temperature of the solder is one of the most basic wavesoldering parameters. The general suitability of 250 °C/480 °F for most wavesoldering tasks has been mentioned already. Close adherence to this value is less critical than is often assumed, an accuracy of ± 2–3 °C/4–6 °F being quite sufficient. It is much more important to guard against a slow, unnoticed upward or downward drift of the solder temperature away from its set value. The temperature readout on the control panel of the machine, together with its warning signals, may be misleading: software or functional errors are not unknown. The safest way to guard against this danger is to check the actual solder temperature halfway through every working shift by checking the solder temperature with a reliable, preferably occasionally re-calibrated, handheld temperature measuring instrument, with its sensor placed in the solderwave about 5–10 mm/0.25–0.5 in below the crest.

The height of a wavecrest is directly linked to the speed of the solder pump, which with most good machines has a slip-free, tachometrically controlled drive which is protected against variations in the supply voltage. The waveheight and its consistency across the whole width of the wave can be checked very simply by sliding a piece of plain FR4, with gradations marked on it, across the length of the wavenozzle while the pump is running (Figure 4.29). It is advisable to carry out this

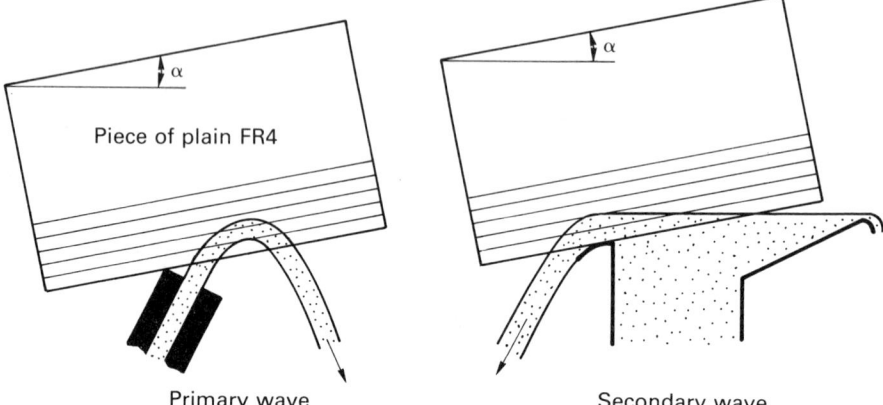

Piece of plain FR4

Primary wave   Secondary wave

Figure 4.29   *Checking the wave height. α: Conveyor angle*

simple check at the beginning of every shift. Some computer-controlled machines are fitted with a sensor-operated surveillance of the height and integrity of their solderwave(s).

The depth of immersion of a board into the crest of the solderwave is normally equivalent to the thickness of the board. It is therefore important that the underside of the board is strictly parallel to the line of the wavecrest to well within this measure. This is easily checked by letting a piece of plain FR4 without copper lamination, as wide as the largest board, run across the wave and stop briefly over the wavecrest. The flattened wavecrest will be clearly visible through the translucent FR4. If the board is parallel to the wavecrest, the width of the band formed by the flattened wave will be the same across the whole breadth of the testboard (Figure 4.30).

If the board is not parallel to the wavecrest, the whole conveyor must be tilted sideways until a parallel position is achieved. Provisions for carrying out this adjustment are, or should be, a feature of every wavesoldering machine. As an alternative to the FR4 board, many machine vendors can supply a plate of heat-resistant borosilicate glass which carries a pattern of parallel lines to make it easy to check the width of the wavecrest across the plate. To make sure that the glass plate does not crack during this manoeuvre, it is advisable to pass it over the fluxer and the preheater before arresting it over the wave. With FR4, this is not necessary.

Uneven or rough running of the solderwave, such as fluttering of the waveheight, can be a sign that deposits of dross or flux residues have formed on the pump impeller or the solder ducts, often as a consequence of an unduly low solder level in the machine (see above). With all jetwaves, even quite small accretions of dross or flux residue in the exit slot of the wavenozzle can ruin the smooth profile of the solderjet. The wavecrest becomes ragged. One big dross particle can depress it by quite large amounts, naturally leading to serious soldering defects. Regular cleaning of the nozzle aperture is an important requirement with all jetwaves. It is

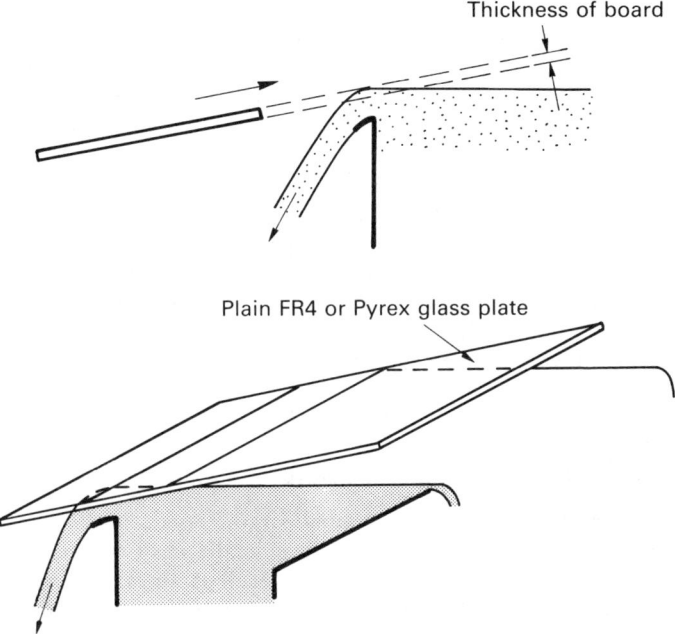

Figure 4.30   *Checking depth of immersion and horizontal alignment between wavecrest and board*

best carried out by drawing a scraping tool, made from soft steel or PTFE, along the whole length of the nozzle aperture. Most vendors can supply suitable implements for the purpose. A scraper made from aluminium, brass, copper or hardened steel should on no account be used.

With all double-wave machines, the second wave is of the 'asymmetrical' type (see Section 4.4.4). On the exit side of this kind of wave, the board lifts off from a horizontal pool of solder, whose surface moves in the same direction and ideally at the same speed as the board on its conveyor for reasons which have been explained already (see Figure 4.14). The match between these two speeds can be checked simply by floating a small steel ball, for instance from a ball bearing, on the solder surface and comparing its movement with that of the board conveyor.

Conveyor speed

The conveyor speed is a critical wavesoldering parameter. On the one hand, the heat received by a board is inversely proportional to the speed at which it travels through the preheating unit at a given setting of the heater panels. On the other hand, the maximum practicable soldering speed of a wave machine is governed not only by the ability of the solderwave to get the necessary amount of heat into the board within the time available for this, but also by the complexity of its pattern and the density of its population of components. Furthermore, multilayer boards with high heat capacity must travel more slowly than simple single-layer boards. Boards with closely set SMDs and fine-pitch multilead components must travel over the wave more slowly to give the solder a chance to flow into the narrow gaps between neighbouring components, and to drain away from the fine pattern of leads.

With most soldering machines, the set and the actual conveyor speed are displayed on the control panel. Nevertheless, it should be part of the machineminder's task to check the actual against the displayed conveyor speed of the machine once every day, with the aid of a stopwatch and a simple marker, travelling on the chain conveyor over a measured distance marked on the conveyor rail. This very simple test can save hours of expensive rework of boards, should the conveyor speed have drifted from its set value, or should the machine control have started to malfunction.

Computer-controlled soldering machines

The large number of interlinked operational parameters makes wavesoldering a natural subject for computer control, which generally has two tasks.

The first task is the monitoring and stabilizing of all parameters, which will have been established as optimal and stored in the program. It is worth saying again that this automatic pilot does not relieve the operating personnel from periodically verifying that the machine does in fact run correctly. The functions which must be watched are those which are difficult if not impossible to monitor by sensors, such as the behaviour of the foamwave, the spraypattern of the fluxer and the correct behaviour of the solderwave.

The second task relates to parameters which can be adjusted to suit a given type of circuit board. The main parameter here is the conveyor speed, which can be raised with boards of simple pattern and modest thermal requirements, or which may need lowering for complex or multilayer boards. Linked to this is the intensity of preheat. Low-temperature emitters respond slowly to a change in heater current (see Section 4.3.2), and this factor must be considered in the program. Alternatively, the inclusion of one or more high-temperature emitters in the preheating unit will permit a much faster response to the commands of the computer. Some types of machine allow for a choice between foamfluxing and sprayfluxing to suit different types of board.

With computer-controlled soldering lines, each board normally carries a barcode which calls up the correct parameter as it enters the machine. Nevertheless, it is not at all advisable to run a soldering line with a random mix of different types of board. It is advisable to gather them in as large batches as possible.

4.7.3   Optimizing machine parameters

The following strategy for starting up a new machine, or changing to a new type of board, has proved its worth in practice:

1. Check that the conveyor angle is near 7°, and that the conveyor is laterally horizontal. Place a plain piece of FR4, of the same thickness and size as the boards to be soldered, in a board carrier or into the chain conveyor and move it forward into the fluxer. With a foamfluxer, adjust the air pressure so that the wave can hold the required height with a good margin. With a sprayfluxer, set the width of the spraypattern to suit the width of the board.

2. Move the board forward to the solderwave. With double-wave machines, set the primary wave, whether it is of the turbulent type or a jetwave, as high as is possible without causing the solder to push through apertures in the board and flood the top surface. With a jetwave of the type where the solder flows in the direction of the travelling board, make sure that the baffle at the trailing edge of the board fixture is high enough to prevent the solder from flooding the top of the board as it leaves the wave. Single jetwaves of the counterflow type need a safety baffle at the leading edge of the board. Again, set the wave as high as is possible without flooding the board.

    The secondary wave is always of the asymmetrical laminar type. The board is moved forward so that its leading edge is just in front of the wavecrest. Adjust the waveheight so that the crest comes approximately level with the top surface of the board. This means that thicker multilayer boards dip deeper into the wave, spend more time in contact with it, and in consequence receive more heat.

    Having done this, move the board forward into the wave and check whether the board is parallel with the wavecrest, as shown in Figure 4.30. If you find that the board is not parallel with the wavecrest, do not try to adjust the setting of the whole solderbath or of the wavenozzle, but tilt the conveyor, as has been described already.

3. Next, set the conveyor speed at half the value which the vendor gives as its maximum speed, unless operational requirements make it necessary to work faster than that. It is worth remembering that it is a fact of life in engineering that the failure-rate or fault-rate of any given equipment or process begins to rise exponentially as it is driven at a rate or speed approaching its designed maximum. (Compare the number of pit stops during an Indiannapolis race with the service requirements of a family car.)

    Run a board of the pattern which is to be soldered on the machine, without components, through the fluxer and the preheating stage and check its temperature on leaving the latter (Section 4.3.4). If it gets too hot, reduce the setting of the heaters. If on the other hand the heater, even at its maximum setting, does not get the underside of a heavy multilayer board hot enough, fit a top reflector to the preheater if none is provided. If the board is still too cool, reduce the distance between the heaters and the conveyor. Only if all else fails (and in this case the design of the machine must be at fault) lower the transport speed or, better, modify the heating stage yourself.

4. Having balanced the setting of the heating stage and the conveyor speed against each other, proceed to solder about ten fully assembled boards, and check the soldering quality of each carefully. If this is satisfactory, enter the set of working parameters into the machine computer or your production

control manual. If faults persist in an erratic pattern, check the stability of the fluxer and wave setting; as a last resort, lower the conveyor speed and reduce the setting of the heating stage accordingly. If faults persist systematically with one or more given components, check their solderability or the suitability of the layout. For details of the systematic analysis, interpretation and elimination of soldering faults and defects, see Chapter 9.

4.7.4 Machine maintenance

Daily

Clean the wavenozzle at the end of the shift, and if necessary also at the mid-shift break. Turn on the solderpump and check whether the wavecrest is level and stable. If you are not satisfied, switch off the pump and scrape the inner walls of the solder conduit with an *annealed* hacksaw blade. An annealed hacksaw blade will not snap and constitute a potential danger, nor will it damage the conduit. Restart the solderpump and skim dross and flux residues, which may now be flushed through the wavenozzle, from the solderbath. This maintenance is especially important with jet nozzles.

At the end of the day, clean splashes of solder and flux from the top of the machine and the rims of the solderbath. With endless-chain conveyors, check the condition and functioning of the automatic chain cleaners. With board carriages, remove excessive flux buildup from the holding jaws.

This schedule can be greatly relaxed with soldering units which work under a nitrogen atmosphere (Section 4.5.2).

Monthly

Lift the wavenozzle assembly and the pump impeller from the solderbath and remove all adhering dross and flux residue. With foamfluxers, renew the air filter and clean the foaming stone. With sprayfluxers, clean spraynozzles, if any, and remove buildup of dried flux, if any. Renew flux in fluxer, unless in the case of a foamfluxer the scheduled flux renewal has taken place earlier. Clean the exhaust system of the fluxing unit. Clean and, if necessary, renew air filters.

Annually

Carry out a complete overhaul of the soldering machine. Check all board carriages, if any, for correct setting and alignment.

This schedule applies for soldering lines which are in constant use. With machines which are used only sporadically the schedule will, of course, be stretched accordingly.

4.7.5 Check-analysis of the solderbath

Depending on the utilization of the machine, the solderbath should be checked at least once a year for its tin content and its impurity levels. Most solder vendors are able to carry out this analysis for their customers. Unless your own organization maintains a central analytical laboratory, and sometimes even then, it is better and quicker to employ the services of an outside specialist. For the interpretation of the analytical report, and the measures to be taken if it is unsatisfactory, see Section 3.3.3.

An analytical laboratory requires a sample of solder weighing 100–200 g/3–6 oz, in the form of a small ingot. In order for this sample to be meaningful and representative of the contents of the solderbath, the following sampling procedure should be followed. The wave is switched on and kept running for 1–2 minutes. The sample is then taken from the over-run of the wave with a small stainless steel ladle, which must be absolutely dry. This is best assured by preheating it in a blowflame. The sample is then poured into a simple mould fabricated from heavy steel or stainless steel sheet, as sketched in Figure 4.31. This mould too must be absolutely dry, but it ought not to be too hot because the sample should solidify reasonably quickly.

4.7.6 Dealing with dross

The nature of dross and the manner of its formation is discussed in Section 4.4.5. The layer of dross which forms on the surface of the solderbath (unless the machine is run under nitrogen) must be removed periodically to prevent it from being sucked into the pump inlet. Skimming the dross twice daily is sufficient for this purpose (Section 4.4.5). The simplest and best method is to gather the dross into one corner of the bath surface with a simple stainless steel implement, and then to lift the lump of dross out of the bath with a flat stainless steel spatula such as is obtainable in any hardware shop. Tilting the spatula after lifting the dross allows most of the clean solder trapped in it to drain back into the bath. The rest is then put into a steel container which is provided with a lid. Most solder vendors are

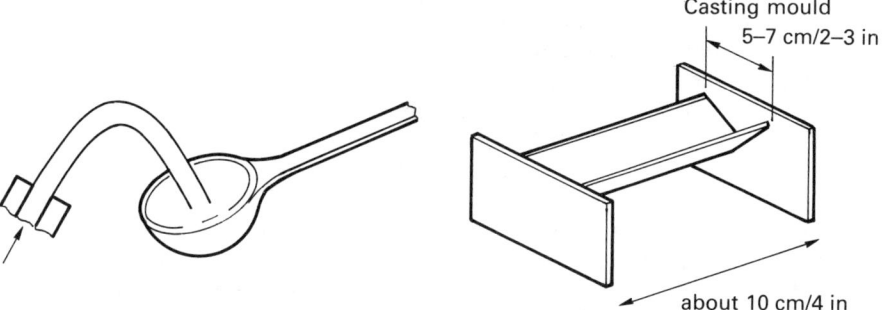

Figure 4.31 *Taking a solder sample*

prepared to take back solder dross from their customers once a sufficient quantity has accumulated, and will credit them for a portion of the clean solder contained in it. Depending on the circumstances, it may be advisable to demand an analysis of the returned dross for its metal content from the solder vendor.

Skimming the dross from the solderbath twice a day should be enough. Frequent skimming in order to make the solderbath look attractive, and maybe to impress visitors or the management, is not only unnecessary but also increases the amount of dross which forms on the machine. An existing layer of dross helps to protect the bath from further oxidation.

4.7.7 Hygiene and safety

Lead and its toxic nature

Solder contains about 40% lead, and lead is toxic. However, if treated and handled with common sense, there need be no danger to any person working with solder in any of its many forms, such as solderwire, solder ingots, molten solder or solderpaste, provided a few basic facts are recognized.

Lead can be absorbed into the human body only through the digestive system, while skin contact is harmless. Put crudely, the basic rule is therefore 'Do not eat lead, in any of its forms.' In practice, this means strict observation of a number of simple rules.

Don't smoke, eat or consume drinks on the job. Having handled solder or dross, wash hands thoroughly before smoking, eating or drinking. The reasons for these rules are obvious: handling a cigarette or food with solder-contaminated fingers carries the danger of ingesting lead-containing solder. Even small amounts matter, because lead is a cumulative poison which is not excreted by the normal bodily functions. Quite apart from that, soft drinks should not under any circumstances be consumed near any part of an electronic assembly line. Fruit juices and fruit sugar form reaction products on metallic surfaces which severely affect solderability, and which are difficult to remove. Aerosol, formed for example by a fizzy soft drink, can be fatal for the solderability of a circuit board.

An often neglected danger point is the habit of chewing fingernails. The spaces under the fingernails are notorious collectors of dirt and dust, picked up from everything that is being handled or touched (as any forensic scientist knows). Habitual nailbiters should therefore on no account be given jobs which involve the handling of solder in any of its forms.

Dross must be handled with caution and common sense: it contains a proportion of powdery lead oxide, which is more dangerous than metallic lead because it is absorbed more readily into the digestive system. Hence the rule of placing dross skimmings into a metal container which is fitted with a lid. Dross must be handled gently, so that it does not form a cloud of dust. It is a sensible precaution to issue a dust mask to all operators who have to handle dross in larger quantities. On the other hand, there is no reason for wearing a dust mask when skimming dross from a soldering machine, because in this form the oxide is trapped within the bulk of the metal and its adhering flux residues.

Handling molten solder

Molten solder is quite hot and must be treated with respect. The main danger when handling it arises from the fact that it will spit and splatter violently when it comes in contact with a wet or even slightly damp surface. This spitting is caused by the explosive evaporation of any surface moisture trapped under the molten metal. Hence the strict rule, already mentioned, that every implement which comes into contact with molten solder must be meticulously dried by preheating. By contrast, small amounts of liquid spilled onto the surface of molten solder will hiss away quietly without spitting.

Drops of molten solder on the skin can be painful, and cause small but relatively harmless local burns. To stop the pain quickly, touch a cold metal surface, or run cold water onto the burn. Never apply oil or grease, which will only make matters worse. Application of a small amount of burn-ointment, which normally contains picric acid, stops the pain, promotes quick healing and prevents blistering. It is useful to keep a tube or tin of it handy near any machine or bench where molten solder is handled.

On the other hand, even a minute drop of molten solder which reaches the eye can fatally damage sight. It is therefore important to issue all operators who have to handle molten solder, for example when taking a sample of solder from the solderwave or when emptying a solderpot, with safety goggles. There is, however, no need to wear goggles when removing the safety screen to watch a board passing over the solderwave, or when skimming a solderpot (provided the skimming tool is dry).

Wearing protective gloves is a wise precaution when sampling the solder or cleaning the wavenozzle. When handling larger amounts of molten solder, such as when emptying a solderpot, it is advisable to wear an apron or protective clothing. Solderdrops clinging to clothing are easily removed by touching them with a small soldering iron set at a low soldering temperature, provided the material is entirely of natural fibre such as wool, cotton or linen. With synthetic fibres, this method would not work, and scraping or plucking the solder off is the only way.

When faced with the task of handling larger amounts of molten solder, it is best to plan one's strategy in advance: decide what you want to do, and how best to do it, before you start. Have all implements and receptacles dry and ready in their proper places. Do not hurry, and move slowly and with deliberation.

4.8 The role of adhesives in wavesoldering

SMDs must be anchored to the board before they are wavesoldered because they have no leadwires or legs with which to hang on to the substrate. Adhesive joints have been found to provide the best answer to the problem. Their mechanical properties are adequate for the task, and they can be broken without undue force if necessary.

4.8.1   Demands on the adhesive and the glued joint

The glued joint must be strong enough to hold the component securely to the board during any handling operation which may precede the soldering process, for example the insertion of wired components with a 'mixed' board, and above all during the wavesoldering procedure itself. These mechanical loads are only modest, at most of a magnitude of a few newtons. It is important, however, that the joint does not distort or disintegrate under the influence of the flux solvents during the preheating stage, and especially during the passage through one or two solderwaves. It should also stand up to the chemical and mechanical stresses of any subsequent cleaning operation, which may involve the action of hot solvents or water, and possibly powerful jets or ultrasonic agitation.

Should the removal of a glued and soldered component become necessary because on inspection it has been found to be faulty or wrongly placed (Section 10.2), the joint should be capable of being broken without undue force and consequent damage to the substrate (Section 4.8.5). Finally, during the life of the assembly, the glued joints should not give off or leak any substance, particularly not one of an ionic nature, which could lower the surface resistance of the board or interfere with the function of the assembly.

4.8.2   Storage and handling behaviour of adhesives

Adhesives for SMDs are of the reactive, single-component epoxy or acrylic type. Solvent-containing adhesives and two-component reactive adhesives, which require mixing before use, are unsuitable for industrial SMD wavesoldering.

Single-component adhesives are a mixture of two ingredients, a polymer–resin and a hardener, which are capable of reacting with one another, forming a rigid structure of crosslinked molecules. This reaction requires a trigger to set it off, which may be a rise in temperature, or exposure to light in the visible or the UV range, or both these triggering agents, acting simultaneously or in sequence.

A good SMD adhesive must satisfy a number of specific requirements:

1. During storage, resin and hardener should not of course react with one another. With some adhesives, this may require storage in a refrigerator at about 5 °C/40 °F, to ensure a storage life of up to one year, which is what the industrial user expects. With many modern adhesives, refrigeration is no longer necessary, and storage times of up to one year at room temperature (say 25 °C/78 °F) are not unusual.
2. A drop of adhesive, as dispensed onto the board, may have to bridge a gap between 0.01 mm/0.4 mil and 0.3 mm/12 mil in height (the standoff height of the component) while, depending on the geometry of the layout, its base may have to fit into a very narrow gap between two footprints (about 1 mm/40 mil with a micromelf) (Figure 4.32). This requires the dispensed adhesive to retain its shape without sagging or 'slumping'. Any sideways spread of the dispensed drop would not only lower its height, so that it might fail to contact and hold the component, but it could also spread over the adjacent solderpads, totally

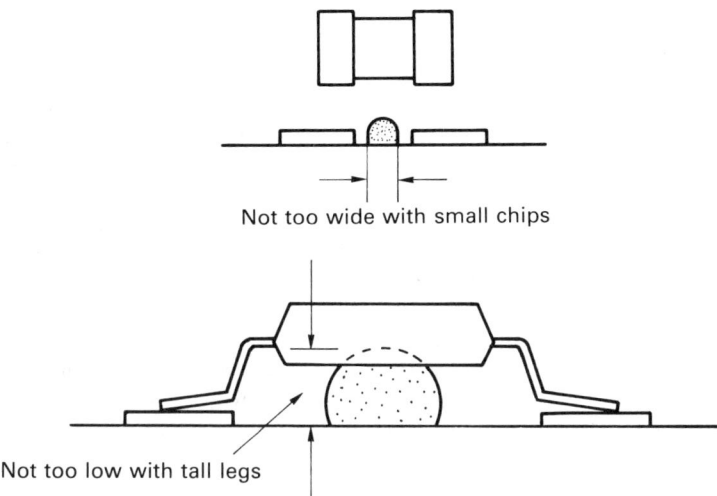

Not too wide with small chips

Not too low with tall legs

Figure 4.32 *Demands on the adhesive spot*

and possibly irreparably ruining their solderability, and thus the whole circuit board. The type of behaviour in which semiliquid substance retains its shape is called 'thixotropy'. For similar reasons, the solderpastes which are used in reflowsoldering must also exhibit thixotropy, which is discussed fully in Section 5.2.1.

Apart from a sideways slump of the adhesive drop, it would be equally fatal should one of the more mobile constituents of the adhesive leak out sideways from the drop and contaminate an adjacent solderpad. Finally, the adhesive must separate neatly from the dispensing nozzle or placement pin, without forming a tail or thread which might tip over and fall on a solderpad.

The flow behaviour of an adhesive is necessarily temperature dependent, making it more mobile at highter temperatures. Most manufacturers have succeeded in reducing this temperature dependence to a minimum. However, since a very precise dosage of the dispensed adhesive drop is of the essence, especially with very small melfs and chips, the dispensing ampoules on some placement systems are heated to a standard temperature.

As a rule, the adhesive does not sit directly on the FR4 of the board, but on the solder resist. This places certain demands on the adhesion and the surface properties of the latter which are discussed in Section 6.1.

Very often one or more conductors will pass between the solder pads of a component. It is important that these conductor tracks are not covered with a layer of solder, as might be the case with boards made by a 'subtractive' process. If they are, the solder will melt underneath the solder resist as the board passes through the solderwave. Because the solder resist starts to crinkle as the solder on which it sits melts, the result is called the 'orange peel effect'.

An SMD glued to the solder resist loses its safe anchorage when the solder underneath the resist melts, so that it is in danger of being washed off in the wave. For this reason, boards for wavesoldering SMDs should preferably be of the 'solder mask over bare copper' or 'SMOBC' type (see Section 6.1).

3. After an SMD has been placed on its adhesive dot, it must stick to it strongly enough to prevent it from shifting its position or falling off, while the board is handled between the placement of the components and the curing of the glued joints. This holding power of the uncured adhesive is called 'green strength'. Furthermore, an adhesive must be able to maintain its thixotropic behaviour and its green strength for at least 24 hours between being taken from its container, or discharged from its dispenser, and its being hardened or cured prior to soldering (open time).

4. Last, but not least, the adhesive should have a distinctive and conspicuous, perhaps luminous, colour, so that missing or misplaced dots are easily spotted. Orange or bright red seem to be the preferred shades.

4.8.3 Applying the adhesive

The precision of both the placement coordinates and the size of every individual dot of adhesive are critically important, especially with small melfs and chips: a misplaced dot, or one which is too large and becomes squeezed out during placement, is liable to cover a solderpad and make it unsolderable. Removing cured adhesive from a pad surface is one of the most costly and hazardous operations in corrective soldering (Section 10.1.1). On the other hand, a dot of insufficient height may not connect with the component it is supposed to hold.

In most situations adhesive dots have to be of varying height, for reasons explained above. Screen or stencil printing (for details, see Section 5.2.2) is therefore rarely suitable for putting them on the board, because both these methods produce deposits of uniform height. Instead, industry has a number of alternatives to choose from.

Sequential application of single dots

Dispensing the adhesives from the nozzle of a cartridge or ampoule is widely practised. Most vendors offer adhesives in air-pressure operated ampoules, which can discharge the content in a controllable manner. For manual placement, the pressure impulse in the hand–held ampoule is controlled by the operator through a footpedal or a press-button. The operation is simple, and misplaced adhesive can be wiped off, with solvents supplied by most vendors.

Dispensing adhesive from ampoules can be mechanized in two ways:

1. For putting down dots of adhesive onto boards before the components are placed, highly developed automatic equipment, which is capable of being programmed, is on the market. With these machines, the dispensing ampoule

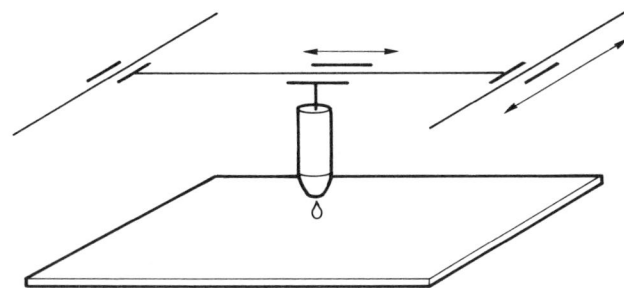

Figure 4.33 *Dispensing gantry*

is mounted on an xy movable gantry (Figure 4.33). The precision and repeatability of dosage and placement are of a high enough order to meet the requirements of adhesive application to modern, closely populated boards.

The distance between the nozzle tip and the board can be varied and programmed, not only to suit the size of individual dots, but also to enable two dots to be placed on top of one another (piggy-back) to cater for exceptionally high standoffs. A 'suck-back' at the end of a delivery impulse prevents the formation of a dangerous string of glue when the nozzle is lifted off. The dispensing program can be derived from the software of the board layout. The speed is limited by the various manoeuvres which have to be performed between discharges, such as starting and stopping the dispensing head, adjusting its position in the vertical z axis, actuating the displacement mechanism, and executing the 'suck-back'. Within these limitations, vendors claim maximum achievable dispensing rates of up to 17 000 dots/hour, i.e. over 4 dots/second.

A new version of this dispensing concept was recently introduced, which uses technology derived from ink-jet printing. The duration of an individual discharge is as short as 0.001 seconds, which makes it possible to use 'on-the-fly' dispensing, where the dispensing head does not stop moving. Thus, dispensing speeds of 20 dots/sec, i.e. 72 000 dots/hour, are possible. Jet dispensing units can be integrated in-line with high-speed pick-and-place equipment. The major adhesive vendors are able to supply adhesives for jet-dispensing.

2. Many automatic sequential pick-and-place machines are or can be fitted with an adhesive dispensing station, which puts down measured dots of adhesive and precedes the component placement station. The dispensing details are the same as with the dispensing gantry. With simultaneous pick-and-place machines, the adhesive is often placed directly on the underside of each component in the time interval between pickup and placement (see Section 7.3.2).

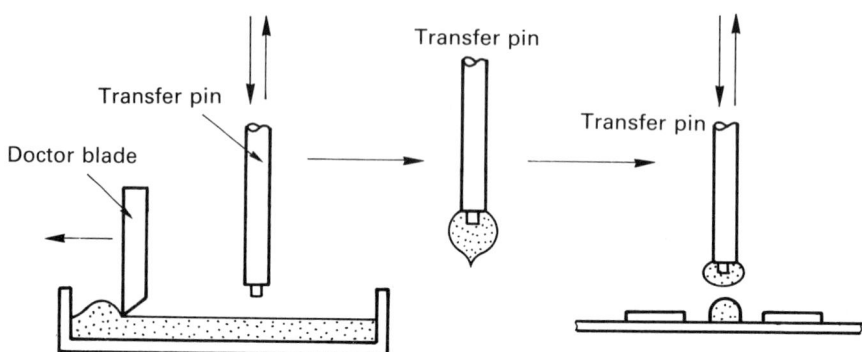

Figure 4.34 *Pin-transfer of adhesive*

Simultaneous placement of all dots on one board

The pin-transfer principle (Figure 4.34) allows the deposition of all adhesive dots on a board in one single operation. Each pin collects a quantity of adhesive from a tray which is covered with a layer of adhesive. The thickness of this constantly replenished layer is controlled by a doctor blade. The amount of adhesive picked up and transferred is governed by the diameter of the transfer pin. The pins themselves are mounted on a tooling plate, being fixed in holes which have been drilled in the appropriate positions. The tooling plate slides between the pickup tray and the board, being lowered onto the board when it has reached the correct position. This simple method is particularly suitable for short runs and small-scale production.

4.8.4   Curing the adhesive joint

After all components have been placed on the 'green' adhesive, the joints must be cured. Curing transforms the adhesive from a viscous mass into a firm, solid body. This is achieved by triggering a reaction between the two constituents of the adhesive, which crosslinks the individual mobile polymer molecules of the resin constituent into a coherent, semicrystalline mass. It is important that the joint does not shrink or crack during curing.

With most epoxy and acrylic adhesives, heat provides the required trigger. As is natural with any chemical reaction, the higher the temperature, the shorter the time in which the crosslinking reaction proceeds throughout the joint (Figure 4.35). With a modern adhesive, all joints on a board are sufficiently hard for wavesoldering after two to three minutes at a curing temperature of 120 °C/250 °F.

Curing is normally carried out in an infrared oven, similar to or identical with a reflowsoldering oven (Section 5.4.4). It must be remembered, however, that with the exception of small melfs and chips, the adhesive joint is shielded from the

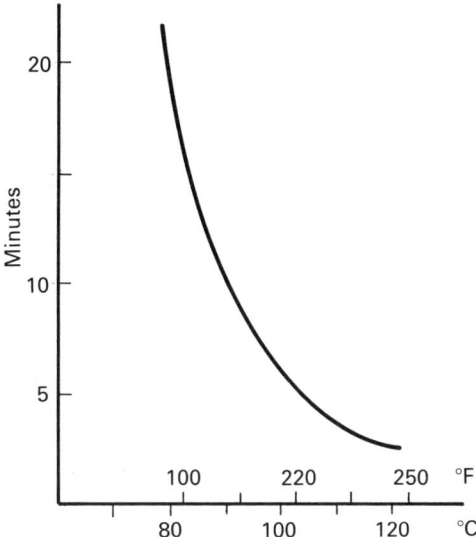

Figure 4.35 *Hardening curve of an epoxy adhesive*

radiation by the sometimes large and thick component. For adhesive curing, therefore, an efficient convection oven is best. A nitrogen atmosphere in the oven is useless in this context, because nitrogen does not pick up the infrared energy from the emitters and therefore it remains cool (Section 5.4.2).

As an alternative to heating, or possibly in addition, exposure of the adhesive to ultraviolet light is also effective in promoting the curing process. This mechanism will only work if the adhesive, or at least some of it, is 'visible' to the light source and not covered by the component (Figure 4.36). Simultaneous exposure to heat and light can shorten the curing time by up to 50 per cent.

Recently, a type of adhesive has become available which contains a light-

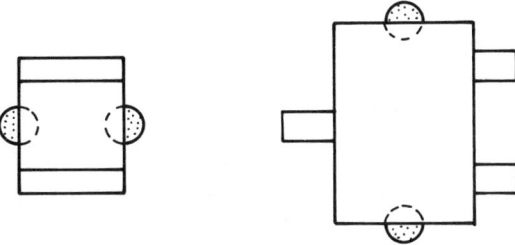

Figure 4.36 *Location of adhesive spots for ultraviolet hardening*

sensitive initiator. This has the effect of adding a 'hairtrigger' mechanism to the crosslinking reaction: exposure of the dots of adhesive to a dose of visible light for one-half to four minutes (depending on the intensity of the light source), before the components are placed, firms up the adhesive and increases its green strength without starting the crosslinking proper. Once the components are placed, which must be done within 30 minutes after the light exposure, full curing of the joints can be completed in a much shorter time or at a much lower temperature than is the case with conventional adhesives.

4.8.5 The glass transition temperature

No assembly process is entirely free from faults, and occasionally glued and wavesoldered components must be removed from a board. However, while soldered joints can be unsoldered, a glued joint cannot be 'unglued' but must be broken. The so-called 'glass transition' mechanism of crosslinked polymers makes it possible to break cured joints without damaging the board. Above a certain temperature, which is called the glass-transition temperature (T_G) the crosslinked bonds between neighbouring molecules begin to open and the molecules start to regain their mobility. For a cured joint this means that the adhesive loses its rigidity and begins to behave like a highly viscous substance. Depending on the type of adhesive and on its curing history, its T_G lies between 35 °C/95 °F and 80 °C/175 °F. This means that after heating it to its T_G, the joint can be separated without much force. In desoldering practice (Section 10.2) this is done by twisting the SMD. The resulting shearing force will readily break the joint. Pulling the joint apart would require a much greater force than twisting, and is liable to damage any conductor tracks which may pass underneath the glued joint.

References

1. Brit. Pat. 798 701, 1956, Fry's Metals, *Improvements Relating to Soldering Components to Printed Circuits.*
2. Brit. Pat. 639 178, 1943, Eisler and Strong, *Manufacture of Electric Circuits and Circuit Components.*
3. Kirby, P. L. and Pagan, I. D. (1987) The Origin of Surface Mounting, *Proc. Europ. Microelectronic Conference, Bournemouth, UK.*
4. Klein Wassink, R. J. (1989) *Soldering in Electronics, 2nd ed.*, Electrochemical Publications, Ayr, p. 489.
5. Smernos, S. and Strauss, R. (1988) Low Temperature Soldering. *Electronic Communications*, pp. 148–151.
6. Klein Wassink, R. J. (1989) loc. cit., pp. 498–500.
7. Oates, W. A., Todd, D. D. (1962) Kinetics of the Reduction of Oxides. *J. Austral. Inst. Met.*, **7**, pp. 109–114.
8. Leibfried, W. (1979) Soldering without Flux. *German Min. for Res. & Technol. (BMFT), Rep. T.*, pp. 79–164 (in German).

9. Keeler, R. (1990) New Fluxless Soldering Process. *El. Pack & Prod.*, **30**, 10, p. 15.

10. Albrecht, J., Scheel, W., John, W., Wittrich, H., Grasmann, K. H. and Liedke, V. (1992) Fluxless Wavesoldering. *DVS Report 141*, Duesseldorf, Germany, pp. 90–99 (in German).

5 Reflowsoldering

5.1 The reflow concept

As has been said in Section 3.1, the making of a good soldered joint needs the right amount of solder, flux and heat, in the right place, and at the right time. With wavesoldering as with handsoldering, the flux always comes first, and solder and heat together come afterwards. With all reflowsoldering methods, the heat always comes last.

To begin with, solder and flux are placed on one or both joint surfaces, either together in the form of a solderpaste or separately, first the solder in the form of a metallic coating and then the flux at a later stage. Subsequently the joints are put together. The important point is that all this happens at room temperature though, with some procedures, the solder may have been predeposited on one or both joint surfaces by a hot-tinning method.

With all reflow strategies, the assembled joints are finally heated to a temperature high enough to melt the solder, and for long enough to let it tin the joint surfaces and fill all the joint gaps. As soon as this has been achieved, heating is discontinued and the solder is allowed to solidify, the faster the better.

5.1.1 SMDs and reflowsoldering

Reflowsoldering is a much older process than wavesoldering, going far back into antiquity; under the name of 'sweatsoldering' it is used in plumbing to this day. With the advent of hybrid technology more than thirty years ago, sweatsoldering was recognized as the logical way of joining SMDs, which were specifically developed for hybrids, to the metallic conductor pattern of the ceramic substrate. Rosin-based soldercreams were already in existence, though not as yet screen-printable, and the assembled hybrid circuits were mostly soldered on simple hotplates. Professional reflowsoldering equipment and printable solderpastes became commercially available by the early seventies, when SMDs had begun to be used on conventional circuit boards.

SMDs and reflowsoldering are ideal partners. With wavesoldering SMDs, the molten solder needs help to find its way to the joints (Figure 4.14). With reflowsoldering, both solder and flux are already in place before the joints are heated by one of the several options available. Thus, for a board surface populated entirely with SMDs, reflowsoldering is the natural choice. Boards with a mixed population of SMDs and inserted components must be wavesoldered, because inserted through-joints cannot be reflowsoldered, at least at the present state of the art. However, this is only true if the SMDs share the same board surface with the through-joints of the inserted components (board construction I(a), Figure 5.2).

Figure 5.1 shows the procedural options of reflowsoldering.

Figure 5.2 shows the various ways of arranging SMDs and wired components on a board. The choices of possible soldering strategies for the different types of board construction are shown in Table 5.1.

Overhead soldering

Reflowsoldering boards with SMDs on both sides involves 'overhead soldering', for which two different strategies are possible:

Two-pass soldering

This strategy is the more common one. One side is dealt with first, involving paste print-down, placement of the SMDs, and soldering by vapourphase, infrared, or hot-air/gas convection. The second side (side two) is then treated in the same way. With vapourphase soldering, the joints on side one, which are now underneath, will melt again, with the components hanging onto the board with the help of the surface tension of the molten solder. Practical experience shows that they will not drop off unless they are heavy and have only a small number of joints, such as power transistors. With vapourphase two-pass soldering of double sided boards, such components should therefore all be placed on side two.

It is safer to use an infrared or convection oven for the second pass, because the temperature of side one, which is now underneath, can be kept below 183 °C/360°F, so that its joints do not melt again. Two-pass soldering in a vapourphase- or a protective-atmosphere oven has the advantage that the footprints of side two cannot oxidize, and possibly lose some of their solderability while side one is being soldered. Laser- or impulse-soldering of side two is of course another possible strategy for double-sided pure-SMD boards.

Single-pass soldering

The second option, of putting the components on both sides and then soldering the whole board in one single pass through the vapourphase or oven, is practicable with boards which carry a deposit of solid solder on their footprints (Section 6.3.2). With such boards, the components on side two are anchored to the board by a dot of air-hardening or curable adhesive, which is applied prior to the placement of the components (Section 4.9.3). These glued joints need not be as strong as those which hold the SMDs to the board during wavesoldering since they only have to prevent them from falling off.

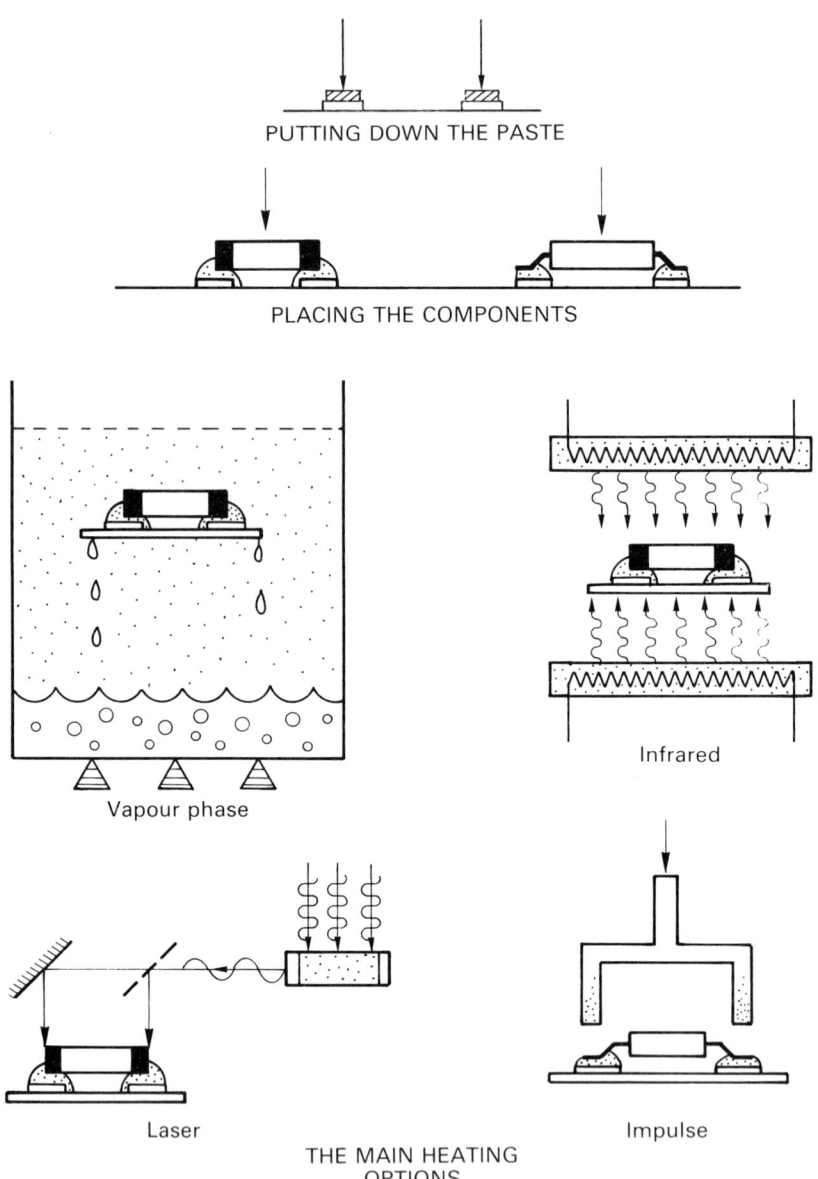

Figure 5.1 *The reflow soldering options*

Single-pass reflowsoldering with solderpaste joints on both sides is not often practised because of the problem of printing solderpaste and placing components on both sides of a board before soldering the joints. Soldering itself is only possible

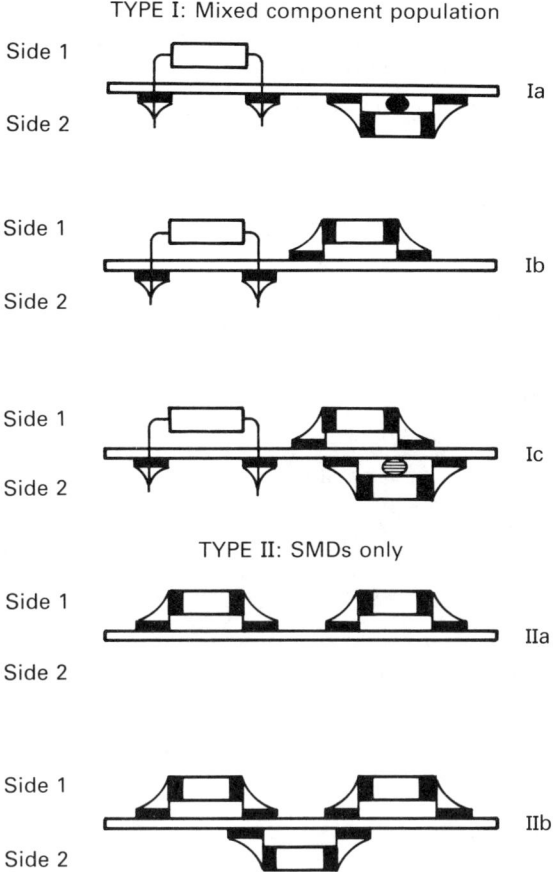

Figure 5.2 *Options of board construction*

if the components on side two are very light, and if the solderpaste has sufficient 'wet strength' to stop them from falling off before the solder melts and its surface tension takes over. Feasibility trials will certainly be required, possibly with some benchtop equipment (Section 5.4.4) before using this strategy.

Visual inspection

Visual inspection between the application of adhesive and placing SMDs, and between placing SMDs and curing the adhesive, is well worth while: it is easy to correct missing or misplaced adhesive spots, or misplaced, misaligned or missing SMDs before the adhesive is cured. Corrections after curing are difficult, often time-consuming, may damage the board, and above all, can be very expensive.

Table 5.1   *Board constructions and their soldering strategy options*

Type I: Mixed component population

Board construction I(a) – Strategy I (Recommended by the USA Surface Mount Council (SMC), 1988)

Side 1:   Insert wired components, crimp lead ends
             (Visual inspection)

                                                        Turn board over

Side 2:   Apply adhesive,★ place SMDs
             (Visual inspection)
             Cure adhesive

                                                        Turn board over

             Wavesolder side 2
             (Inspect, clean, test)

Board construction I(a) – Strategy II

Side 2:   Apply adhesive,★★ place SMDs
             (Visual inspection)
             Cure adhesive

                                                        Turn board over

Side 1:   Insert wired components, do not crimp leads†)
             (Visual inspection)
             Wavesolder side 2
             (Inspect, clean, test)

Board construction I(b) – Strategy I (Recommended by USA SMC, 1988)

Side 1:   Print solderpaste
             (Visual inspection)
             Place SMDs
             (Visual inspection)
             Reflowsolder
             Insert wired components††
             Wavesolder side 2
             (Inspect, clean, test)

Board construction I(b) – Strategy II (If the board carries only a few QFPs or TABs)

Side 1:   Insert wired components
             Wavesolder
             Impulse-solder SMDs to their previously and
               suitably pretinned solderpads (Section 5.9)

Board construction I(c) (Recommended by USA SMC, 1988)

Side 1:   Print solderpaste
             (Visual inspection)
             Place SMDs
             (Visual inspection)
             Reflowsolder

                                                        Turn board over

Table 5.1 *Continued*

Side 2:	Apply adhesive,‡ place SMDs	
	(Visual inspection)	
	Cure adhesive	
		Turn board over
Side 1:	Insert wired components	
	Wavesolder side 2	
	(Inspect, clean, test)	

Type II: SMDs only

Board construction II(a) (Recommended by USA SMC, 1988)

Side 1:	Print solderpaste
	(Visual inspection)
	Place SMDs
	Reflowsolder
	(Inspect, clean, test)

Board construction II(b) (Recommended by USA SMC, 1988)

Side 1:	Print solderpaste	
	(Visual inspection)	
	Place SMDs	
	Reflowsolder	
		Turn board over
Side 2:	Print solderpaste	
	(Visual inspection)	
	Place SMDs	
	Reflowsolder‡‡	
	(inspect, clean, test)	

★Adhesive applied by pin–transfer or gantry–mounted programmed syringe (Section 4.9.2). The crimped wire ends of the inserted components make screenprinting or stencilling of adhesive impracticable.

★★Screening or stencilling of adhesive practicable.

†The underboard crimping tool of the insertion machine may knock neighbouring SMDs off the board.

††Leadwire ends may be crimped but take care that the head of the insertion machine does not knock neighbouring SMDs.

‡Any method may be used, including screening and stencilling.

‡‡Reflowsoldering a board a second time, with previously reflowsoldered SMDs hanging underneath (overhead soldering) is practicable with certain provisos (see text).

5.1.2   Reflowsoldering versus wavesoldering

The respective merits and disadvantages of wavesoldering and reflowsoldering, specifically of SMD joints, are best compared under two separate aspects of making soldered joints, the metallurgical and the operational ones.

Metallurgical aspects

Wavesoldering

With wavesoldering, all joints are filled from a solder reservoir which holds from 40 kg (90 lbs) up to over 200 kg (450 lbs) of molten metal, held at an accurately controlled and monitored temperature (normally around 250 °C/480 °F). The composition of the solderbath changes slowly over the course of time: the solder can and will pick up copper from the solderpads and other metals from the component leads; its tin content will gradually drop, because tin oxidizes more quickly than lead, and because the joint surfaces pick up more tin than lead. This means that there is need for occasional check analyses of the metal bath (Sections 3.3.3 and 4.8.5).

All SMD joints are fully exposed to the washing action of the solderwave, which removes the greater part of the intermetallic products of the reaction between the solder and the substrate and returns them to the solderbath, where they are greatly diluted. Thus the solder content of every joint can be regarded as a very small part of a very large metallurgical system.

With wavesoldering, the 'confrontation interval', by which is meant the length of time during which the joint surfaces confront molten solder and react with it, is short, from two to five seconds (Sections 3.2 and 5.4.1). Therefore, the brittle intermetallic layer, which is the product if the reaction between the two, is thin ($<1\ \mu m/0.04$ mil).

A wavesoldered joint solidifies quickly: its solidification interval, i.e. the time span between its passage through the solderwave with its temperature of about 250 °C/480 °F and its reaching the solidification temperature of 183 °C/361 °F, is short too, two seconds at most, and this makes for a fine-grained structure.

As a consequence of all this, a wavesoldered SMD joint, whether on the ends of a melf or chip, or under the lead of an IC, has a favourable metallurgical structure: it is finegrained with few, if any, intermetallic solids dispersed in it, and the brittle layer of intermetallic compound between the solder and the substrate (Section 3.2) is thin. The tubular joints in the throughplated holes do not concern us here; in any case, they are not washed by the solderwave directly, but are filled by capillary action.

Reflowsoldering

With reflowsoldering, matters are drastically different: every individual joint forms a separate, closed metallurgical system with all the products of the solder/substrate reaction trapped in it. This has a pronounced effect on the joint structure, particularly since a very small volume of solder interacts with a relatively large surface area of substrate.[1]

Methods of supplying the soldering heat

Equilibrium heating methods

The various reflowsoldering methods differ mainly in the manner in which the joints are heated. Some heating methods present a so-called equilibrium situation, in which all joints reach the well-defined and constant temperature of the heating medium after a certain time. With vapourphase soldering, this is the temperature of the saturated vapour of the working fluid, i.e. its boiling point. With convection soldering, in a convection oven or a hot gasjet, it is the temperature of the heating medium. With impulse soldering, it is the temperature of the soldering tool. In an equilibrium situation, overheating is impossible (Section 5.5.1).

The rate of heating and cooling varies widely between the different methods. With vapourphase soldering, the joints approach their end-temperature asymptotically, in a matter of minutes, but not all joints necessarily at the same time. Consequently the confrontation interval too is not the same for every joint; it may extend to between 30 seconds and one minute. The speed of the subsequent solidification depends on the design of the equipment. With modern systems, it can be less than 5 to 10 seconds.

With convection soldering, the rate of heating is faster, depending on the type and the efficiency of the equipment, and the same is true of the solidification speed. The confrontation interval can be kept from a few to up to 30 seconds, depending on the scale of the operation.

With impulse-soldering, where only the joints themselves but not the whole board are heated, both the confrontation interval and the solidification period can be controlled and kept within one or two seconds.

Non-equilibrium heating methods

Infrared and laser soldering present non-equilibrium situations: the temperature of the heat source is well above the intended soldering temperature, and the thermal energy is transferred to the joints by heat radiation. The temperature reached by the joints depends principally on the duration of their exposure to the heat source. The higher the temperature of the heat source, the more critical is the timing of that exposure. With infrared soldering, heating rates depend on the temperature of the emitters, and are mostly relatively slow; confrontation intervals range between 10 and 30 seconds. Solidification periods can be kept within about ten seconds. By contrast, with laser soldering both confrontation intervals and solidification periods are counted in milliseconds.

Whatever the soldering method, the metallurgical structure of a joint determines its reliability and life expectancy. The finer the microstructure, and the thinner the brittle intermetallic layer between the solder and the substrate, the better both of them will be. As long as confrontation intervals can be kept well below a minute, or better 30 seconds, and the solidification periods below ten seconds or so, all will be well.

Figure 5.33 gives a comparison between the temperature profiles of wavesoldering and all current reflowsoldering processes.

Operational aspects

Any comparison between different soldering methods ought to take the following requirements into consideration:

1. Minimum percentage of faulty joints, minimum rate of rework
2. Joints with optimum metallurgical structure and operational reliability
3. Lowest possible thermal and mechanical stress for boards and components
4. Soldering results independent of complexity of the board, SMD-population density and component shape
5. Suitability both for integration in CIM in-line production, and for coping with lots-of-one short runs
6. Low production costs
7. Environmental acceptability.

Wavesoldering

A great advantage of wavesoldering is its status as a mature and well understood technology of more than thirty years' standing. Its main drawback is the critical dependence of its soldering efficiency, as measured by a consistently low joint-defect rate, on the correct setting, high stability and consistent monitoring of a large number of operating parameters, many of them interlinked, and involving the management of a large mass of moving molten solder. The list of parameters is formidable: the density, the freshness and the water content of the flux, and the nature of the flux cover; the temperature of the solder and its purity; the speed of the solderpump; the shape and stability of the solderwave; the temperature of the preheating stage; and the speed of the conveyor, to name the main ones. It is not surprising that the number of microprocessor-controlled wavesoldering machines which are on offer is growing steadily.

While these are aspects of wavesoldering in general, wavesoldering SMDs raises further problems: SMDs must be glued to the circuit board and the adhesive must be cured before soldering. The board layout must take the 'shadow effect', which an SMD exerts on the wave, into consideration and many board configurations have only one optimal direction of travel over the wave. With wavesoldering, it is difficult to influence the amount of solder in a joint, especially with melfs and chips; thus the control of the joint profile becomes problematical. Multilead SMDs with a joint pitch of below 1.27 mm/50 mil become difficult to wavesolder in a normal atmosphere without bridging. A pitch of 0.75 mm/30 mil is presently regarded as the ultimate limit of wavesolderability. Wavesoldering under a cover of nitrogen, and with less than 50 ppm of oxygen present, has pushed this limit to 0.5 mm/20 mil and below.

Reflowsoldering

The great attraction of all reflowsoldering methods from the operational point of view is the relatively low number of soldering parameters which need to be set and monitored: with vapourphase soldering, it is the duration of dwell in the vapour chamber; with infrared ovens, the temperature of the heating elements and the conveyor speed; with laser-soldering, the duration of the energy pulse, and with impulse-soldering the temperature of the soldering tool and the duration of its dwell.

It is a general rule that the fewer the operational parameters of a manufacturing process, the smaller the chance of something going wrong, and the higher the probability of keeping the fault rate and the amount of corrective soldering down. The ideal 'first time right, every time right' is easier to realize with one of the reflowsoldering methods than with wavesoldering.

The great attraction of reflowsoldering is that in every deposit of solderpaste, both flux and solder are factory-fresh. A growing problem is the tendency of the footprints, which must be covered with solderpaste, and the distance between them (their pitch) to get progressively smaller, especially narrower. This is beginning to push the technologies of solderpastes, and of printing them, towards the limits of technical and commercial feasibility. Systems by which depots of solid solder can be placed on SMD footprints by the board manufacturer (Section 6.3.2) instead of the board user may provide an answer to this particular problem.

Until the advent of the nitrogen-filled wavesoldering machine, the answer to the wave-versus-reflow dilemma was simple: do not use wavesoldering, unless you must. Even then, if you are faced with a board construction of type I(a) (Figure 5.2), it may be worth considering changing the board architecture to type 1(b) and avoiding the need to wavesolder SMDs. The trade-off of lower reject rates against costs of redesign, loss of board real estate and maybe having to buy a reflowsoldering installation may still be favourable. Since wavesoldering under nitrogen has pushed the limits of wavesolderability into a region where solderpaste-printing is beginning to get problematical, and wave-or-reflow dilemma must be solved on its respective merits in each particular case. A checklist on which this choice could be based will be found in Section 5.9.

5.2 Solderpaste

Solderpaste is the most widely used medium for putting down both solder and flux in one single operation on a footprint for subsequent reflowsoldering. The success of a reflow process which involves solderpaste depends in the first place, and critically so, on the quality of the paste which is used.

It is essential that every footprint on every board of a production run receives without fail its solderpaste deposit of an identical and predetermined outline and thickness. On each one of them, and on every board, the solderpaste must be of known and unvarying composition, with a consistently satisfactory soldering behaviour. Otherwise, faulty or open joints or solder bridges between neighbouring joints are bound to result, and the soldered board will not be able to function. This means that 'soldering success' as defined in Section 9.1.1 will not be achieved, with all the consequences in corrective soldering, additional costs and loss in quality which this failure entails.

5.2.1　Operational requirements

It follows that the properties and the behaviour of the solderpaste must be closely defined, maintained, and monitored to ensure that at all times it meets the following long list of requirements:

- The paste must be easily dispensable or printable in precisely predetermined and repeatable quantities.
- It must be of a stable consistency, without variation from batch to batch. The paste must not separate out in storage nor during use. These properties should not be unduly sensitive to the working environment such as temperature and relative humidity.
- The paste printdown or deposit must firmly remain where it was put down, retain its shape and have sufficient 'tack' to hold every component firmly in place between placement and soldering.
- Its soldering behaviour must be impeccable: the solder in the paste must melt at the correct temperature, and instantly and perfectly wet solderpads and component connectors.
- All the solder from the paste printdown or deposit must finish up in the soldered joint, even though some of the paste may have been squeezed out of the joint during placement. None must be allowed to stay behind outside a joint in the form of a bridge between neighbouring footprints, or as individual prills or globules.
- Finally, the flux residue which remains after soldering must, if necessary, be removable by one of the presently available and acceptable methods.

This complex list of requirements has turned the formulation and the manufacture of solderpastes into a science, if not an art, which involves such diverse disciplines as metallurgy, physics, colloid chemistry and printing technology, to name just a few.

The successful and economic running of any reflowsoldering line which employs solderpaste depends critically on the correct, predictable and constantly repeatable performance of that paste. To assure the user of its quality and to enable him to check it, several standard specifications have been formulated in cooperation between paste users and producers, two important ones being ANSI/IPC-SP-J-STD 005 (USA) and DIN 32513 (Germany).

5.2.2　Standard specifications

Solderpastes contain finely powdered solder with particles of closely defined shape and size-distribution. Their composition conforms to one of the standard specifications for solders for electronic use which are outlined in Table 5.2. The solder particles are dispersed in a viscous medium which contains soldering flux, a solvent or carrier, and various additives which control the physical behaviour of the paste during storage, dispensing or printing, in the period between deposition and soldering, and finally during the exposure to the heat of soldering.

The flux portion of the liquid phase conforms to one of the standard specifications for soldering fluxes for electronic assemblies, though recent and continuing developments in soldering and cleaning technology make it difficult to fit some fluxes into existing standard specifications (see Sections 3.4.3 and 5.2.4). The standard specifications also inform the user how to measure and verify such parameters of consistency and printing behaviour which affect him or her.

Table 5.2 *Solderpaste parameters covered by US and German standard specifications*

Parameter	IPC-SP-J-STD 005 (USA)	DIN 32513 (D)
Solder:		
Composition	Melting point below 427 °C/800 °F	Solder to DIN 1707 60% Sn, 63% Sn, 62% Sn/2% Ag
Particle:		
size[2]	4 ranges	3 ranges
shape	spherical or elliptical	spherical only
surface	no 'satellites'	not specified
Flux: (including solvents, etc. see Section 3.4.2)	to IPC-SF-818 or QQS 5625	to DIN 8511, part 2
Paste:		
Metal content	75–92% weight	no weight % range specified
Viscosity	in centipoises no values specified measured with a Brookfield pump viscometer	in centipoises no values specified RVTD or Malcolm spiral
Shape retention (slump)	circular testpattern	testpattern of fine-pitch pads
Wetting test on Cu	required, not quantified	required, quantified
Tack test	quantified	not quantified
Solderballing	halo of discrete balls, <75 microns diameter allowed	no discrete balls, but up to 3 adjacent 'satellites' allowed
Flux residue	as specified in IPC 818	Test for effect on surface insulation resistance specified

5.2.3 Solderpowder

Metal percentage and its effects

Standard specifications, as well as paste manufacturers, usually give the metal content of a solderpaste in per cent weight. The user, on the other hand, is more interested in the volume percentage of the solder in his paste, because he has to fill a joint gap or make a solder fillet, and both these tasks demand a definite volume of solder. Therefore the user needs to know how thick the paste printdown has to be or how much paste the syringe dispenser must discharge to give the volume of solder needed. Table 5.3 converts the weight percentage of the solder content of a paste into volume percentage, using the density of 63% Sn solder (8.5) and assuming a density of the flux portion of 0.9 g/ml.

The table shows, for example, that in order to produce a solder depot 0.2 mm/8 mil thick, a paste with a metal content of 90% weight will need a printdown thickness (wet-thickness) of 0.4 mm/16 mil, while with an 80% paste the wet-thickness will have to be 0.6 mm/24 mil.

Generally speaking, the higher the metal content of a solderpaste, the higher is its viscosity, which in turn determines its suitability for a given method of application. However, other factors such as the particle-size distribution and the nature of the flux also affect the printing or dispensing behaviour of a solderpaste. It is therefore advisable to follow the recommendations of the vendor regarding the best grade of paste for a specific method of application.

Solder composition

Reflowsoldering, like all soldering methods used for electronic assemblies, demands tin–lead solders of eutectic or near-eutectic composition. The eutectic (Section 3.3.1) with 63% tin/37% lead melts at 183 °C/361 °F, the lowest melting point within the tin–lead series. A small addition of silver to the eutectic tin–lead solder shifts the composition towards the ternary tin–lead–silver eutectic and lowers its melting point to 178 °C/352 °F.

Apart from this obvious advantage, there are two more reasons why such tin–lead–silver solders find increasing use in SMD soldering. First, the silver addition slows down the rate of attack of the molten solder on silver-based substrates such as the metallized faces of certain chips (Section 3.3.3). Secondly, silver has been found to improve the strength and fatigue resistance of soldered joints.[3]

Table 5.3 *Weight per cent versus volume per cent of solder*

Weight % solder	Volume % solder	Best suitable for
90	50	Stencil and screen printing
85	38	Screen printing, Syringe dispensing
80	30	Syringe dispensing
75	24	Syringe, Pin transfer

In international practice, the choice of solder compositions for use in solderpastes has narrowed down to three basic types, which are all covered by respective national standard specifications (Table 5.4):

I.	60% Sn, 40% Pb	melting range 183–188 °C/361–370 °F
II.	63% Sn, 37% Pb	melting point 183 °C/361 °F
III.	62% Sn, 2% Ag, 36% Pb	melting point 178 °C/352 °F

Most manufacturers offer pastes for special purposes, containing solders with compositions different from the above three. For example, indium-containing solders slow down the leaching action of the molten solder on the gold-based metallization of LCCCs.

Pastes with solders of melting points other than the Sn–Pb eutectic are available for special soldering tactics. For example, a sequence of pastes with graded melting points can be used for 'sequential soldering', where an assembly is reflowsoldered more than once: the first joints are made with a high-melting solder, later ones with solders of progressively lower melting points. This prevents the de-soldering of the first joints, while the later ones are being made. A typical sequence of alloys would be the following:

57% Sn, 43% Bi	melting point 139 °C/282 °F
63% Sn, 37% Pb	melting point 183 °C/361 °F
96.5% Sn, 3.5% Ag	melting point 221 °C/430 °F
10% Sn, 2% Ag, 88% Pb	melting range 302–310 °C/576–590 °F

Sequential soldering has, for example, been proposed for reflowsoldering boards of Type II(b) (Section 2.2), with SMDs on both sides, soldering the first side with a higher-melting solder than the second one. Practice has shown that both sides can be soldered with the same type of paste without the components of the first side falling away from the board while the second side is being soldered, provided they are not too heavy (Section 5.1.1). Sequential soldering needs precise temperature control, and may require two separate soldering systems. As a rule, its use is only worthwhile with special tasks or assemblies, where costs are a secondary consideration.

Table 5.4   *Corresponding solder standards*

Type	UK	USA	Germany
	BS 219	QQ-S-571E	DIN 1707
I.	Grade KP	Sn 60	L-Sn 60 Pb
II.	Grade AP	Sn 63	L-Sn 63 Pb
III.	Grade 62S	Sn 62	L-Sn 63 Pb Ag★

★DIN 1707 specifies an Ag content of 1.3%–1.5%. There is a suspicion that with 2% Ag, which is slightly above the silver value of the ternary eutectic, brittle crystals of Ag_3Sn may form in a slowly solidified joint, such as may occur in vapourphase soldering.[3] See also Section 3.2.3.

Normally, pastes are made with fully alloyed solder powder, which means that all particles have the same composition and melt at the same temperature. At times, the attempt has been made to slow down the melting of the solder by adding a mix of pure Sn and 20% Sn/80% Pb powder to 63% Sn/37% Pb powder, maintaining the overall 63/37 composition. The aim was to avoid the 'tombstone' or 'Manhattan' effect, which causes chips to stand up vertically on one end during soldering. However, a far better way to avoid this inconvenient and costly soldering fault is to pay proper attention to the correct layout of the solderpads and to good solderability of the components while working with a standard solderpaste (see Sections 6.2.3 and 9.1.1).

Shape and size of solder particles

The shape and the size-distribution of the solder particles, as well as their surface condition, have a pronounced effect on the behaviour of a solderpaste during and after printing or dispensing, and during soldering.

It is now generally agreed that all solder particles should be as nearly spherical as possible, though IPC-SP-J-STD 005 admits elliptical shapes with a maximum mismatch between length and width of 1.5 : 1.0 DIN 32513 is adamant about sphericity, though it does not quantify it. The problem with non-spherical particles is their tendency to block the apertures of dispenser nozzles and the ever-narrower slots in stencils for fine-pitch work, or to clog fine-mesh screens. A few non-spherical particles locked against one another can soon start a 'traffic block', even if individually they are small enough to pass through the critical aperture. As a rule, the diameter of the largest particles in a given paste should be less than one-third of the width of the narrowest aperture through which the paste has to pass.

Another malformation of solder particles is known as 'satellites'. These are very small solder globules which adhere to the surface of individual solder particles, and they are mainly caused by unfavourable working parameters of the powdering process. The circumstances under which they can form are known, and are avoidable.[2] Satellites too can lead to blocking of nozzles or screens. On automatic production lines and with close-pitch layouts, a few blocked apertures can cause a large number of costly soldering faults.

The solder particles themselves should be free from surface oxide: IPC-SP-J-STD 005 wants them to be bright and smooth, because there must be no impediment whatsoever which could prevent neighbouring solder globules from fusing together as soon as they have started to melt. Solder particles free from surface oxide demand powdering procedures which operate in an oxygen-free atmosphere, and which by now have become routine.

If the particles do not unite, or are slow about it, they may be left behind as small prills or globules, while the bulk of the molten solder is drawn into the joints. This is the mechanism behind the so-called 'solderballing', a serious functional defect of a solderpaste which can cause malfunction of soldered boards in service. It is a recognized cause of rejection of any batch of paste which, when tested by the user, is found to be affected (see Section 5.3.3).

The size distribution of the particles is very important. Particle sizes are always given in microns (μm) diameter, one micron equalling 0.04 mil. IPC names sieving, sub-sieve size analysis, microscopy or light beam scatter as suitable methods of size determination. DIN prescribes microscopic measurement.

IPC specifies four types of powder:

Type 1: >80% weight of the powder 150–75 microns
Type 2: >80% weight of the powder 75–45 microns
Type 3: >80% weight of the powder 45–20 microns
Type 4: >80% weight of the powder 38–20 microns

At the top end of the scale, a 1% weight overshoot of size is allowed. At the bottom end, a maximum of 10% weight of the particles may be smaller than 20 microns.

DIN specifies three types of particle size distribution:

Type 1: >85% in number 75–125 microns
Type 2: >85% in number 45–75 microns
Type 3: >85% in number 20–45 microns

At the bottom end, a 10% overshoot is allowed for particles smaller by a defined amount than the bottom limit for the bulk, and below that a further 3% allowance for still smaller particles. At the top end of the scale, there is a 3% allowance for particles larger by fifteen microns for type 1 and five microns for types 2 and 3. No particles are allowed larger than that.

While size measurement is mainly the province of the paste manufacturer, the practical user has two particular concerns: at the top end of the particle-size range, he does not want to find any large 'rocks', even one of which could block the nozzle of his dispenser or the mesh of his screen.

Of equal practical importance is the absence of fine dust in the range below 10 microns, because this dust can be a dangerous contributor to the solderballing effect. Here, IPC with an open-ended lower margin of 10% weight gives less protection than DIN with an open-ended 3% allowance in numbers. In this context, counting is better than weighing: a simple calculation shows that 28 000 solder particles of twenty micron diameter, 225 000 particles of ten microns, and 1.8 million particles of five microns all weigh one milligram. However, whatever standard specifications prescribe, in practice it will be found that manufacturers of quality solderpastes are fully alive to the significance of these circumstances. Vendors realize the importance of supplying their customers with a product which will not let them down during printdown and soldering.

The advent of fine-pitch technology (FPT) has made new demands on the properties and the performance of solderpastes. In this context, pitch denotes the centre-to-centre distance between the leads of a component. It has been said that what surface-mounting technology (SMT) was to through-hole technology, FPT is now to SMT, as far as reduction in weight and size of the assembly, closer component placement and QFPs with ever more numerous, more closely spaced leads are concerned.[4]

A spacing of 1.25 mm (50 mil) has by now assumed the role of 'standard pitch'. Fine pitch starts with 0.875 mm (35 mil), progressing via 0.750 mm (30 mil) and 0.625 mm (25 mil) to 0.5 mm (20 mil) and 0.3 mm/12 mil (ultra-fine pitch). With FPT, the gap between two neighbouring footprints is half the pitch distance, and so is the width of the footprints themselves. The impact of fine-pitch technology on the task of printing the paste onto such closely spaced, narrow solderpads is discussed in Section 5.3.2.

Remembering that IPC powder Type 4 allows particles of up to 38 micron (1.9 mil) diameter, and that the narrowest aperture through which such powder has to pass during printing should be at least three times as wide as the largest particle diameter, it becomes clear that with a footprint 150 μm (6 mil), FPT powders of smaller particle size are required to ensure printdowns with consistent solder contents. Size ranges of 20–30 micron (1.25–1.5 mil) are reported to give promising results. In consequence, the 'fine-pitch' pastes of the major vendors are based on this type of powder. Obviously, such narrow size ranges make high demands on the technology of both powder and paste manufacturers.

However, particle size is not the only parameter which characterizes an FPT paste. Other relevant factors will be discussed presently.

5.2.4 The flux and its residue

In the context of the residue left by a solderpaste, it is logical to consider the whole of its non-metallic portion, which contains one or several solvents, thickeners, thixotropic additives and possibly other ingredients, apart from the soldering flux itself. When considering the residue left by a solderpaste after soldering, it will be convenient to call this whole cocktail 'flux', and its residue 'flux residue'.

Until recently, the flux in a classical solderpaste conformed to one of the standards for electronic fluxes, such as MIL-F-14256 RA or RMA, BS 5625 Classes 4, 5a or 5b, or 7, or ISO DP9454 Classes 1.1.3, 2.1.3, 2.2.1 or 2.2.3 (Sections 3.4.2 and 3.4.3). These fluxes were designed to be left on the board after soldering without causing corrosion or lowering the insulation resistance, in a normal, benign environment, and to be removable with a standard CFC solvent.

The recent turmoil in the world of cleaning (Sections 8.3.4–8.3.6) has changed the attitude of the electronics industry not only towards cleaning, but also towards the fluxes whose residues can make cleaning necessary. The growing importance of reflowsoldering has focused attention on solderpaste residues, and in turn spurned the paste manufacturers to develop new types of pastes with residues that meet these new demands. Since the residue from reflowsoldered solderpaste is strictly localized, and not likely to interfere with ATE testing with needle adaptors, the physical presence of a coating of flux residue is not very important. What matters is the effect of the flux residue on insulation resistance, especially with fine pitch, high pin-count components, and on the conformal coating or lacquering of a finished board.

As a result, new pastes have reached and continue to reach the market. At the time of writing (1993) they fall broadly into three classes.

The first class covers low-residue pastes, containing fluxes, which belong to the same family as the low-solids fluxes used in wavesoldering (Sections 4.2.1 and 4.2.2). Some recent ones obviously derive from the fluxes used with wavesoldering in low-oxygen atmospheres (Section 4.6). In consequence, they demand reflowsoldering in a similar atmosphere, either in a nitrogen-filled infrared or convection oven (Section 5.6.5 and 5.7.1) or by a vapourphase technique, where the vapour of the working fluid is suitably low in oxygen.

Secondly, rosin-free pastes with a fully water-soluble residue are on offer. These permit simple water or modified-alcohol cleaning. This is gaining favour in situations where a cleaning process is obligatory, irrespective of the type of flux used.

Finally, some solderpastes have reached the market which are claimed to be truly without residue. One of them requires infrared reflowsoldering in a specific atmosphere which vaporizes the residue (USP 4.960.236).

5.2.5 *Printing and dispensing properties*

The consistency of a solderpaste determines its behaviour during and after printing or dispensing. Both methods of putting the paste down on a board make precise and quantifiable demands on the paste during its deposition, and on its immobility once it sits on its pad. The apparent contradiction between these two requirements is taken care of by a characteristic flow behaviour, which solderpaste shares with many dispensable substances from decorators' paint to mayonnaise, known as 'thixotropy'.

Specifically as applied to solderpaste, the term 'pseudoplasticity' has been proposed as more fitting.[5] In practice, both terms mean that the effective viscosity of the solderpaste depends on the rate of shear or the force of deformation which acts upon it. In simple terms, it moves easily when it is pushed: the harder it is pushed the more easily it moves. When left alone, it stiffens and retains its shape. In the terminology of solderpaste usage the lack of the latter property is called 'slump' or 'spread'. The slump behaviour of a given paste is naturally of great significance with fine-pitch work.

Viscosity and its measurement

Viscosity is an often quoted and certainly important property of a given solderpaste, but thixotropy makes its measurement difficult. Viscosity is defined as the resistance of the paste to movement. Measuring viscosity involves measuring the force needed to move a sample of paste in a defined way. Thixotropy means that the force needed to move the paste depends on the rate of movement of the viscometer, a Catch 22 situation which can only be resolved by accurately defining the test conditions. Even so, the viscosity of a solderpaste is given by most vendors within very wide margins, hardly ever as a definite value.

Generally speaking, the higher the metal content of a paste the highers its viscosity, other things being equal. In practice, they hardly ever are. With finer solderpowder, the paste becomes more mobile and its viscosity is lower for a given metal content. At the same time, it shows less slump, which is not surprising: it is easier to make a good mudpie out of sand than out of gravel.

The contribution of the flux portion of the paste to its flow behaviour and the shape-retention of the printdown plays an important part in its suitability for fine-pitch work. It has been shown that, if a paste printdown remains immobile during the heating stage of reflowsoldering, lateral spread of the melting solder is much reduced, and with it the danger of bridging between closely spaced solderpads.[6]

Federal specification QQ-S-571E lists a range of solderpaste viscosities set against their metal content and recommended methods of application (Table 5.5).

IPC-SP-J-SDT 005 does not prescribe or recommend viscosity ranges for a given method of paste application; it only specifies how to measure the viscosity of pastes within a given viscosity range, using a Brookfield RVTD Viscometer. DIN 32513, recognizing the complex nature of paste viscometry, prescribes measurement at three different rates of shear, and gives a choice of test methods, including one using a Brookfield RVTD and another one which simulates syringe dispensing. Again, no absolute viscosity values are suggested.

The aim of standardizing viscometry is to enable vendors and purchasers to speak the same language. The man on the shopfloor, who has to produce consistently faultless prints without too much trial and error when he opens a fresh tin, wants to know whether today's paste will behave in the same way on his printing frame or his machine as yesterday's paste.

A Brookfield RVTD or the Malcolm spiral pump viscometer will tell him that, within a margin of $\pm 10\%$. This instrument, operated in the 'Helipath' mode or one of its equivalents, is the best choice for bench viscometry: the test can be carried out without having to take any of the paste out of the tin as long as the tin is at least 5 cm/2 in high and wide. Thus, little if any of the expensive paste is wasted. Before the test, the tin is stirred with a clean stainless steel or plastic spatula to counteract

Table 5.5   *Solderpaste viscosities and methods of application (according to Federal Specification QQ-S-571E)*

Code no.	Nominal metal content (% weight)	Range of viscosities (centipoises)	Examples of application
2	90	800.000–1.000.000	Stencilling
2	88	600.000–800.000	Screenprinting
3	86	400.000–800.000	Screenprinting/Syringe-dispensing
4	81	300.000–400.000	Screenprinting/Syringe-dispensing
5	75	200.000–300.000	Pin-transfer

any settling out of the metal – it is as well to do this in spite of the advertised antisettling properties of a paste. After stirring the paste, it is advisable to wait a few minutes to let the paste recover its true viscosity before measuring it.

Viscometry requires a certain amount of skill and experience – but so do screen and stencil printing. The results are bound to show some scatter; in particular they are very operator-dependent, and it is therefore best to entrust viscometry always to one particular operator (Dr M. Warwick, Multicore, UK, private communication).

A further flow property of solderpaste has recently been introduced in the form of a 'yield limit'. This quantifies the shear force above which a body of paste begins to move, and which appears to have no direct relation to the measured viscosity of the paste.[7] A high yield limit is said to make the paste appear 'gooey' and stick to the squeegee, too low a yield limit causes the paste to form a 'wave' rather than a 'roll' in front of the squeegee, and to increase the slump of the paste printdown. Measuring the yield limit involves determining the shear force between a stationary and a rotating disc, the gap between them being filled with the solderpaste.

5.2.6 The solderball test

The success of a reflow process depends critically on the way in which the paste behaves during soldering. One important aspect of this is the capability of a paste printdown to gather all its individual solder particles into one coherent volume and into the joint which it is meant to fill, without leaving any stray solder prills behind. This property can be readily checked with the 'solderball test', a simple procedure which was originated by Philips many years ago. It is now incorporated in every relevant standard specification for solderpaste (see Table 5.2).

ANSI/IPC-SP-J-STD 005 prescribes the following procedures. Using a suitable metal stencil, three circular deposits of the paste to be tested, 0.65 cm in diameter and 0.25 mm thick, are produced on a non-tinnable support such as a thin sheet of stainless steel or a ceramic wafer. The specimen is then heated to a temperature 25 °C/45 °F above the liquidus temperature of the solder alloy in the paste (near enough 183 °C/361 °F for normal electronic solderpastes), which means a test temperature of 208 °C/406 °F. After 20 seconds, the solder in all three deposits should have melted and each have formed a single ball of solder with perhaps a 'halo' of a few individual small balls of <75 microns/3 mil at some distance. The German specification DIN 32513 is stricter, permitting 3 'satellites' adhering to the solderball itself.

This test is so simple to perform that it can and should be used on the shopfloor every time a fresh tin of paste is opened, or before paste which has been recovered from the printing frame at the end of a run is used again for printing. Some vendors of paste can supply stencils for the solderball test. Alternatively, a stencil can be punched from a sheet of metal. More simply, punching a hole through a self-adhesive paper label folded double upon itself with a normal office paper-punch will produce a stencil aperture for a paste printdown with a diameter of 5.5 mm and a thickness of 0.2 mm, which is adequate for a reproducible practical test result (Figure 5.3).

The test specimen can be heated by placing it on a hotplate, though for preference it should be floated on a small solderbath which is thermostatically held at the test temperature. In actual paste-printing and reflowsoldering practice, test temperatures tend to differ somewhat from the prescriptions of ANSI/IPC-SP-J-STD 005: if the paste is to be used for reflowsoldering in an infrared oven, a test temperature of 250 °C/452 °F is preferred. If reflowsoldering is carried out in a vapourphase installation, the test temperature will be the same as the vapour temperature, i.e. 215 °C/419 °F. It is normal practice to commence using the tested paste for production as soon as the solderball test has been carried out and proved satisfactory. It is wise, however, especially when starting with a new delivery of paste, or when testing an alternative product, to set some test specimens aside and heat them after the maximum time which will elapse between printing down the paste and soldering the fully assembled boards on a given production line. The German DIN standard prescribes waiting times of one and seventy-two hours as a regular test procedure.

The solderball test checks whether the flux in the paste is capable of retrieving all the solder particles from that portion of the printdown which has been squeezed out beyond the confines of the solderpad during the placement of an SMD, so that they do not form stray solder globules. Oxidized solderpowder, old or insufficiently active flux, deterioration during storage, or loss of volatile but essential flux-constituents during previous use may all contribute to the formation of stray solder globules. As is discussed in Section 11.2.2, such globules are a disqualifying soldering fault with most classes of electronic assemblies.

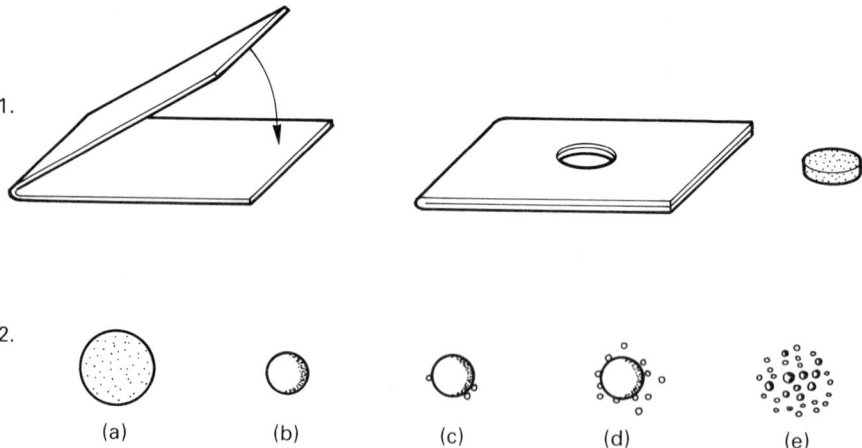

Figure 5.3 *The solderballing test. 1. Making the stencil from a self-adhesive label; the paste printdown made with it; 2. Results: (a) paste printdown; (b), (c) acceptable results; (d), (e) unacceptable results*

With multilead fine-pitch components, the solderpaste is sometimes not printed individually on every solderpad, but instead it is applied in the form of a 1.0–1.5 mm/40–60 mil wide stripe across all the leads on each side of the component. This makes fewer demands on the fine-pitch behaviour of both paste and stencil, but the paste must have the required soldering behaviour to permit this procedure. Some solderpastes for fine-pitch work remain very stiff during heating up to the soldering temperature and so will not permit stripe deposition; such pastes are unsuitable for strip-printdown.

To test the suitability of a given paste, a soldering trial with a stripe deposit must be carried out on a bare circuit board which carries some fine-pitch footprints. The stripe of paste can be put down with a hand-operated syringe dispenser, or with a special stencil with stripe apertures, which is available from some paste vendors or supply houses. The trial board is then passed through the same infrared or vapourphase reflowsoldering equipment which is used in routine production. If the molten solder has gathered on the solderpads without leaving any globules or bridges between the pads, the paste is suitable for stripe printing.

Predrying

Until recently, it was considered necessary to predry circuit boards between placement of the components and reflowing by whatever process, at a temperature between 80 °C/176 °F and 100 °C/212 °F for about 30–60 minutes. This was in order to 'precondition' the paste so that it should not misbehave during soldering: spitting and causing solderballs, or allowing the components to 'swim' or to stand upright, forming tombstones, for example. Most modern pastes do not require circuit boards to be predried, and ANSI/IPC-SP-J-STD 005 does not prescribe predrying in its testing schedule, though the German DIN 32513 does.

5.3 Putting the solderpaste on the board

The basic task here is to put the right amount of paste into exactly the right place. There are two ways of doing this. First, with sequential placement or dispensing, single or, with some methods twin, circular dots of paste of controlled size are deposited on their footprints, either manually or mechanically. Secondly, printing, either through a screen or a stencil, puts paste on every footprint on a board simultaneously in one operation. With printing, the shape of the paste deposit matches the outline of the footprint on which it is placed, with certain provisos which will be discussed later. Printing requires a flat board surface, free from any obstruction such as the projecting ends of connecting wires or leads of components inserted from the other side of the board (see Section 5.1.1). Whenever paste has to be put down on a board surface which is not strictly two-dimensional, printing is impracticable and a dispensing method must be used.

5.3.1 Single-spot dispensing

Repair work is one main field for single-spot dispensing, when either an open joint has to be filled with solder or a replacement component has to be soldered in position. For these tasks, the paste is dispensed either from a hand-held syringe, which may be operated with compressed air controlled by footpedal or finger action, or from a hand-operated dispenser gun. With all of these methods, the dispensing tool can be set to discharge a fixed, constant amount of paste at every stroke, or else every discharge can be operator-controlled. Many vendors supply dispensing tools or guns suitable for clip-on paste cartridges.

Some automatic pick-and-place machines are fitted with twin syringe dispensers fed from paste cartridges. They put down metered amounts of solderpaste on the solderpads of bipolar components like melfs and chips prior to their placement. Metered syringe dispensing demands a paste of constant viscosity. This means either a stable temperature in the workroom, or a paste with a reasonably temperature-insensitive viscosity. For accurate dispensing, the paste in the cartridge must be absolutely free from trapped air bubbles. Otherwise, accurate metering becomes impossible and, what is worse, the sudden bursting of an airbubble, as it reaches the tip of the dispensing nozzle, scatters small drops of paste in the neighbourhood, leading inevitably to a multitude of solder prills.

The exact put-down location for the paste depends on the type of joint. With melfs and chips, the paste deposit must touch the metallized ends of the component but it should not be squashed underneath its body, since this can cause stray solder globules left underneath. For components with flat legs or leads, the paste is deposited in the middle of the footprint. The amount of paste put down should provide just enough solder to completely fill the joint, while the edges of the lead remain visible, or to give the solderfillet at both ends of the melf or chip a concave profile, but no more than that. Naturally, the metal content of the paste by volume (see Table 5.2) must be borne in mind when working out the dosage (Figure 5.4).

Recently, equipment has become available for the mechanized, processor-controlled and programmable placement of a pattern of paste dots of controlled size on one or a run of circuit boards. One suggested use is the placement of solderpaste on small experimental or production runs of boards, where the preparation of a special screen or stencil would be uneconomical or too slow.

With these applicators, a mechanically or pneumatically actuated dispenser cartridge is mounted on a gantry-type xy plotter, which straddles the board (see Figure 4.33). The software which controls the location and size of the individual dots of paste can be derived from the board layout, or created by teach-in and stored. The lateral accuracy of the placement coordinates is reported to be within 0.2 mm/80 mil, using automatic sighting of a fiducial reference mark on the board. Metered dispensing with a screw-feed mechanism is the preferred method of discharge, to prevent settling-out of the solder from the paste and consequent nozzle blocking through repeated pneumatic or piston impulses during operation.

Put-down rates of 13 000–16 000 dots/hour are quoted for single-nozzle applicators, and up to 25 000 dots for twin nozzles. A further field of application for this type of equipment is the placement of solderpaste on boards on which the

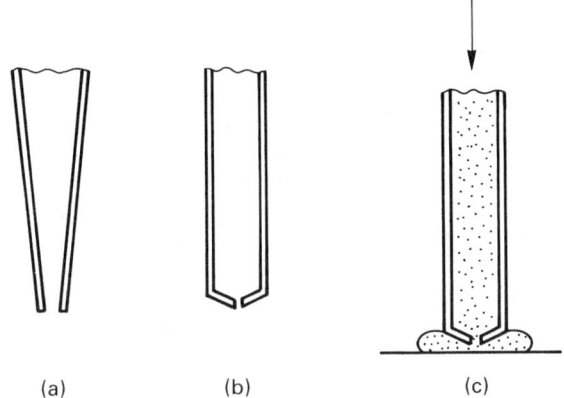

Figure 5.4 *Placement of solderpaste by dispensing. (a) Unsuitable nozzle shape; tends to block; (b) recommended nozzle shape; (c) nozzle too close to footprint; squashing of paste deposit*

joints are located on different levels, or where obstructions on the board surface prevent the use of a screen-printing or stencil-printing method. To meet this need, the location of the paste discharge nozzle in the vertical z axis is variable and programmable. The same type of equipment can be used for putting down drops of adhesive for anchoring SMDs to a board prior to wavesoldering (Sections 4.9 and 5.1.1).

5.3.2 Stencilling and screen printing

Stencil versus screen

Stencilling and screen printing are the most widely used methods for putting solderpaste on printed circuit boards. Both demand a free, unobstructed and flat board surface. Flatness is important for a precise printdown without smudging or lateral squeezing out of the paste. The solder resist too should be of equal thickness over the whole board, because the surface of either acts as a gasket against the screen or stencil, and prevents lateral squeeze-out of the paste.

As far as the choice between stencilling and screen printing is concerned, most industrial users tend to opt for stencilling unless the company concerned has in-house screenmaking facility and expertise. The distinctive virtue of screen printing is its ability to create ring-shaped patterns like an 'O', being able to support the central dot on the mesh of the screen. Since solderpaste is almost always printed in solid squares and rectangles, there is no compelling need for a screen.

High print quality and precision can be achieved with either method. Metal stencils generally cost less and making them requires less specialized skill. Stencils are easier to store, more forgiving towards mishandling and, if properly treated, will last longer than screens. The thickness of the printdown equals the thickness of the stencil, while with screen printing both the nature of the mask and of the screen determine the thickness of the printdown. Above all, with fine-pitch technology, footprint dimensions and distances are getting ever smaller, pushing screen-printability to its limits, if not beyond.

Stencilling allows local reduction of printdown thickness, by thinning the stencil by local etching. This is useful when individual fine-pitch components require less paste deposit than the rest of the board population. However, there is a penalty involved: to allow the squeegee to drop down to the lower level of the etched-back area, this area must extend by 3–5 mm (120–200 mil) beyond the fine pitch footprints, which wastes valuable real estate. The preferred alternative is to make the apertures in the stencil shorter than their corresponding footprints in order to reduce the amount of paste on the fine-pitch footprints.[8]

Thickness and dimensions of the printdown

From the soldering point of view, it is crucial that every pad receives the correct amount, i.e. volume, of solder needed to fill the joint. What interests the printer is the thickness of solderpaste deposit (called the wet-thickness) which must be put down on the board. The ratio wet-thickness/solder-thickness depends on the metal content of the paste and can be derived from Table 5.3. Table 5.6 is based on these values and lists the relationship between wet-thickness and solder-thickness for the two types of paste normally used for stencilling.

When deciding on the amount of paste which footprints are to receive, it is perhaps wise to err, if at all, on the generous side. In subsequent inspection and correction it is easier to detect and remove the occasional solderbridge than to find and fill a solitary empty joint, especially with fine-pitch work. Another problem with the thin deposits of fine-pitch printing is the high degree of coplanarity of the legs of multilead components which a thin printdown demands. Typical coplanarity tolerances are 25–75 microns (1–3 mil).[9] Lack of coplanarity obviously increases the risk of open joints. The above-mentioned shortening of the length of the printdown and use of a thicker stencil is a simple way out.

Table 5.6   *Thickness of solderpaste printdown and of soldercoating*

| Stencil thickness = wet-thickness of paste | | Thickness of solder cover with paste | | | |
| | | 90% wght/50% vol of solder | | 95% wght/67% vol of solder | |
mm	mil	mm	mil	mm	mil
0.30	12	0.15	6	0.2	8.0
0.25	10	0.125	5	0.17	6.7
0.20	8	0.10	4	0.13	5.4
0.15	6	0.07	3	0.10	4.0
0.10	4	0.05	2	0.07	2.7

Stencils and stencil printing

Stencils for paste printing are made from sheet metal, usually hard-rolled brass. For demanding work and fine-pitch printing, beryllium copper, nickel-chromium or stainless steel are preferred. Stencil thickness ranges from 0.75 mm/30 mil to 0.1 mm/4 mil for ultra-fine pitch work. There are a number of ways of creating the printing apertures.

For short runs or prototype work, the stencil openings can be produced by drilling instead of etching. Stencil thickness and hole diameters must of course be suitably chosen to provide the required amount of solderpaste for each pad. Drilling requires a precision drilling machine with optical registration or a numerically controlled circuit board drilling machine.

With the majority of stencils, the printing apertures are created by etching, normally from both sides. For that purpose, both sides are pre-coated with a photomechanical etch resist, often by the vendor of the stencil sheets. The stencil pattern can be derived from software for the artwork for the board. To allow for possible misregister between stencil and the circuit board pattern, it is customary with standard pitch work to make the linear dimensions of the openings in the stencil somewhat smaller than those of the corresponding footprints, but the etch resist pattern must also make allowance for the undercutting of the stencil sheet around the outline of the apertures during etching.

Double-sided etching is often carried out in such a way that the apertures are wider towards the underside of the stencil, which faces the circuit board. This aims to reduce the risk of paste sticking to the sides of an aperture. This can be a danger with fine-pitch work, where the area on the footprint to which the paste must stick comes close to the area of the sidewalls of the stencil aperture, which should neatly slide away from the printdown as the stencil is lifted from the board after printing. For this to happen without fail requires not only a correctly etched stencil, but also a solderpaste with finely adjusted stickiness and drying behaviour.

With ultrafine-pitch work, this measure is no longer enough. Stencils with laser-cut straight-walled apertures are available from several vendors. Obviously, the cost of a lasercut stencil is proportional to the number of apertures rather than the size and complexity of the pattern, as with etched stencils. Nickel-plated brass stencils or molybdenum stencils, available in the US, are said to give particularly clean lift-off from ultrafine pitch paste-prints.[10]

Stencil printing

Stencils in sizes of up to 170 mm–250 mm (7 in–10 in) can be used in simple hand-operated stencil printers. The stencil is held in a hinged frame which can be lowered onto the board, which itself is held on a vacuum table and located against movable locating pins. Within this size range, high-precision printing can be obtained. Larger formats should only be printed on such equipment if the print pattern is a simple one and high precision is not required. Because the stencil is held in the frame without being tensioned, larger formats tend to sag. This leads to inaccurate deposition and can cause smearing of the paste on the substrate. Larger stencils should be used on a regular screen-printing unit.

The accurate register between the stencil and the footprint pattern on the board is critical. Whatever the pitch of the footprint pattern, all of a paste-print must be deposited within the confines of its respective footprint. With an ultrafine pitch of say 0.3 mm/12 mil, the footprints are only 1.5 mm/6 mil wide, which means that the stencil must be aligned relative to the print pattern on the board to an accuracy of within 0.1 mm/4 mil in the x and y axis. This makes high demands not only on the precision and repeatability of the board pattern, but also on the skill of the paste printer. With in-line printing machines, automatic alignment systems based on video recognition of fiduciary marks are used. With manual printing frames, alignment should be verified by setting up the stencil with a ×10 magnifier for every print.[11]

An in-depth discussion of the operational details of stencil printing goes beyond the confines of this book. A number of excellent publications and books are available to the practitioners of paste printing.

Recently, equipment has become available which allows a stencil, suitably perforated on two opposing sides, to be tensioned in the printing frame without being glued to a supporting mesh screen. Care must be exercised here so as not to overtension and deform the stencil, thus causing misregister between printdown and board pattern on large boards.

Most practitioners of stencil printing agree that the stencil should be placed in direct contact with the circuit board, without the 'snap-off' which is used with screen printing. This measure ensures maximum precision and avoids smearing of the paste over the edges of the printed areas. A hard rubber squeegee within the range of 75–95 shore with a diamond cross-section is often recommended. The rigidity of the hard diamond edge reduces 'scoop-out' of paste from the larger stencil apertures. Where stencils with locally reduced thickness (see above) are used, a blade-shaped rubber squeegee instead of a diamond will provide the elasticity required for the blade edge to follow the contour of the stencil. An increasing number of practitioners, on the other hand, prefer a steel blade as a squeegee.

With fine-pitch work, it has been reported that narrow rectangular stencil apertures which are parallel with the direction of travel of the squeegee fill well with solderpaste, but right-angle ones fill poorly. Since this situation arises with square components such as quadpacks, it may be preferable to place these diagonally on the board and accept the loss of valuable board surface rather than a low yield on soldering.[12] In any case, with fine-pitch work, a low squeegee speed is recommended. Speeds of 1 cm–4 cm (0.5 in–1.5 in) per second have been mentioned, while with standard pitch 5 cm–10 cm (2 in–4 in) per second are normal.

With stencils, both the forward and the return travel of the squeegee are used for printing. At the end of a stroke, the squeegee is lifted over the remaining paste and travels back again, pushing the paste before it. The printing pressure should be such that the stencil surface is wiped clean by the advancing squeegee. Though stencil pastes are more viscous and stiffer than pastes for screening, the stencil apertures are not obstructed by the mesh of a screen and therefore pressures need not be higher than for screen printing. Excessive pressure leads to smudging under the stencil and to early wear or deformation, especially of the narrow bridges between neighbouring apertures in fine-pitch work (coining). At the start of a stroke, the squeegee is set down on the stencil, not on the surrounding screen, and the same is true for the end of the stroke. Equally, the length of the squeegee should be less than the width of the stencil. Both measures ensure that the screen which supports the stencil is not damaged or strained, which would affect the register between printdown and the circuit board.

In conclusion, it can be said that stencilling, which can cope with practically every task of solderpaste printing, is a technique which demands manual skill and a conscientious approach, but which can be mastered by in-house training.

Screens and screen printing

In contrast to stencil printing, working with professional screen printing equipment demands a good deal of experience and skill and requires the control of a considerable number of parameters which influence print quality. Silkscreen printing is a profession in its own right which demands an extended apprenticeship. The use of screen printing as a means of putting down solderpaste on circuit boards is decreasing. Screen printing is not really suitable for fine-pitch work: the presence of the screen wires in the printing apertures complicates their geometry and interferes with the clean transfer of the paste from the apertures to the footprints which is so essential with fine-pitch and ultrafine-pitch technology. Therefore only some basic factors which distinguish screening from stencilling need be mentioned here. For a detailed account of the technique, one of the many excellent books on silkscreening should be consulted.

Printing screens are fabrics, woven from a large variety of materials such as polyester or polyamide. Stainless steel fabrics are sometimes recommended for solderpaste printing because of their superior wear resistance against the solder particles. They are, however, unforgiving towards mishandling such as kinking or creasing and should be used only by professionals. The fabric is tensioned and bonded to the screen frame so that the threads run diagonally across the frame. This is to ensure the required precision and definition of the contours of the apertures, which run normally parallel to the sides of the frame. Screen fabrics are characterized by their mesh number, that is the number of openings per linear inch (or cm), and by the thickness of the thread. Both of them taken together determine the width of the mesh opening, which can range from 400 micron/16 mil down to 72 micron/2.9 mil. As a rule, the diameter of the largest solder particles in a given solderpaste should be not larger than one-third of the mesh opening of the screen with which the paste is being printed.

The printing pattern on the screen is created by coating it with a light-sensitive layer of photopolymer emulsion or bonding a photopolymer-film of a given thickness to it in such a manner that, in printing, the squeegee bears against the photopolymer, not against the screen fabric. The polymer is then exposed to strong ultraviolet light, using a pattern derived from the circuit board artwork as a mask in the same way as has been described above for the etch resist pattern of a stencil. The unexposed portions of the photopolymer are water soluble and are washed out in water. When no longer required, the photopolymer mask can be washed from the screen, which can be recoated and used again.

In contrast to stencilling, the screen is held at a small distance ('snap-off') from the board which is being printed, 1.0 mm–1.5 mm (40 mil–60 mil) being normal. Because of the elasticity of the screen, the downwards pressure of the squeegee overcomes the snap-off and creates a line-contact between screen and board, which traverses the board. Along this line, the paste, having been pressed through mask and mesh, is deposited on the board. Behind the moving line of contact, the screen lifts off the board, leaving the paste printdown on the board (Figure 5.5). The thickness of the paste printdown does not have the simple relationship with the thickness of screen-plus-mask that it has with the thickness of the stencil in stencilling, because the threads of the screen take up some of the space in the printing apertures. Its exact value will have to be determined by trial.

With screen printing of solderpaste, as with stencil printing, the squeegee is lifted over the left-over paste, additional paste being added when necessary, and the next board is printed on the return stroke. The 'flooding stroke' between two printing strokes, which is customary for screen-printing with fluid inks and which serves to redistribute the ink after a printing stroke, is not normally practised with solderpaste printing. Certain adjustments may have to be made to the register of the second board to accommodate the slight shift in the position of the screen mask due to its elasticity.

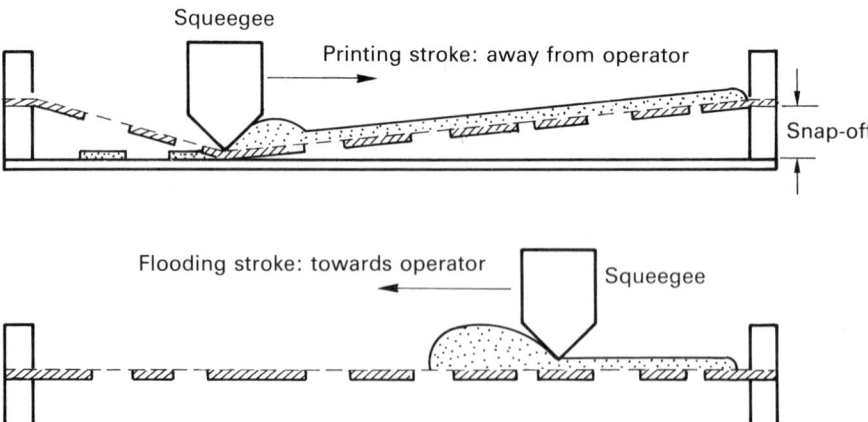

Figure 5.5 *Screen printing by hand*

Care of stencils, screens and paste

At the end of a printing run, the stencil or the screen must be cleaned at once. Dried paste in an aperture reduces its area and thus starves the corresponding footprint of paste and therefore solder, which puts the formation of the joint at risk. If this type of fault passes undetected during the next printing run, locating and correcting the fault may become very costly.

For cleaning, the stencil or screen is taken from the machine and laid flat on a bench covered with paper. Cleaning the screen or stencil *in situ* is bad practice, because paste and/or solvent are liable to drop into the machine or on to the vacuum bed. The bulk of the left-over paste is removed from the stencil or screen with a flexible, blunt spatula; the remainder can be removed by hand using a squeegee as a scraper. Much of the paste remaining in the apertures will be pulled out of them when the paper which covered the bench is peeled away. The stencil or screen is then washed with a solvent, often specific to a given paste and obtainable from the paste supplier. The use of dedicated stencil or screen washing equipment is recommended, because stencils, especially thin ones, are delicate, easily damaged and difficult if not impossible to repair. Their replacement costs not only money, but also valuable if not critical time.

Paste left over after a printing run or a shift need not be dumped, but it should on no account be returned to the tin or jar it came from, which still contains fresh, unused paste. Left-over paste goes into a separate, marked container, and must undergo the 'solderball' test (Section 5.2.6) before re-use. As soon as possible, lids must be replaced on tins or jars, which must never be left standing around open, nor in sunlight even when closed.

After removing paste from a tin, or after putting left-overs back into one, any paste adhering to the sides is scraped back into the bulk, which should sit neatly in one coherent mass. If necessary, the contents of a tin can be compacted by tapping it down against the benchtop. Thin smudges of paste spread around the inside of a container will dry, and having dropped back into the paste, can block a stencil or a screen.

After opening a paste container, it is good and safe practice to briefly and gently stir the contents for a few seconds with a clean spatula, preferably made of plastic, before loading the paste onto the stencil or screen. Even the best of pastes may form a thin layer of solvent, or at least of more dilute paste, on top of the bulk after prolonged standing. Long or violent stirring must be avoided: the thixotropic or pseudoplastic nature of the paste means that the shearing force of stirring it lowers its viscosity. A paste needs a certain recovery time, which varies from one to the other, before it regains its inherent flow properties and printing behaviour. The stronger the shearing force, the longer will the loss of viscosity persist.

Solderpastes which must be stored in a refrigerator, at, for example, 4 °C/39 °F, so as not to settle out or to deteriorate are becoming rare. In any case, a container taken from the refrigerator must never be opened before its contents have been given enough time to attain room temperature. Normally this means taking the next day's supply of paste from cold storage on the evening before, and putting it on a bench in the printing shop. If a container is opened while still colder than the ambient air, atmospheric moisture will condense on the paste and as likely as not spoil its printing and soldering behaviour. Lastly, the label of every paste container carries, or should carry, a use-by date. This date is disregarded at the peril of a run's, if not of a day's, wasted production.

One last warning: a solderpaste should not be tampered with. Even if the printer feels that a few drops of thinner or a pinch of fine solderpowder would make the print come out even better, there is every likelihood that the soldering behaviour is going to be worse. If the print does not come out as it should, alter the printing parameters, but do not 'improve' or 'adjust' the paste. Solderpastes are the result of a careful balance between the demands of printing and of soldering, and this balance is a delicate one.

5.3.3   Depots of solid solder

As the distance between footprints drops below 0.5 mm/20 mil, putting down depots of solder paste with sufficient accuracy, and soldering without unacceptable numbers of bridges or misses, becomes progressively more difficult and expensive. Separating solder and flux from one another and applying first the solder, then the flux, may seem a retrograde step after the simple concept of applying both of them together: however, in recent years, several proposals have been made which emphasize the merits of this approach.[13, 14]

The essence of these proposals is the creation of flat depots or pads of solder, each of controlled and even height and of predetermined volume, fused to every footprint. Flatness is important, because the domed profile of the solder depots on hot-air levelled circuit boards is unacceptable for the safe and precise placement of the thin, flexible legs of close-pitch multilead components (Figure 5.6). Creating the flat solder depots is part of the manufacturing process of the circuit board, and thus the responsibility of the board manufacturer (Section 6.3.2). Boards with preplaced solder as described in that section have meanwhile become commercially available in Germany.

The flat preplaced solder depots relieve the board assembler of the tasks of handling solderpaste and of screen-printing or stencilling it on to the boards, with

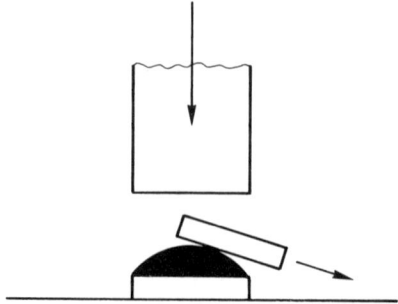

Figure 5.6   *Danger of domed solder depot*

all the headaches that this technology implies. He has only to flux the boards, or selectively the footprints, and then to place the components before reflowsolder- ing. The flux must remain sufficiently sticky after being applied to a board or footprint, so as to hold the placed component securely before soldering. For selective application to the individual footprints it must also be screen or stencil printable. Fluxes which meet these demands are on the market. For impulse- soldering or thermode-soldering of individual multilead components on the other hand, the flux, which should be quick drying, is normally applied by hand (Section 5.9).

5.4 Vapourphase soldering

5.4.1 The basic concept

Vapourphase soldering (or condensation soldering, as it was called originally) makes use of an elegant concept of heating the joints. When a cold body is placed in the saturated vapour of a boiling liquid, the vapour instantly condenses on its surface, giving up its latent heat of condensation in the process. This continues until the body has reached the same temperature as the vapour. About twenty years ago, very stable and chemically inert fluorocarbon fluids with boiling points above the melting temperature of electronic solders, i.e. above 183 °C/361 °F, became commercially available and thus made 'condensation soldering' of electronic assemblies a viable possibility.[15] In the light of present knowledge they do not add to the depletion of stratospheric ozone, being free of chlorine. Their contribution to global warming is probably negligible because of their low volatility (Section 8.3.5).

The working principle of vapourphase soldering is delightfully simple and elegant, but its physics are somewhat complex. Also, the high-boiling working fluids are not cheap, with prices in the neighbourhood of £70–80 or over $100 per kg. Vapourphase soldering represents an 'equilibrium situation' (Section 5.1.2): the items to be soldered attain the same temperature as the source of heat, i.e. the vapour in which they are immersed. With the usual working fluids, this temperature is 215 °C/419 °F. There can be no overheating, and at the end of the soldering cycle all parts of the circuit board have reached more or less the same temperature.

Furthermore, no or extremely little atmospheric oxygen is present during the heating and soldering cycle, which thus amounts to soldering in a controlled atmosphere with all the consequent advantages (Section 5.6): the absence of oxygen makes it possible to use solderpastes with less active, low-residue fluxes, and that reduces the need for cleaning the soldered boards and makes cleaning easier. Furthermore, metallic surfaces do not oxidize while they are being heated to soldering temperature. This helps the solderpaste to climb up the vertical end faces of chips and melfs and make joints with good, 'lean' profiles (Section 9.3) (Figure 5.7).

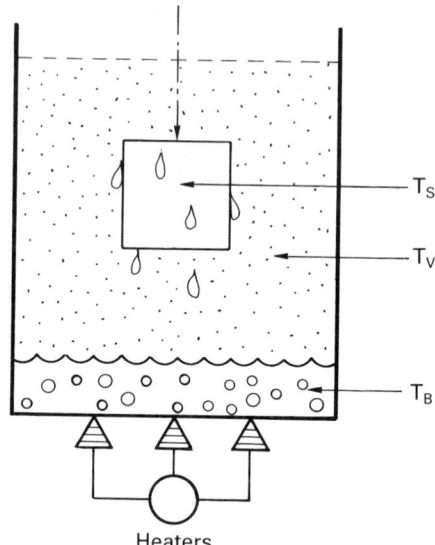

Figure 5.7   *The working principle of vapourphase soldering.* $T_B = T_V =$ *Boiling temperature of the working fluid;* $T_S =$ *Temperature of the item to be soldered* $\leqslant T_V$

5.4.2   Vapourphase working fluids

At the time of writing (1993), vapourphase working fluids are made and marketed by three vendors (Tables 5.7 and 5.8).

While the fluids themselves are non-toxic, certain precautions must be observed when using them in soldering installations. These will be discussed later.

5.4.3   The physics of vapourphase soldering

The thermodynamics of heat transfer by condensation

The rate at which thermal energy is transferred from the working vapour to the circuit board depends on the temperature difference between the two. The heat transfer at any given surface area on the board stops when it is as hot as the surrounding vapour. This fact has two consequences:

1.   The rate of heating is initially high, up to $40\,°C/70\,°F$ per second, but it gets slower the nearer the surface temperature of the board, and the components and solder joints on it, get to the temperature of the vapour.
2.   Solids with a good heat conductivity and low specific heat (for instance metals like copper or solder) heat up more quickly than substances with low heat conductivity and high specific heat, such as ceramics, FR4, rosin, and the working fluid itself.

Table 5.7 *Description of working fluids*

Vendor	Name	Description	Constituents	Boiling point or range
3M	Fluorinert FC-70 Fluorinert FC-5312	Perfluoro- triamylamine	C, F, N	215 °C/ 419 °F
Montefluos	Galden LS 230★	Perfluoro- ether	C, F, O	225 °C– 235 °C/ 437 °F– 455 °F
Rhone- Poulenc	Flutec PP11★★	Fluorocarbon	not available	215 °C/ 419 °F

★Grades with higher boiling ranges are available.
★★Grades with lower and higher boiling points available.

A further factor to be considered is the thermal diffusivity of solid bodies, which governs the heat flow from the surface into their interior. The thermal diffusivity behaviour of a body depends on its thermal conductivity, its specific heat and its density, and it determines how quickly the bulk of the body assumes the temperature of a hot medium in which is immersed.[16]

Table 5.9 shows the practical consequences of thermal diffusivity, as far as they are relevant to vapour phase soldering.

The condensation process

The condensate from the working vapour, which collects on the surfaces of the board assemblies on entering the working vapour, is a bad conductor of heat. Since

Table 5.8 *Important properties of working fluids and other relevant substances*

Substance	Specific gravity	Thermal conductivity $watt/cm/°C$	Density of vapour
FC 70	1.94	0.0007	not available
FC 5311	2.03	0.0005	13 × air
LS 162	1.82	0.0007	16 × air
Flutec PP1	2.03	not available	not available
Freon (R113) SF-2-I	1.57	0.0007	6 × air
Cu		3.9	
FR4		0.002	

Table 5.9   *Heat diffusion in vapourphase soldering. Lapse of time before the centre of a 1.6 mm/63 mil thick plate, immersed in a heated medium, has approached its end temperature to within 90%, calculated in Kelvin*[16]

Substance	Lapse of time
Copper	0.006 seconds
Solder	0.019 seconds
AlO$_2$ (ceramic)	0.074 seconds
Glass	1.1 seconds
FR4	5.6 seconds

a board is in a roughly horizontal position during soldering, the condensate forms a 'puddle' on its upper surface, which impedes the transfer of further heat. Using plates made from copper, it could be shown that holding the plate vertically in the working vapour, the rate of heat transfer was 1.4 times better than with the plate in a horizontal position.[17] Reflowsoldering circuit boards in a vertical position is of course a practical impossibility, but the experiment does demonstrate the puddle effect. Designing the holding fixtures of the boards so that they tilt slightly towards one corner can help the condensate to run off.

Practical consequences

Not all joints on a board will reach full soldering temperature at the same time, for reasons of their thermal properties and their accessibility to the working vapour. A given board will have to stay in the vapour until the slowest joint has been properly soldered. This dwell time must be determined experimentally in preliminary soldering runs. Naturally, it should not be longer than necessary, because prolonged 'confrontation' times between molten solder and substrate affect joint quality (see Section 3.2). Once the slowest joint is fully soldered, the whole assembly is cooled to below the solidification temperature of the solder (183 °C/419 °F) as soon and as quickly as is practicable.

The peculiarities of heat transfer by condensation can cause some operational problems. The initially fast rise in temperature can put some components, such as ceramic condensers, at risk from internal cracking. It may also rupture the housings of large ICs through the 'popcorn effect' (see Section 1.4). Finally, the 'tombstone effect', which causes chips to stand upright, has been attributed to the fast heat-up in vapourphase soldering, though solderability problems and unsuitable layout are probably the main culprits here (see Section 9.1.1).

The puddle-effect, combined with the different temperature rise of metallic and non-metallic surfaces, is responsible for another problem met with in vapourphase soldering, the so-called 'wicking' of PLCC legs. In wicking, the solder climbs up the legs of PLCCs, thus starving the bottom bend of solder, with the risk of an open joint (Figure 5.8).

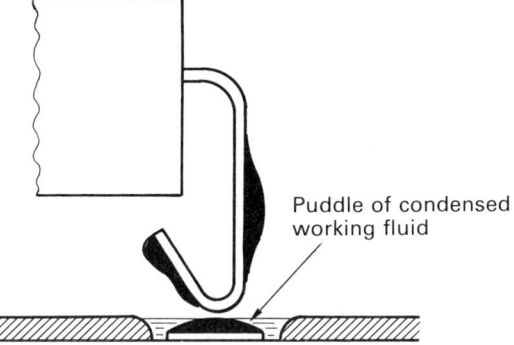

Puddle of condensed
working fluid

Figure 5.8   *Wicking*

The mechanism of wicking is briefly as follows. The J-legs of a PLCC, being vertical and made of metal, warm up more quickly than the solderpaste on which they sit. The paste is a bad conductor of heat, and is soon covered with condensate, much of it having run down the J-leg. This puts the footprint underneath at a disadvantage as far as heat transfer is concerned, and it heats up more slowly than the J-leg. As soon as the solder begins to melt, it is attracted by the hottest available metal surface, which is the J-leg. The strong forces of interfacial tension (see Section 3.2) cause it to climb up the leg, and starve the joint between leg and footprint sufficiently to cause an open joint.

Recently developed vapourphase soldering systems, which combine infrared preheating with a vapourphase soldering station, and which will be described presently, have helped to overcome the consequences of the initial fast rise in temperature.

Boiling behaviour and safety aspects of working fluids

The heating arrangements for keeping the working fluid on the boil must be engineered with great care, because perfluorinated liquids must on no account be overheated lest they form the very toxic decomposition product perfluoro-isobutylene (PFIB), a colourless, odourless gas which can form at temperatures above 300 °C/572 °F. Its occupational exposure limit is one part in 108, which is very near its limit of detectability.[18]

During boiling, local overheating could occur if a vapour blanket is allowed to form between a heating surface and the working fluid. To prevent this from happening requires correctly designed and well maintained equipment, as well as reliable safety measures, such as the monitoring of local temperatures, the replenishment, filtering and monitoring of the condition of the working fluid, and periodic removal of paste particles and other foreign matter which might have dropped into the boiling sump.

5.4.4 Vapourphase soldering equipment

Benchtop equipment

Simple laboratory vapour immersion apparatus, made in glass, is on the market. This apparatus can handle boards up to about 10 cm/4 in square, the boards being placed horizontally on a suitable wire mesh support and lowered into the vapour volume (Figure 5.9). This type of apparatus is also useful for checking large ICs for their tendency to crack under the popcorn effect during vapourphase or infrared soldering, for testing the soldering performance of fresh or alternative supplies of solderpaste, or for carrying out pilot soldering tests on experimental boards. Since the apparatus is made of glass, the progress of soldering can be readily observed in all its stages. Though it is, or should be, protected against overheating, it is advisable to operate this type of laboratory vapourphase apparatus in a fume cupboard.

Industrial equipment

Until the end of the nineties, industrial vapourphase soldering equipment was of two different types: the batch immersion plant and the in-line continuous soldering plant.

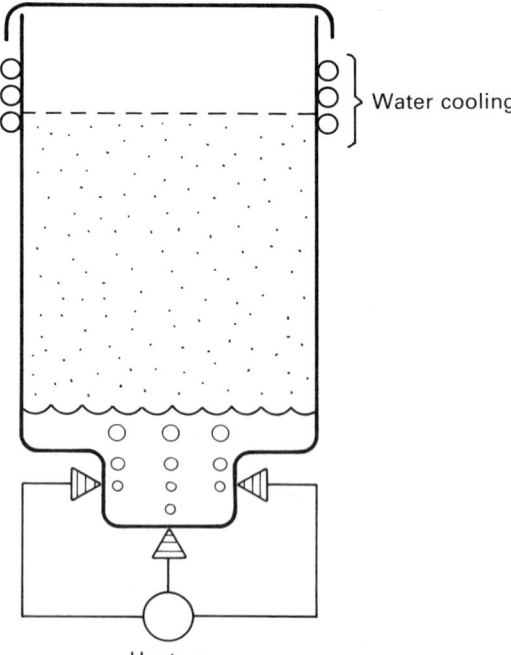

Water cooling

Heaters

Figure 5.9 *Bench top vapourphase soldering equipment (Multicore)*

The batch immersion machine is basically an open-top vessel, with a boiling sump of the working fluid carrying either internally or externally placed heaters (Figure 5.10).

Since this type of equipment is in the process of being phased out, it needs only a brief description. Its principal feature is a 'lid' of secondary vapour, which floats on top of the heavier working vapour (for the vapour densities, see Table 5.7), and prevents the escape and loss of the vapour of the expensive working fluid. Before the environmental danger of the chloro-fluoro-carbons (CFCs) had been recognized (Sections 8.3.5 and 8.3.6), CFC 113 (e.g. Freon), which boils at 47.6 °C/117.7 °F, was universally used for the secondary vapour blanket. Since then, a safe alternative, chlorine-free perfluorocarbon, is marketed by 3M under the designation of SF-2-I. It boils between 38 °C/100 °F and 40 °C/104 °F, and has a zero ozone-depletion potential, though it does possess a small global-warming potential.

The levels of working and secondary vapour are controlled by cooling coils, located at the appropriate levels on the inside walls of the vapour chamber. Sensors monitor the existing vapour levels and activate pumps to replenish the working or secondary liquid when required.

The boards to be soldered are placed horizontally on trays, or stacks of trays, made of stainless steel wire. It is good practice to stack the boards on their carriers in such a way that the condensate from boards higher in the stack does not drip onto the soldering areas of boards below. The board trays are mostly lowered and raised mechanically. As has been said already, their dwell time in the vapour must be

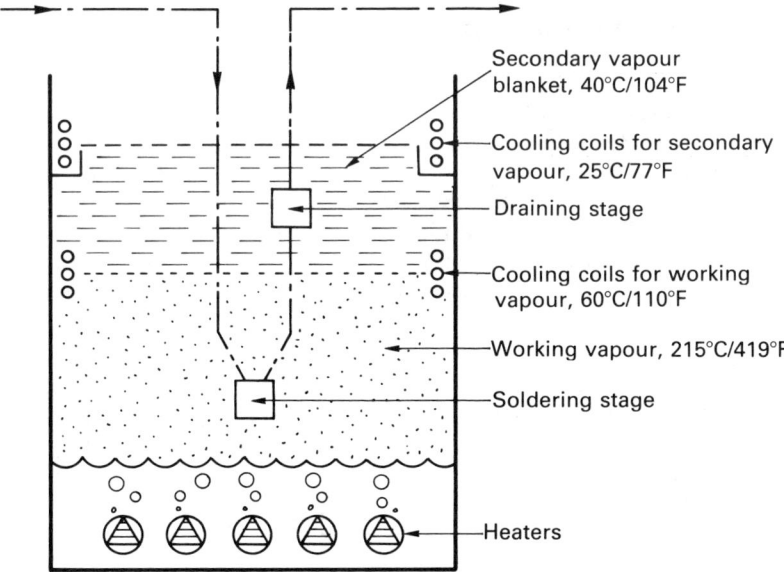

Figure 5.10 *Vapourphase batchsoldering principle*

established experimentally for each type of board. In descending into the vapour and on emerging from it, the boards pass through the blanket of secondary vapour. To minimize drag-out of the expensive working fluid, the boards may rest in the secondary vapour blanket for a short-time during emerging to allow condensate from the working vapour to drain back into the sump.

Like all vapourphase soldering equipment, batch-immersion plants must be fitted with all manner of monitoring and fail-safe equipment. This guards against overheating, circulates, filters and replenishes the working fluid, monitors its acidity (pick-up of water might cause the formation of HF in the liquid), monitors and maintains vapour levels, and guards against failure of the cooling-water supply, to name its main tasks. Recent batch-immersion machines no longer have an open top, but are fitted with a lid, with the boards to be soldered entering the machine sideways.

In-line vapourphase soldering systems are larger in size, but simpler in design and they need no vapour blanket. The boards travel on an endless stainless steel belt of open wiremesh or link-belt construction, entering and leaving the vapour chamber through downwards-sloping and upwards-sloping watercooled tunnels which prevent the escape of the heavy working vapour from the machine (Figure 5.11). Given the length of the vapour chamber, the conveyor speed is governed by the dwell time which the slowest joint on the board needs to solder properly. It follows that with mixed batches of board the slowest joint governs the rate of production.

5.4.5 'New-generation' vapourphase soldering systems

The preheat concept

During the late eighties, a fresh concept of vapourphase soldering was introduced to the market, under the general name of 'new-generation' vapourphase soldering. Its main feature is a preheating stage, through which the boards pass before they enter the vapour chamber. The 'pre-heat' raises the board temperature to between approximately 125 °C/260 °F and 150 °C/300 °F, at a heating rate of not more than about 5 °C/9 °F, per second. The boards are then held at that temperature for one and a half to two minutes, before they enter the working vapour with its

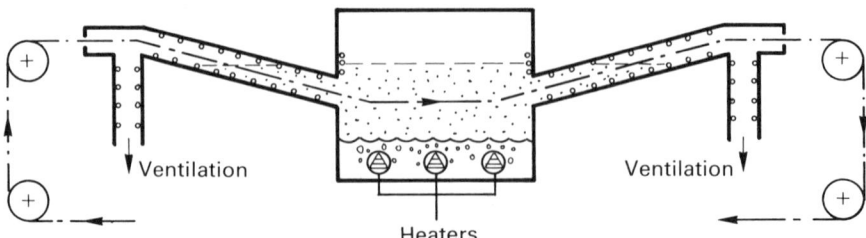

Figure 5.11   *In-line vapourphase soldering system*

temperature of 215 °C/390 °F.[19] The resulting heating profile of the boards is shown in Figure 5.12.

Raising the temperature of the boards up to soldering heat in two steps brings a number of advantages:

1. The initial sharp temperature rise of straight vapourphase heating is eliminated, together with its dangers of spluttering paste, tombstoning, and damaged components. Preheating is also reported to prevent wicking of the J-legs of PLCCs.[20] When the preheated boards enter the vapour, their initial temperature rise is about 20 °C/36 °F per second, but by that time the volatile constituents of the solderpaste have disappeared and the components are much less likely to suffer internal cracks or to burst, having been preconditioned and dried.
2. As a board enters the vapour, the whole assembly is at a uniform temperature, only about 30–60 °C/55–100 °F away from the melting point of the solder. This means that it melts soon afterwards at more or less the same time in all the joints. In consequence, the dwelltime of a board in the vapour can be considerably shortened. Confrontation periods between the molten solder and the joint surfaces down to 15–30 seconds have been reported, with a consequent improvement of the metallurgical structure of the joints.

It might be held against the preheating concept that it heats the assembled board in the presence of atmospheric oxygen. This is probably not a serious impediment to the soldering process which follows, the preheating temperature being too low to seriously oxidize the metal surfaces or to degrade the solderpaste.

Two schemes of operation of vapourphase machines with preheat are shown in Figure 5.13.

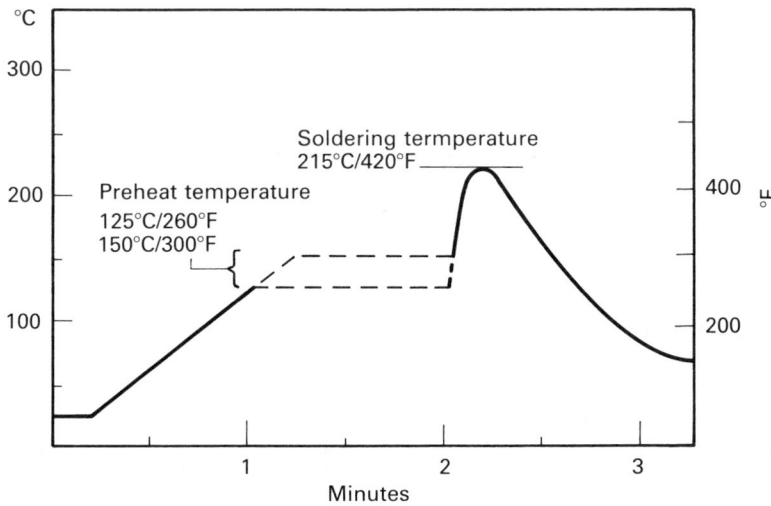

Figure 5.12 *Temperature profile of vapourphase soldering with preheat*

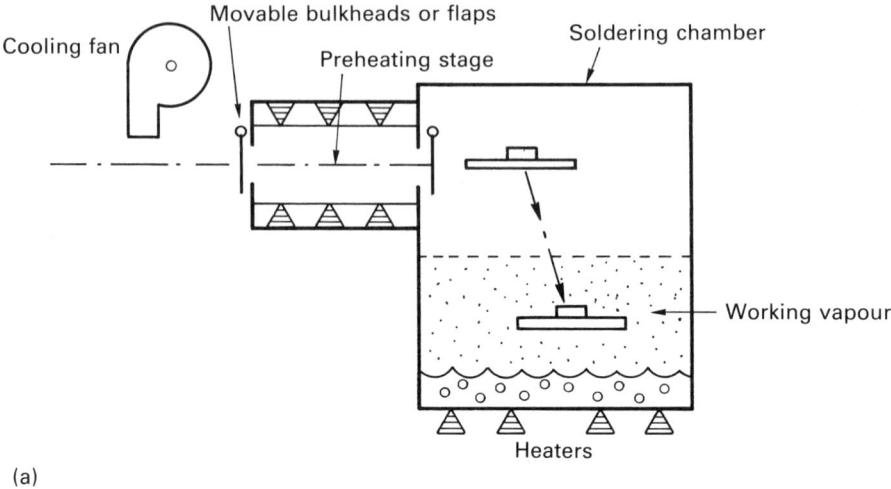

(a)

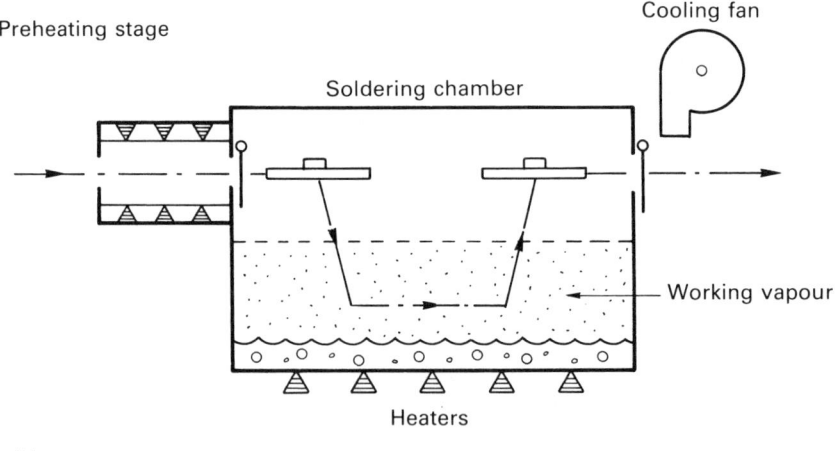

(b)

Figure 5.13 *Vapourphase soldering with preheat. (a) Batch system; (b) in-line system*

Vapourphase soldering equipment of the new generation has reached a level of sophistication well above that of the machines based on the original concept. With most of the new machines, the boards are carried on pallets, which are lowered into the vapour space in such a manner that they remain horizontal at all times. With a batch system, this makes it possible to fully close the vapour chamber, for example with hinged flaps, before and after a pallet has entered it, thus reducing vapour loss and fluid consumption. With in-line systems, it does away with the sloping entry and exit tunnels, and thus shortens the length of the machine. Again, the vapour chamber can be fitted with vapour locks to minimize the loss of working fluid.

Preheating is effected by infrared emitters, such as internally heated ceramic panels, which of course must be operated below the decomposition temperature of the working vapour. One manufacturer has chosen to preheat the boards indirectly with hot air to pre-empt objections on that score. Postcooling with a forced airstream, to shorten the confrontation period and to speed up the solidification of the solder in joints, has become universal practice.

One make of machine is fitted with an exhaust system, which draws air inward through the entry and exit ports of the board conveyor towards the space above the vapour chamber. In this way, any backward diffusion of working vapour into the preheating section and with it the risk of PFIB being formed is avoided. A water scrubber retrieves solvent from the vapour which may have been carried away with the exhaust.

Depending on the type of machine, fluid losses as low as 50 g/hour are claimed to be achievable. With large in-line machines the losses can be higher, but at least one vendor offers vapour-recovery units which can be retrofitted to keep fluid loss and operating costs within the acceptable limits of the smaller batch systems.

5.5 Infrared soldering

5.5.1 Working principle

At first sight, infrared (or IR) soldering appears even simpler than the vapourphase concept: passing the boards through a tunnel fitted with infrared emitters seems to be all that is required, and the early IR soldering ovens did just that. However, practical experience soon showed that a thorough understanding of the physical principles which form the basis of infrared soldering, together with sound engineering, were required to make infrared soldering the reliable and largely prevailing production method it now is.

Unlike vapourphase soldering, infrared soldering represents a 'non-equilibrium situation', because the temperature of the heat source is considerably higher than the temperature which the soldered joints are intended, and can in fact be allowed, to reach. It will be useful to restate at this point that every soldering method represents either an equilibrium or a non-equilibrium heat-transfer situation (Section 2.1), and to detail the implications of the difference between them.

In an equilibrium situation:

1. The temperature of the heat source is identical to the soldering temperature which the joints must reach.
2. The speed with which the joints reach that temperature depends on their thermal coupling with the heat source. Process control regulates the duration of exposure of the joints to the heat source.

3. There is no danger of overheating the joints. Faulty process control can result in two possible faults: cold joints and overcooked joints. Cold joints happen when they are not given enough time to reach their full soldering temperature. In consequence, the joint is weak or non-existent. Overcooked joints happen when they remain at the soldering temperature too long (>30 secs). This results in a thickened intermetallic layer, so that the joints are brittle, with lowered life expectancy.
4. Examples are wavesoldering, vapourphase soldering, hot-air or hot-gas convection soldering, impulse or thermode soldering.

Non-equilibrium situation:

1. The temperature of the heat source is higher than the soldering temperature which the joints must reach, as well as the maximum temperature which they can be allowed to reach. In most cases, the heat source is not in direct physical contact with the joints, but transmits its thermal energy by radiation.
2. The rate of temperature rise of the joints depends on the temperature-difference between the heat source and the joints, the power of the heat source, its distance from the joints, and the heat capacity of the joints themselves.
    Process control regulates the temperature of the heat source and the time of exposure of the joints to it. Depending on the intensity of the heat radiation, exposure times range from milliseconds to minutes.
3. The demands on the precision and reliability of process control are in direct proportion to the speed of the temperature rise of the joints. Cold joints occur when the exposure time is too short. Overcooked joints are the result of too long exposure to the heat source. The consequences are as in the equilibrium situation.
4. Examples are infrared soldering and laser soldering.

It follows from this that precise control of the operating parameters is essential with infrared soldering. Confusion often arises from the fact that the oven interior itself has no defined temperature of its own. It is traversed by a field of infrared radiation which is on its way from the emitters to the circuit boards. A considerable part of this radiation, 30% or more, is absorbed by the oven atmosphere before it reaches its target. This absorption factor depends not only on the distance between the emitters and the boards, but also on the rate at which the oven atmosphere is circulated in the oven interior. Thus, it makes no real sense to speak of an oven temperature as such.

The temperature of the emitters is known and controllable. So is the rate of circulation of the air in the oven, and with it the efficiency with which it passes the heat, which it has absorbed from the radiation, by convection to the boards.

This is a complex state of affairs, from which it follows that the success of an infrared soldering operation depends critically on the setting, the stability and the monitoring of the operating parameters, on a meaningful and accurate monitoring of the flow of heat in the oven, and on a sufficiently quick response of the system to automatic or manual controls. This aspect will be dealt with in Section 5.5.4.

Most industrial IR soldering installations are in–line tunnel ovens, with arrays of infrared emitters placed above and below a conveyor on which the circuit boards travel (Figure 5.14). More rarely, a batch process is used with one or more boards being soldered in a front-loading oven which is fitted with emitters and a circulating system for the oven atmosphere.

5.5.2 The physics of heat transfer by radiation

The three laws of radiation

In order to understand what is going on in an infrared soldering oven, and how to run it in an efficient manner, three physical laws which govern the transmission of heat by radiation, and some facts about heat transfer by convection, must be discussed. This can be done in quite simple language and practically without mathematics.

The temperature of a body is a measure of the thermal energy which it contains. The greater its thermal energy, or heat content, the higher is its temperature. Heat emitters are of course so designed that the surfaces which face the circuit boards are as hot as the heat source inside them, while their back surfaces reflect as much heat as possible back in to the emitter body.

In principle, there are three ways in which a hot body can transmit its heat to its cooler surroundings: conduction, convection and radiation. Conduction we should be able to discount in a well constructed oven. In a well designed IR oven, emitters should not lose much heat by convection either: virtually all the energy put into them should be given off as radiation. With true convection ovens, the heat sources are located out of sight of the circuit boards and they are designed as heat exchangers, which transfer all of their thermal output to the oven atmosphere by convection (Section 5.6).

Thus, radiation is the main vehicle which transfers the heat needed for soldering from the emitters to the boards and the joints on them. The invisible infrared heat radiation differs from visible light, and in fact from any other radiation, be it cosmic, X-ray or radio, only by its wavelength, taking like all radiation the form of electromagnetic waves. (For the purpose of this discussion we shall not go into quantum physics and photons.)

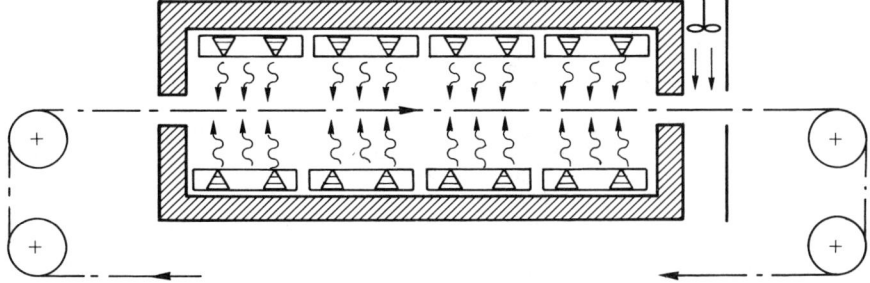

Figure 5.14 *Working principle of an infrared soldering oven. △ Heaters*

Electromagnetic waves, like all waves, are characterized by their wavelength and their amplitude, i.e. their energy content. In principle, infrared radiation can be emitted not only by a solid body, but also by a hot gas (e.g. from a flame) or a hot liquid. In infrared soldering, we are concerned with the radiation given off by the surfaces of hot solid bodies, our emitters.

The spectrum of wavelengths and the intensity of the radiation given off by any hot body follow a number of physical laws, of which the following three are important for us.

The Stefan–Boltzmann law★

This law states that the energy emitted by a hot body is proportional to the fourth power of the absolute temperature★★ of its surface. In consequence, a small increase in the temperature of the heat source, i.e. the surface of the emitter, produces a large increase in emitted thermal energy, which means in its heating power.

Table 5.10 gives the numerical values of heat emission per unit of surface area, within a range of temperatures which includes those in which emitters for IR soldering normally operate. These values assume that the emitter surface converts its thermal energy into radiation with 100% efficiency. In practice, most industrial emitters have an efficiency of 95%, the other 5% being reflected back into the emitter body ('grey factor').

Table 5.10   *Relationship between surface temperature and emitted thermal energy. Values taken from NASA Document TT-F-783*

| | Surface temperature | | Emitted thermal energy |
K	°C	°F	(watt/cm²)
373	100	212	0.1
473	200	392	0.3
573	300	572	0.6
673	400	752	1.2
773	500	932	2.0
873	600	1112	3.4
973	700	1292	5.1
1073	800	1472	7.7
1273	1000	1832	15.0

★Josef Stefan (1835–1893), German physicist; Ludwig Boltzmann (1844–1906), German physicist.
★★Absolute temperatures are measured on the Kelvin scale, in which a temperature of x Kelvin is the same as $(x - 273)$ °C. (The Kelvin scale is named after Sir William Thomson (1824–1907), later Lord Kelvin of Largs, British physicist.)

For example 0 K, the absolute zero, -273 °C (it makes no sense to convert this into °F). Reckoning in absolute temperatures means, for example, that at room temperature, say 20 °C/68 °F = 293 K, solder is within 36% of its melting point of 183 °C/331 °F = 456 K. Hence the comparative ductility and low mechanical strength of solder at normal temperatures (Section 3.3). At very low sub-zero temperatures, solder becomes quite strong and brittle.

Planck's law*

Planck's Law describes how the radiation energy given off by the surface of a body is distributed over the wavelength spectrum at any given surface temperature. Figure 5.15 shows how the maxima of these energy distribution curves, which are quite sharp at high temperatures, move towards longer wavelengths and get flatter as emitter temperatures get lower.

On the scale of wavelengths, the visible radiation, which could of course also be called thermal radiation, lies between $0.38\,\mu$m (violet) and $0.76\,\mu$m (red).

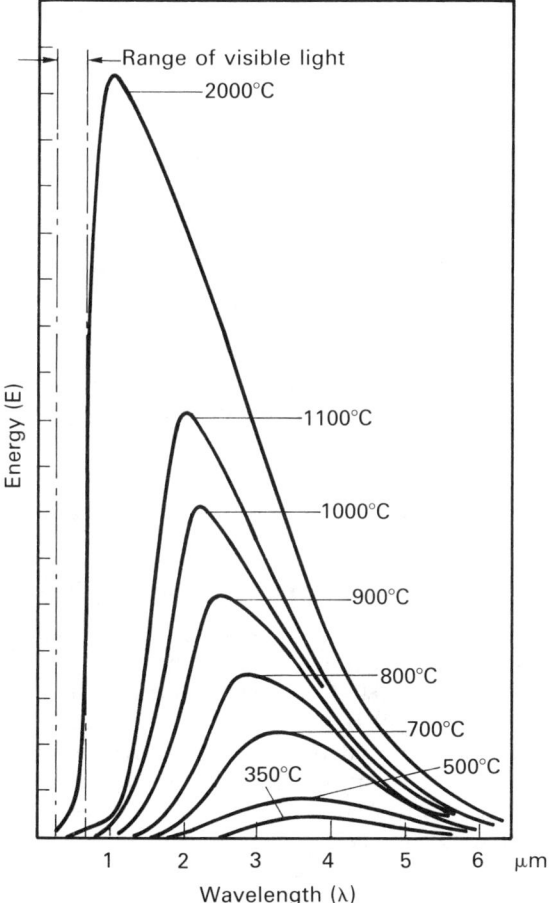

Figure 5.15 *Distribution of emitted energy over the wavelength spectrum for different emitter temperatures*

*Max Planck (1858–1947), German physicist.

Ultraviolet begins below 0.38 μm, while above 0.76 lies the range of infrared (the prefix 'infra–', meaning 'below', refers to the wave frequency of the radiation, not its wavelength).

(The surface temperature of the sun is approximately 5500 °C/9900 °F, at which temperature the Planck curve has its maximum at 0.52 μm. The wavelengths of the visible range of thermal radiation, i.e. of the light which our eyes can see, is spread neatly on either side of that maximum. This is as it ought to be, since our retina which perceives this radiation had millions of years to evolve under the light of our sun. One might add that many animals, whose requirements for survival differ from ours, can perceive radiation with rather different ranges of wavelengths.)

For practical convenience, infrared radiation has been subdivided into three wavebands:

0.76–1.5 μm	Near infrared
1.5–5.5 μm	Middle infrared
>5.5 μm	Far infrared

Wien's law★

Wien's law gives a formula for the location of the maximum of the energy distribution curves as they shift towards longer wavelengths with lower emitter temperatures: the wavelength of the maximum is inversely proportional to the temperature of the emitting surface. As has been said already, these maxima get progressively flatter with lower emitter temperatures. Below 500 °C/930 °F, the precise location of the energy maximum has no longer any practical significance. Table 5.11 illustrates these relationships.

The significance of the temperature difference between emitter and board

The emitters used in reflowsoldering ovens generally fall into two categories:

1. Low-temperature and medium-temperature emitters usually take the form of internally heated metal panels or ceramic bodies. Their operating temperatures range from 250 °C/480 °F to 400 °C/750 °F for the metal panels and 300 °C/570 °F to 500 °C/950 °F for the ceramic ones. Their operating characteristics can be compared to those of an electric hotplate: slow to get hot, and slow to respond to changes in the heating current.
2. High-temperature emitters are as a rule quartz tubes with a tungsten or tungsten alloy heating coil running down the middle. Quartz (SiO_2) is chosen for this purpose because it is transparent to infrared. The back half of the tube may carry an external film of gold to act as a reflector. High-temperature emitters operate at temperatures between 700 °C/1290 °F and 850 °C/1560 °F. Their operating characteristics are similar to those of an electric light bulb in their fast response to being switched on or off, or dimmed.

★Wilhelm Wien (1864–1928), German physicist.

Table 5.11 *Emitter temperatures and wavelength pattern of the emitted radiation*

Temperature					
°C	250	300	350	500	700
°F	452	572	662	932	1292
Emitted energy:					
watt/cm^2	0.43	0.60	0.89	2.0	5.1
watt/in^2	2.58	3.60	5.34	12.0	30.6
kw/m^2	4.3	6.0	8.9	20	51
Wavelength of the energy maximum: λ_{max}	5.5 μm	5.0 μm	4.6 μm	3.7 μm	2.4 μm
Emitted energy at λ_{max} in relation to energy at $\lambda_{max/300\,°C}$	0.6	1.0	1.5	4.4	13.9

As would be expected, the temperature difference between an emitter, the heat source, and a circuit board, the heat sink, governs the rate at which thermal energy flows from the one to the other. In fact, because Planck's law is involved, the fourth power of the absolute temperatures of both source (T_1) and sink (T_2) enter into the heat-flow formula:

$$Q = C_{1,2}[(T_1/100)^4 - (T_2/100)^4]$$

in which $C_{1,2}$ represents the emission and absorption characteristics of the surfaces involved. The radiating efficiency of most practical emitters is about 95% (see above). The heat absorption of the surfaces involved in reflowsoldering will be discussed later. Table 5.12 gives numerical values, computed from the formula, for the relative rates of heat flow between a low-temperature and a medium-temperature emitter and a circuit board at different points of its travel through an IR oven.

Many emitters, for example gold-backed quartz tubes, are designed so that their back reflects the heat from the emitter interior back into it, in order to direct as much as possible of its thermal output towards the circuit boards and not the casing of the oven. Failing that, heat reflectors must be mounted between the emitters and the oven wall.

Several important practical conclusions can be drawn from the figures computed in Table 5.12:

1. With a low-temperature emitter of 300 °C/572 °F, a temperature variation of ±50 °C/120 °F drastically affects the rate of heat transfer from the emitter to the board, by up to ±40%. This demonstrates the importance of accurately controlling and monitoring the temperature of that type of emitter.
2. As the board gets warmer during its travel past an array of low-temperature emitters, the efficiency of heat transfer from emitter to board drops drastically: by the time the board has reached soldering temperature (183 °C/361 °F), the

Table 5.12   *Comparison between heat transfer efficiencies between low-temperature and high-temperature emitters and a circuit board at various stages of heating.* T_1, *Temperature of the emitter;* T_2, *Temperature of the circuit board and the solder joints*

Comparative heat-transfer efficiencies Q with a low-temperature or medium-temperature emitter

T_1		250°C/482°F	300°C/572°F	350°C/662°F
		Q	Q	Q
T_2	20°C/68°F	0.64	1.00	1.47
	100°C/212°F	0.49	0.84	1.27
	150°C/302°F	0.31	0.75	1.14
	183°C/361°F	0.21	0.65	1.08

Comparative heat-transfer efficiencies Q with a high-temperature emitter

T_1		750°C/1382°F	800°C/1472°F	850°C/1562°F
		Q	Q	Q
T_2	20°C/68°F	0.82	1.00	1.21
	100°C/212°F	0.81	0.99	1.20
	150°C/302°F	0.80	0.98	1.19
	183°C/361°F	0.80	0.98	1.18

rate of heat transfer has dropped by 35%. This factor obviously lengthens the time which a board must spend in the final heating stage, where the solder melts. This in turn lengthens the confrontation interval during which the solder remains molten in the joint, with its consequent undesirable effects on its metallurgical quality.

3. With a high-temperature emitter (800°C/1472°F), a variation of emitter temperature has less drastic consequences on the amount of heat emitted than with a low-temperature emitter. More importantly, the rate of heat transfer to the board is hardly affected while the board warms up and approaches its soldering temperature.

It follows that while there may be a case for low-temperature emitters in the preheating stage of an infrared oven, where the heating rate of the board should not exceed about 5°C/10°F per second, high-temperature emitters are certainly more suitable for the last stage of the soldering cycle, where a final burst of thermal energy has to raise the board temperature from about 100°C/210°F–150°C/300°F quickly and briefly beyond the melting point of the solder to between 220°C/430°F and 250°C/480°F. This time-span should be short and the maximum temperature be readily controllable for metallurgical and operational reasons which have been discussed already.

By reason of their construction, the surface temperature of low-temperature emitters takes several minutes to respond to changes in the heating current and even longer for the emitters to reach their operating temperature when switched on from cold. By contrast, high-temperature emitters reach their full operating temperature and respond to heating-current changes in well under one second.

Directionality of the emitted radiation

A hot body emits heat radiation in all directions, with a falling off at angles close to the tangent to the surface. This means that in the interior of an oven heated by low-temperature and medium-temperature emitter panels, the pattern of infrared radiation is diffuse so that there is no danger of sharp shadows. Nevertheless, any joint which is totally out of sight of any emitter surface can only rely on such heat as convection from the hot oven atmosphere can bring to it. This applies for example to badly placed PLCC J-legs.

On smooth, heat-reflecting surfaces with gold-backed quartz tubes, IR radiation follows the laws of optics. The IR output is directional and this must be taken into account when designing the spatial disposition of quartz emitters.

Reflection and absorption of incoming heat by the board and the joints

When thermal (or any other) radiation encounters the surface of a solid (or a liquid) body, three different effects occur:

1. Some radiation is reflected by the surface.
2. The rest passes into the body where some of it is absorbed as it passes through it. Depending on the transparency of the body towards the radiation, and if it is thin enough, part of the radiation will reach the other side.
3. At the exit surface, some of the radiation is reflected back into the body. The rest emerges and travels on.

Since circuit boards, together with their components and metallic laminates, are too massive for any thermal radiation to pass through, only reflected and absorbed heat need to be considered, both of them adding up to 100% of the incoming radiation.

Figure 5.16 shows the reflection/absorption behaviour of solder and FR4 polyester/glassfibre laminate in the infrared range. With solder, as with all metals, such absorption as does take place occur in a few top atomic layers. FR4, like all organic substances (including the human body, as users of infrared lamps know), is to some extent transparent to infrared radiation, and thus allows a part of it to penetrate to the inner layers. Generally, this is beneficial, as it reduces warping of the boards from one-sided heating, though the virtue of heating the boards equally from above and below remains.

The flux portion of solderpaste, being an organic substance, is also somewhat transparent to infrared radiation, and allows it to enter the paste deposit, where it bounces around between the solder grains and quickly gives up its thermal energy to them.

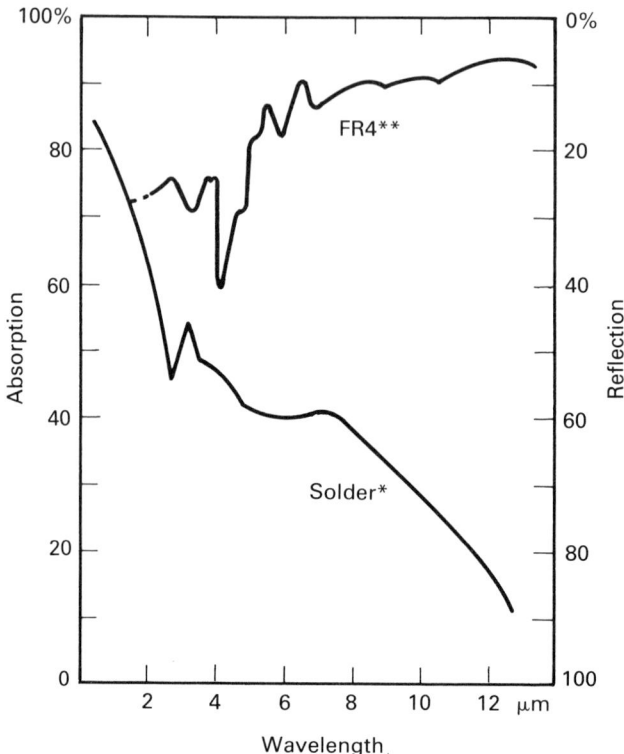

Figure 5.16 *Reflection and absorption behaviour of solder and FR4 in the IR range. * After Lusk, E. F. (1984)* Electronic Packaging Prod., **24***, 3, pp. 98–104.* **After Dow, S. J. (1985)* Brazing and Soldering, No. 8, pp. 16–19*

Solder reflects visible light, but it is an efficient absorber of heat radiation in the very near infrared. This is useful when soldering with a Nd:YAG laser which emits at a wavelength of 1.06 μm (Section 5.7). Absorption drops to 50% and below within the range of wavelengths of high-temperature and low-temperature emitters, which is good enough for all practical purposes and a help with solderpaste, as mentioned above. The steep drop in absorption beyond 8 μm makes CO_2 lasers (emitted wavelength 10 μm) uninteresting for laser soldering.

Effect of the nature of the components on their heat absorption

Ceramics are good conductors of heat and absorb infrared radiation near their surface. This means that there is no danger of thermal stresses and cracking with melfs and chips on that score. However, that does not diminish the risk of internal cracks in ceramic condensers when heating them too quickly from room temperature. On the other hand, the plastic housings of ICs, like FR4, are somewhat transparent to IR radiation. This warms their interior, not sufficiently to put the embedded chips at risk but enough to turn any internally trapped moisture into steam, and thus to trigger the popcorn effect, especially with large ICs.

The colour of an IC housing is quite irrelevant as far as reflection or absorption of heat in an infrared oven, and getting hot or staying cool, are concerned, as has been proved experimentally. What the human eye can see is irrelevant to what happens to infrared radiation on a given surface, as has been explained earlier on. If overheating of an individual large IC should be a worry because of the popcorn effect, a piece of self-adhesive aluminium foil applied to its surface will keep its temperature down.

What does count is the mass of an irradiated component. As Figure 5.17 shows, there is a direct relationship between the mass of a component and its temperature at the end of its passage past the IR emitters: a component weighing 0.1 g may get 60 °C/108 °F hotter than one weighing 4.5 g.[21] The temperature of a component affects the temperature of the component leads and the joints themselves: by the time a board leaves the oven, every joint on every component, including the one with the slowest temperature rise, must have become hot enough for the solder to melt. That means that the board must get hot enough for the solder on the largest components to melt, which may make matters uncomfortable for the small ones.

Effect of the shape of a component

Because of the low heat conductivity of non-metallic bodies, such as IC housings and also circuit boards themselves, their edges and especially their corners get

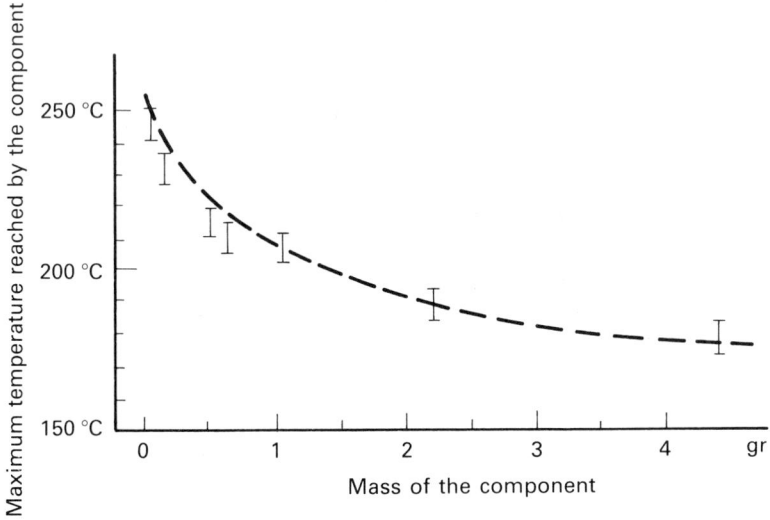

Figure 5.17 *Component peak temperature versus mass during IR soldering*

hotter than their flat surfaces. The reason for this is that in a diffuse, multidirectional heatflow, a given volume of material receives twice as much heat along an edge and three times as much in a corner than in the middle of a flat surface (Figure 5.18).[22]

5.5.3   *The physics of heat transfer by convection*

The role of the oven atmosphere

The role of the atmosphere, which fills the space between emitters and boards, is crucial because infrared radiation is strongly absorbed in normal air, as Figure 5.19 shows.

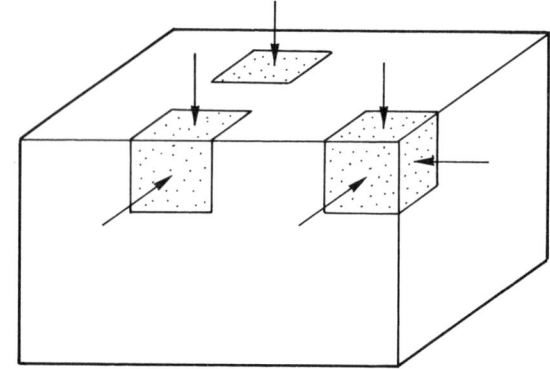

Figure 5.18   *Energy flow into a solid body*

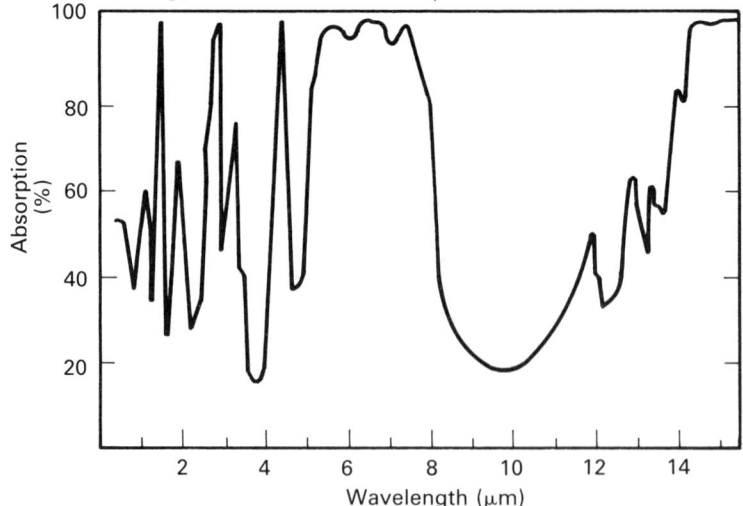

Figure 5.19   *Absorption of infrared radiation in air. After Lea, C. (1988) A Scientific Guide to SMT, p. 224*

If the oven atmosphere also contains some water vapour and solvent vapour, such as might be given off by the solderpaste, additional absorption bands appear in the curve. This absorption pattern means that in an IR oven of normal dimensions, up to 50% of the thermal output of the emitters gets absorbed in and heats the oven atmosphere, not the boards. This fact must be, and is by now, taken into account with all modern industrial IR soldering equipment, with perhaps the exception of some small laboratory reflow ovens.

It is worth noting here that, in contrast to air, no absorption of radiation in the infrared range is reported for nitrogen. This is important when an IR oven is operated with a nitrogen atmosphere instead of air: the boards will receive all of the thermal energy radiated by the emitters, but none by convection unless the nitrogen in the oven is heated by forced circulation through a separate system of heat exchangers.

Today's industrial IR soldering ovens provide for the forced and efficient circulation of the air in the oven in order to make use of its heat content, which would otherwise be lost. Apart from making full use of the energy input into the oven, forced circulation of the hot oven atmosphere helps to smooth out the previously mentioned local temperature differences between SMDs of differing mass, and excess heat on edges and corners. Finally, air circulation improves the heat supply to solder joints located in partial shade, like the J-legs of badly designed or very closely spaced PLCCs.

The physics of convection heating

The physics of convection heating, which governs transfer of heat from the oven atmosphere and the boards with their population of components, is complex. Two factors matter in the present context. The rate of heat transfer from a hot gas to a solid body is directly proportional to the temperature difference between that gas and the surface to be heated, and proportional to the square of the relative speed at which the gas flows over the surface. Also, turbulent flow improves the rate of heat transfer. For that reason, a growing number of atmosphere circulation systems in IR soldering ovens are designed for moving smaller volumes of air at higher speeds, without of course dislodging components from their footprints.

Normal, sometimes termed 'standard', atmospheric air is a mixture of a number of gases, the principal ones being:

Nitrogen	78 % vol.	75.5 % weight
Oxygen	21 % vol	23.2 % weight
CO_2	0.03% vol	0.05 % weight
Hydrogen	0.01% vol	0.001% weight

Some relevant physical properties of these gases, such as their density, their specific heat content per unit volume and their heat conductivity, are given in Table 5.13.

As the table shows, nitrogen is a more attractive heat transfer medium than air since, at a given temperature, a litre of it carries 1.2 times as much heat as a litre of

Table 5.13   *Physical properties of atmospheric gases*

	Standard air	N_2	O_2	CO_2
Density (g/litre)	1.3	1.25	1.43	1.98
Specific heat (cal/g K)	0.24	0.25	0.22	0.20
Specific heat (cal/litre K)	0.31	0.38	0.31	0.40
Heat conductivity (cal/g cm sec 10^{-4})	5.8	5.8	5.9	3.4

air. However, as has been said already, it does not pick up thermal energy from the flow of infrared radiation in the soldering oven, and therefore needs separate heat exchangers.

Carbon dioxide looks even more attractive, and some nuclear reactor designs make use of its good heat transfer properties. Unfortunately, its usefulness for soldering is marred by its tendency to decompose to some degree at elevated temperatures, which leads to the presence of some free oxygen, which ruins its properties as a protective atmosphere, and of some carbon monoxide, which is dangerous.

5.5.4   Operation of infrared ovens

Temperature profile

Definition

The term 'temperature profile' here describes the thermal experience of the solder joints themselves, as the board to which they belong travels through an IR oven. It has become a convention to plot temperature profiles along a time axis, as shown in Figure 5.20. The definition of a temperature profile, as normally encountered in technical and vendors' literature, varies. Often the temperature values represent the temperature of the various groups of emitters which are disposed along the oven, with the horizontal time scale derived from a given conveyor speed. However, the relationship between the temperature of an array of emitters and of the solder joints which travel past it is neither a direct or simple one.

Achieving a given temperature profile

Therefore, with a growing amount of high quality reflow ovens, the temperature values fed to the oven controls and displayed on the control panel of an oven are measured values, determined by temperature sensors disposed along the conveyor track of the oven. Provided the receptor surfaces of the sensors

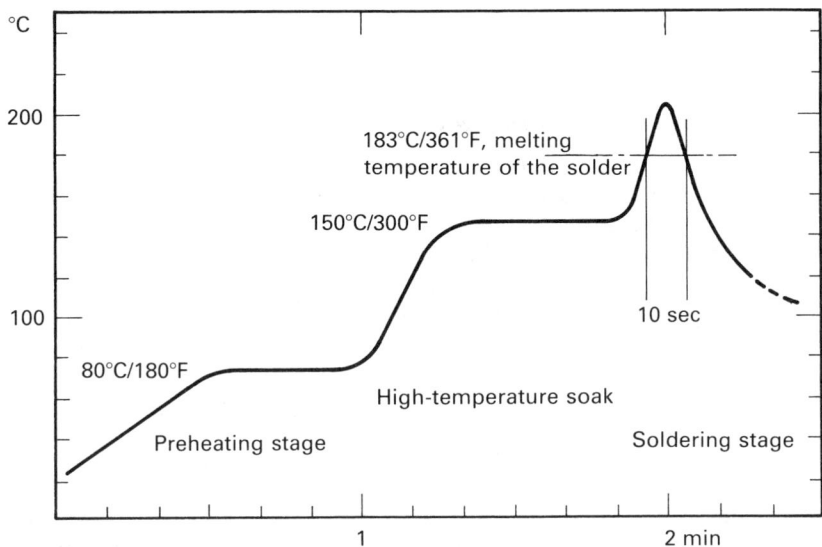

Figure 5.20 *Idealized temperature profile of an infrared reflowsoldering operation*

approximate the reflectivity of solder joints, i.e. the soldering terminals of the components, their readings provide a sufficiently realistic measure of the real temperature profile as it has been defined above, and a better basis for temperature control.

When the only available data are the temperatures of the emitters, the true temperature profile for a given setting of the emitter temperatures can only be established experimentally. A trial board, preferably an assembled board of the type for which the temperature profile has to be set, is sent through the oven, with thermo-elements attached to some typical or critical joints. Often, some sensors are also attached to the housings of some ICs, which may be temperature sensitive. Thermally conductive adhesives for the purpose are obtainable through the trade, if not from the vendors of solderpaste.

Various makes of temperature recorders suitable for travelling through an IR oven alongside the trial board are commercially available (see Section 4.3.3 under temperature control of wavesoldering machines). A recorder can usually accommodate up to six temperature/timetracks, which it can play back to a PC or a visual display after it has emerged from the oven. Others can transmit the temperature data in real time to an outside receiver as they travel through the oven. Such a measured temperature profile, together with the set of emitter temperatures which produced it, forms a reliable basis for setting, controlling and, if necessary, adjusting the real temperature profile not only of boards represented by the trial board, but also of others of a similar type. Once the relationship between a given setting of the emitter temperatures and the actually realized temperature profile of the joints on a board has been established, it should of course be entered and stored in a databank.

The shape of the temperature profile

By a tradition which has only recently been challenged, temperature profiles are subdivided into several separate temperature regions, each corresponding to a separately controlled group or array of emitters disposed along the oven. The number of emitter groups varies from vendor to vendor, but as a rule there are never less than three or more than five.

Until a few years ago, it had been held that the ideal temperature profile started with a preheating period of about one minute, during which the board temperature rises to a plateau of between 90 °C/195 °F and 110 °C/230 °F. This is intended to precondition the paste in the joints so that it does not spit or misbehave in other ways. A second preheating stage, lasting another minute, followed, raising the temperature to about 150 °C/300 °F. The heating rate leading to either plateau was aimed to stay below 4 °C/8 °F per second.

At the end of the second plateau, a rather faster, final spurt of heating, which might be as steep as 6–8 °C/11–14 °F per second, was to take the temperature beyond 183 °C/361 °F, the melting point of the solder. The final temperature peak must be high enough to ensure that the solder has melted and wetted the joint surfaces in every joint. Recommended peak temperatures vary from 215 °C/390 °F–280 °C/540 °F[23] and 230 °C/420 °F–260 °C/470 °F.[24]

The time spent by the joints above the melting point of the solder should be as short as is compatible with assuring that every single joint has reached it, but preferably not longer than 30 seconds. As has been said before, this important parameter on any temperature profile is the 'confrontation period' during which the molten solder in the joint is in contact with the metallic joint members. The shorter this confrontation time, the thinner will be the brittle intermetallic layer between solder and joint members and the better will be the joint (see Section 3.2). To keep it below 15 sec is practically impossible in IR reflowsoldering, but it should be the aim of good soldering practice to keep it within 30 sec or at most 45 sec.

Hardly less important is the speed at which the joint cools from the peak temperature down to about 170 °C/340 °F. Cooling rates between 20 °C/35 °F and 30 °C/55 °F should be the aim, to give the joints the desirable close-grained structure. These can be readily achieved by an efficient and correctly placed aircooling unit at the exit end of the reflow oven. Figure 5.20 shows an idealized version of such a temperature regime.

Recent work has shown that, with today's solderpastes, a simplified temperature profile for the approach to the final peak gives as good results as the stepped heating mode: this temperature profile has now only one temperature plateau, sometimes called the 'soaking' stage. The recommended duration of the soak varies between <1 minute[23] and 2 minutes,[24] with a soaking temperature of about 150 °C/300 °F. The rates of temperature rise between the various heating stages are still within the limits as before (Figure 5.21).

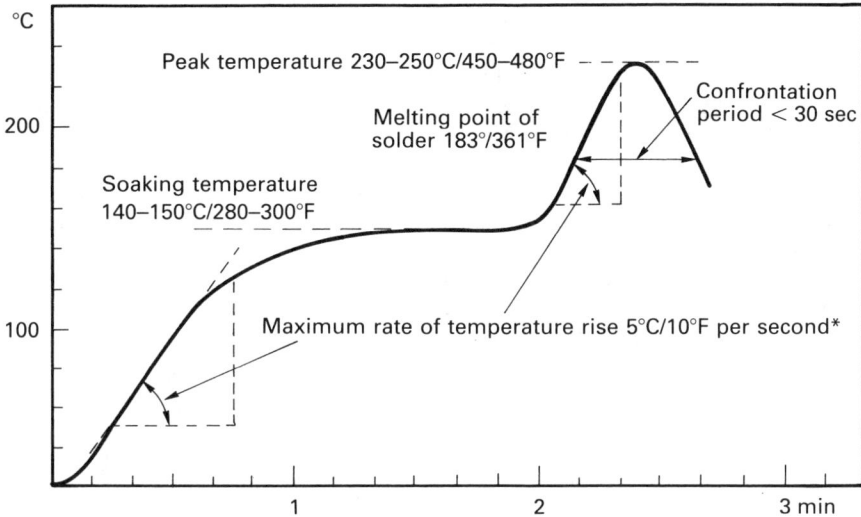

Figure 5.21 *Realistic temperature profile of an infrared reflowsoldering operation. *After Reithinger, M. (1991) Reflow Soldering Technology for Surface Mount Assembly, Electr. Productn., March 1991, pp. 27–37*

5.5.5 Oven design

Size of oven and position of emitters

Once the physical principles of reflowsoldering by infrared radiation are established and understood, designing a suitable oven becomes a straightforward problem of production engineering.

The starting parameters of any oven design are the maximum dimensions of the boards to be soldered and the maximum demanded output. The former governs the width of the oven; the latter, together with the maximum achievable speed of the conveyor, its length.

With industrial high-capacity ovens the conveyor speed may reach 0.7–1.0 m/min (28–40 in/min). Such conveyor speeds demand soldering chambers ranging between 1.5 m/50 in to 2 m/80 inch in length, to which the charging platform at the entry end and the cooling zone and the discharging platform at the exit end must be added. At the time of writing (1993), a high-speed oven, operating with a nitrogen atmosphere, and operating in an integrated production line for domestic video equipment, has been reported as the longest unit to date with an overall length of 16 m/48 ft.

Several vendors are offering ovens which are constructed on a modular basis with each module representing one of the heating zones on the temperature profile. This gives users more flexibility in adjusting their soldering line to the needs of changing circumstances.

Bench-type ovens for small-scale production and laboratory use are often quite short, from 50 cm/20 in to 75 cm/30 in length of chamber. Whether such short ovens are in fact able to put a circuit board through a definable temperature profile is open to doubt, especially as far as the short final melting peak is concerned. With the low conveyor speeds which are a necessary feature of such small ovens, the temperature profile is likely to become blurred unless the boards are very short.

It has become standard practice to place emitters both above and below the board conveyor, mostly with the same type and rating of emitter at both top and bottom. The emitters are controlled in separate groups, normally not less than three and often five or more. Within each group, the top and bottom emitters must be separately controlled, to enable the reflowsoldering of boards with SMDs on both sides (Section 5.1.1). With most makes of oven, emitter settings, once they have been established for a given type of board, can be stored and called up on demand.

Emitters

There are three types of emitters:

1. Internally heated metal panels, mostly with an integral or externally fitted thermocouple, for thermostatic control. They operate within a temperature range between 250 °C/480 °F and 350 °C/750 °F. Their temperature takes well over a minute to respond to a change in heating current.
2. Internally heated ceramic panels, equally fitted with a thermocouple. Their operating temperature extends up to 550–600 °C/1020–1100 °F and their response to a change in heating current is somewhat faster, but still in the minute range.
3. Quartz tubes with an internal tungsten coil extending down their centre. They can be fitted with an integral thermocouple, but seldom are. Tubular heaters are mostly installed at right angles to the direction of the boards. With some high-performance ovens, a number of quartz tubes are also arranged lengthwise along the sidewalls of the oven, to flatten the transverse temperature profile which tends to drop towards the sidewalls due to heat loss. The operating temperature of quartz tube heaters ranges from 700 °C/1300 °F upwards, and their response to a change in the heating current is fast, within less than one second.

Disposition of emitters

With most types of oven, the first two or three groups of emitters are of the metal-panel or preferably the ceramic type. Irrespective of the type and heat absorption characteristics of the boards which travel through the oven, the heatflow needs no fine tuning during the run-up to the soaking plateau on the temperature profile. What matters, though, is that the temperature of the emitters should be well above that of the boards which travel between them. As has been explained in Section 5.5.2, the efficiency of the transfer of heat from a hot to a cooler body is related to the fourth power of the temperature difference between them.

With most industrial infrared ovens, the emitters in the narrow melting zone at the end of the oven are a closely spaced group of quartz tubes, because only this type of heater can provide the short zone of intense, but accurately controllable heat irradiation which is essential to keep the confrontation interval between molten solder and joint surfaces brief, and to avoid overheating.

With some makes of high-performance oven, all emitters are high-temperature quartz tubes. There, the tubes are more widely spaced for the run-up to the soaking plateau, but closely set together in the zone of the melting peak.

Controlling the temperature profile

The control panel of almost every make of oven displays a temperature reading for every group of heaters. As long as this readout is taken for what it is, namely the temperature of the emitters and not of the boards themselves as they travel through the oven, then all is well. This is worth pointing out, because not all makes of oven make this clear. With almost all ovens, the displayed emitter temperature can be set, thermostatically controlled, and often called up from a stored program which has been established as suitable for a given type of board.

With some ovens, a temperature sensor for each group of emitters is located at conveyor level, with a surface which matches the thermal absorption/reflection characteristics of a typical solder joint. The readout from these sensors gives a more realistic picture of the heat to which the boards are exposed as they travel through the oven, though it is not necessarily equal to their actual temperature. Nevertheless, it forms a good basis for a thermostatically maintained temperature profile.

Finally, with some high-performance ovens, an infrared–responsive remote temperature sensor is focused on the surface of the travelling boards themselves. This does indeed give a direct reading of the actual board temperature. This expensive method is normally used only for the final melting zone, where its feedback controls the fast-response quartz emitters, allowing them to be run at high temperatures and consequently high efficiency.

With most other types of oven, the optimum setting of the emitter groups must be established experimentally, preferably with the travelling temperature recorder which has been described already, and then related to the readout on the control panel, and recorded.

It may be worth pointing out again that the fine tuning of the temperature profile of the run-up to the soaking plateau does not matter very much with modern solderpastes, as long as the melting zone is kept short and sharp, with its peak temperature between 200 °C/390 °F and 300 °C/570 °F. It has in fact been reported that a single temperature profile is suitable for most types of board.[25]

Cooling after soldering

As soon as all the joints have been soldered, they ought to solidify as quickly as possible, with their temperature dropping from the peak to below 183 °C/361 °F, this being the melting point of the solder, within a few seconds, certainly not more than about five. This will ensure a fine-grain structure and good mechanical strength.

A stream of air of room temperature, produced by a blower and directed at the boards both from above and below, is enough to achieve this. Obviously, the air velocity should not be such as to cause the components to shift on their footprints before the solder has set. Cooling the emerging boards from both above and below is not always practised, but it does repay the extra effort and cost by improved joint quality.

This cooling section should be located as close to the exit from the melting zone as possible. Suitable baffles must prevent the cooling air from blowing back into the oven.

Circulating the oven atmosphere

Recognizing that about half of the heat produced by the emitters is absorbed by the oven atmosphere before it meets the circuit boards, designers are giving much attention to efficient systems of air re-circulation in their ovens to ensure that this heat is not lost, but transferred to the boards by convection. (The physical laws which govern the transfer of heat from a stream of hot air to the surface of a solid body such as a circuit board are dealt with in Section 5.5.3.) A small and controllable bleed-off must be provided to prevent a build-up of the volatile constituents given off by the solderpaste, and the possible formation of an explosive solvent concentration in the oven.

In order to ensure that the air circulating in the oven does not blur the contours of the temperature profile, most ovens have separate circulation systems for the warm-up, the soaking and the soldering zone. In some designs, the different zones are separated from one another by bulkheads with horizontal slots just high enough to accommodate the conveyor and the boards riding on it. On the other hand, doubts are beginning to appear in some quarters about whether it is essential to strictly separate the first two zones. It is agreed, though, that the final soldering zone, which may be relatively short, must have its separate, accurately controlled temperature regime.

An efficient air circulation system, together with a correct layout of the emitters, will ensure that the temperature profile within an oven is level to within 5 °C/10 °F across the width of the oven. Some makes of oven subdivide the emitters into several separately controllable groups across the width of the oven. This is intended to ensure an even heat supply if the thermal requirements of a given type of board vary substantially from side to side.

Overhead soldering

Many ovens provide for separately controlled heating and circulation systems for the underside of the boards. This is useful when reflowsoldering boards with SMDs on both sides. After placing and soldering the components first on one side, the second side is populated with SMDs and passed through the oven again, with the SMDs soldered in the first pass hanging on to the underside. In vapourphase soldering it cannot be helped that the solder holding them to the board melts again but, as has been pointed out in Section 5.3, the surface tension of the molten solder will not let them drop off unless they are too heavy. Infrared reflowsoldering avoids this risk, because the underside of the board can be kept below the melting point of the solder. Specifically cooling the underside it is not necessary; in fact, this would reduce the efficiency of the soldering process on the top surface. Keeping the temperature of the underside of the board at about $100\,°C/210\,°F$ does no harm to the joints on it and assists the soldering process on the top side.

The importance of placing the cooling zone as closely as possible to the exit from the soldering zone has been pointed out already.

Conveyor systems

Boards travel through the oven on a continuous conveyor, which can be either an endless, flexible open-mesh stainless steel belt, or consist of two endless link-chains which support the two sides of each board. The distance between the chains must be adjustable, either manually or through a programmed servosystem, to suit the width of boards; a stainless steel wire may run between these side supports to prevent wide boards with heavy components from sagging.

A side-chain conveyor is of course obligatory for soldering double-sided boards: the side which was soldered first and carries components cannot be placed face down on a grid conveyor; with a chain conveyor, the central supporting wire must be lowered to clear the highest component. This too can be done either manually or through a programmed servo system.

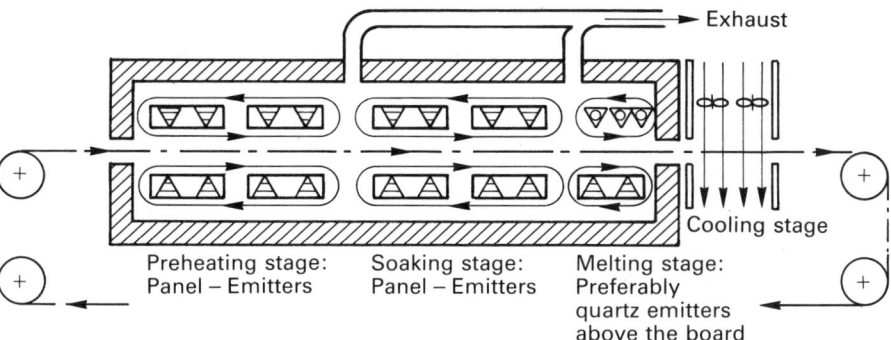

Figure 5.22 *Principal features of an IR reflowsoldering oven*

One vendor equips his ovens with two conveyors running at different speeds. The boards first travel on a relatively slow conveyor through a combined heating-up and soaking zone, where the boards are heated principally by convection. They are then picked up by a faster conveyor, which carries them quickly through the IR-heated soldering zone, keeping the high-temperature peak as short as possible.

5.5.6   Infrared soldering in a controlled atmosphere

The merits of soldering in the absence of atmospheric oxygen have been pointed out already, for both wavesoldering and vapourphase soldering. It has been found that IR reflowsoldering too benefits in several respects from a controlled atmosphere, provided the oxygen level can be kept below 10 ppm.[26] Soldering in an oxygen-free environment allows the use of solderpastes which contain less aggressive fluxes. Their residues can be left on many types of board which would require cleaning when soldered with a paste based on a conventional flux. Because metallic surfaces do not oxidize in an oxygen-free atmosphere, the metallized ends of melfs and chips are more readily wetted by the solder in the paste, the joints present a more attractive contour, and all solder surfaces have a clear and sparkling appearance.

Some striking reductions in the fault rate have been reported following a change from IR reflowsoldering in air to soldering in nitrogen, especially as far as solderbridges, non-wetting and incompletely filled joints are concerned.[27] As with vapourphase soldering, boards which have been IR-reflowsoldered in nitrogen are free from discolouration.

Some further features are specific to oxygen-free IR reflowsoldering in nitrogen. As has been said already, nitrogen does not absorb infrared radiation. Therefore it stays cool and does not contribute to the transport of heat from the emitters to the board. Thus, the entire heating process takes place under the laws of radiation, with convection playing no part, which may affect the temperature distribution on the board. To bring the benefits of convection back into the process, the nitrogen atmosphere must be heated by means of a heat exchanger system, and some makes of oven provide for this. Nitrogen protects the flux portion of the paste from oxidizing and carbonizing, thus making its removal easier when cleaning is required. It also seems to have the effect of promoting the wetting of FR4 epoxy-laminate or other organic surfaces by the flux, especially if it contains rosin. This can mean that flux residue may be found in places where normally it would not have flowed. It has been reported that the addition of 5% H_2 to the nitrogen prevents this from happening.[28]

A 'no-residue' solderpaste is commercially available which, when used for IR reflowsoldering in a reactive, nitrogen-based atmosphere (of unstated composition), is claimed to leave almost no flux residue, the flux portion having been volatilized in the process.[29] This can thus be classed as a genuine 'no-clean' soldering method (Section 5.2.4).

Ovens for IR reflowsoldering in an oxygen-free atmosphere must be designed in such a way that a minimum of the costly atmosphere is lost through inlet and outlet ports or other escape route. A certain amount of 'bleeding' will always be needed to prevent the buildup of the volatile constituents of the solderpaste, such as solvents and carrier substances.

Once all the joints have been soldered, they can be allowed to solidify in ordinary air, since at that stage the presence of oxygen cannot do any harm. Therefore, on emerging from the oven, the boards are cooled by a stream of ordinary air. Equipment which incorporates all these features is commercially available.

5.6 Reflowsoldering with hot air or gas

5.6.1 Convection versus radiation

With infrared heating, there is no simple answer to the plain question of every cook (or soldering engineer) 'How hot is my oven?' It all depends on such factors as emitter temperatures, the absorption properties and shape of the soldering goods and their travelling speeds.

Reflowsoldering by infrared radiation is a non-equilibrium process, where the emitters are much hotter, often very much so, than the joints can be allowed to get. This demands precise and reproducible control of the length of their exposure to the heat flow from the emitters. The higher the temperature of the emitters, the more sophisticated and flexible must be the control system of the oven, with emitter parameters having to be adjusted for different types of board to suit their specific heating requirements. Even so, smaller components get hotter than large ones, and in consequence, temperature differences of up to 30 °C/55 °F can occur between their respective joints, which results in widely differing microstructures.[30] A further cause of uneven heating under infrared radiation is the effect of the geometry of a solid body on its heat take-up, through which edges and corners get hotter than flat surfaces (see Figure 5.17, Section 5.5.2).

5.6.2 The physics of convection reflowsoldering

Convection heating in a moving stream of hot air or gas is an equilibrium situation. The oven temperature is precisely defined and measurable, being that of the circulating atmosphere. Whether the soldered goods fully attain the temperature of that heat transfer medium depends only on whether they remain in it for long enough. In practice, all the joints on a board reach the same end temperature to within ±5 °C/10 °F.

The rate of heat transfer between a flowing gas and a solid surface is mainly proportional to:

- the temperature difference between the gas and the surface
- the square of the speed at which the gas flows across that surface.

Other factors enter into it as well, such as the mode of flow at the interface: the more turbulent it is, the higher is the efficiency of heat transfer. Thus, a densely populated board with a highly complex topography, which produces local turbulence in the hot air flow over it, picks up the heat better than a plain one (the design of every heat exchanger is based on this principle). Thermal conductivity of the oven atmosphere, and its specific heat per volume, also matter (see Table 5.13), with nitrogen having a 20% edge over air as regards heat content per volume.

Flame soldering is a particular embodiment of hot-gas convection heating, and at the same time an extreme example of a non-equilibrium system (Section 3.5.2). Sophisticated microflame soldering equipment is commercially available, with robot-mounted flame heads and a controlled feed of fluxed solderwire on to the solder spot. At the time of writing (1993), these systems are used only for the soldering of non-electronic joints, but that situation may change.

5.6.3   Convection reflow ovens

Design considerations

In recognition of the rules of convective heat transfer, most oven designs prefer to circulate the oven atmosphere at higher speeds and in lower volumes rather than the other way round. The speed must of course not be so high as to dislodge any component from its footprint before the solder melts. Another universal design feature is the direction of the hot atmosphere at right angles to the board surface, so that its lateral spread on impact produces the maximum turbulence, and an even temperature profile across the width of the boards. With most convection ovens, the heating arrangement underneath the board conveyor is symmetrical to that above, so that the boards can travel nose-to-tail, forming a horizontal dividing baffle along the oven without affecting its efficiency. With second-pass soldering of double-sided boards (Section 5.1.1), this way of working makes it easy to keep the temperature of the underside of the boards below the melting point of the solder.

The oven atmosphere is heated by passing it through a system of heat exchangers, which follow the pattern of a well established technology. Because convection-oven soldering represents an equilibrium system, the temperature of the heating air or gas in a given zone is held at the same temperature or only slightly above that which the boards are intended to reach as they pass through it.

Most convection ovens are subdivided into zones, numbering from four up to ten, depending on the intended throughput of the oven. The final soldering stage is, or ought to be, designed so as to generate a steep temperature rise, leading to a narrow peak of 250 °C/450 °F–300 °C/540 °F, followed by a quick temperature drop. Ideally, joints should spend no more than about thirty seconds above the melting point of the solder (183 °C/361 °F), for reasons which have been explained before (Figure 5.21, Section 5.5.4).

Controlled-atmosphere working

Except for a controlled bleed-off to remove the volatile constituents given off by
the solderpaste, the oven atmosphere is recirculated to conserve heat. For the same
reason, the entry and exit ports for the board conveyor are kept narrow. These
features make it relatively easy to design convection-soldering equipment for
ready convertibility between air and nitrogen operation.

The length of a convection oven must take the relatively low rate of heat transfer
between atmosphere and boards into consideration. Overall lengths of between
3.5 m/10.5 ft and 4.5 m/13.5 ft are common. The conveyor systems of convection
ovens are basically the same as with IR soldering ovens (Figure 5.23).

Operating features

The temperature–time profile of convection soldering operations is less sharply
defined and divided into zones than is the case with infrared soldering. In fact, it has
been found possible to solder all but a few boards with unusual temperature
requirements with the same temperature profile, which begins with a steady slope
of 2–4 °C/4–7 °F per second from room temperature to about 140 °C/280 °F–
160 °C/370 °F, followed by a sharp rise to the above-mentioned soldering peak.
This obviates the need for creating and storing a large number of board-specific
oven programmes.[31]

5.6.4 *Development potential of convection reflowsoldering*

Oven atmospheres

Compared with infrared reflowsoldering, the pure convection reflow oven seems
attractive on several counts:

1. The infrared oven is by necessity a mixed system, using both radiation and
    convection for heating the boards, because of the unavoidable heat absorption
    in the oven atmosphere. The convection oven turns the non-equilibrium
    infrared system, with its incidental and uncontrolled role of the hot oven

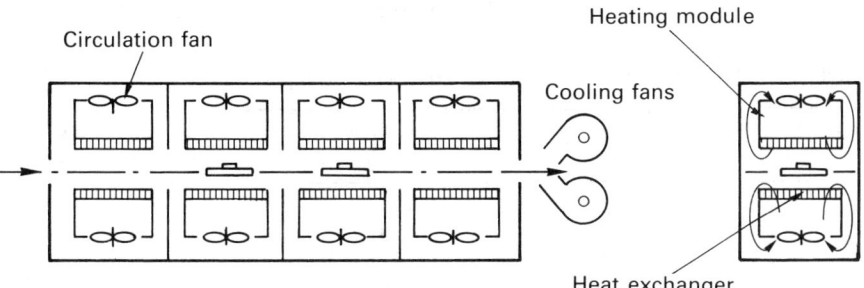

Figure 5.23 *Working scheme of a convection soldering oven*

atmosphere, into a controllable heating mechanism by means of a volume of air or gas raised to a defined temperature by an array of purpose-designed heat exchangers.
2. Convection heating makes it tempting to consider a wider choice of oven atmospheres. What matters in an oven atmosphere for convection heating is its specific heat, i.e. the heat content of a given volume of hot gas, its heat conductivity, which determines how efficiently this heat can be transferred to a circuit board, and above all its chemical interaction with the surfaces of the joints and the molten solder.

In normal air with 21% vol of oxygen, any soldering operation needs a soldering flux. From a chemical point of view, soldering in the chemically inert nitrogen needs no flux, though for physical reasons fluxless soldering under nitrogen has its problems, as has been shown.

Hydrogen starts to become able to reduce tin oxide and other metallic oxides above 350 °C/630 °F, an uncomfortably high temperature for electronic soldering. However, at lower temperatures it already actively assists the wetting of copper and other solderable surfaces by molten solder, but the extreme flammability of pure hydrogen rules it out as an acceptable soldering atmosphere. 'Forming gases' on the other hand, which are mixtures of hydrogen and nitrogen, do not suffer from this disability. A forming gas consisting of 25% vol hydrogen and 75% vol nitrogen has been found to actively encourage wetting of copper and some of its alloys by molten solder.[32]

Table 5.14 compares some relevant physical properties of normal air, nitrogen, hydrogen and 75 N_2/25 H_2 forming gas. As has been said already, nitrogen carries more heat to a circuit board per volume than ordinary air, and protects it and the solder from oxidation. Forming gas carries somewhat less heat per volume, but gives up its heat to the board more readily. Above all, it has been shown to positively promote soldering. Thus, forming gas could be worth considering.

Adding an infrared melting stage

In the conventional IR oven, convection is a basically unplanned adjunct which utilizes the unavoidable heat picked up by the oven atmosphere; this heat would

Table 5.14 *Physical properties of some convection-oven atmospheres*

	Density g/litre	Specific heat cal/(litre K)	Heat conductivity cal/(g cm sec)
Air	1.3	0.31	5.8
N_2	1.25	0.38	5.8
H_2	0.09	0.31	41
Forming gas 75 N_2/25 H_2	0.96	0.36	6.27

otherwise be lost. With a convection oven, it seems tempting to use an IR heating stage as a deliberate addition.

The single unattractive feature of convection ovens is the final melting stage, which must produce a steep and narrow temperature peak, rising from about 150 °C/300 °F up to or above 250 °C/480 °F and then quickly dropping below 183 °C/360 °F. To achieve this with hot air demands a narrowly confined stream of high-temperature air or gas, operating separately from the preceding gradual and more gentle heating regime of the preheating stages. It might be simpler, and perhaps cheaper, to replace this necessarily somewhat blurred high-temperature blast by a readily focused beam of heat radiation produced by one or two closely-spaced quartz emitters. Their ready response to current changes would make them well suited to automatic control by suitably placed sensors. At the time of writing (1993), no convection oven maker seems to have made use of this possibility.

5.6.5 Convection soldering of single components

The process

Hand-held and bench-mounted hot air/gas reflowsoldering tools also represent a form of convection soldering. A gas which could be used instead of air is mainly nitrogen. Whether the expense of nitrogen is justified must be decided in the context of the possibility of using a weaker flux, and thus avoiding or simplifying any subsequent cleaning, should that be called for. Hot air/gas soldering is used mainly for attaching individual components, such as large multilead ICs, to circuit boards which already carry the bulk of their SMDs, having been soldered by an in-line method, either by wave or reflow. Another major use of hot air/gas soldering is the de-soldering and re-soldering of SMDs in repair work (see Chapter 10).

If solderpaste is used as the solder/flux depot, it is mostly applied with a hand-held dispenser gun, which may be manually or pneumatically operated (Section 5.3.1) Alternatively, if the footprints are already covered with a sufficient layer of solder from a preceding hot-air-levelling (HAL) operation or from having passed through a solderwave, a low-solids flux, such as is used for wavesoldering or handsoldering, is suitable. Its choice will be governed by any postsoldering treatment which may be necessary or called for.

Soldering with these tools does not represent an equilibrium situation: the temperature of the hot air or gas is between 350 °C/650 °F and 450 °C/850 °F, and thus well above the final joint temperature of about 250 °C/480 °F. These high gas temperatures are chosen because in this work speed is of the essence. The aim is to keep the confrontation period during which the solder in the joint is molten to within a few seconds, and not more than five if possible. Once the target temperature has been reached, heating is discontinued. Control is either visual, by programmed timing or with a temperature sensor.

With both types of equipment, a coherent stream or jet of hot air or gas of moderate velocity is directed more or less vertically against an array of joints, with the solderpaste deposit or the flux in place. On impact, the stream becomes turbulent, which, as with convection-soldering ovens, is the basis of effective heat transfer to the joints.

Hand-held tools

These are normally in the form of hot-air guns, either fitted with an integral air blower and heater or fed via a flexible hose from a stationary small compressor and heat exchanger. The latter arrangement is preferable for close work, because the unencumbered airnozzle is more manouvrable. The controllable air temperature is normally set between 350 °C/650 °F and 450 °C/850 °F. This is well above the joint temperature aimed at: as has been said already, heating is discontinued as soon as the solder in the joints is seen to have melted and filled them. The jet is conveniently controlled with a footpedal, so as to keep the operator's hand free.

The component is placed on the prepared footprints, normally with a vacuum pipette, and held down while the joints are heated with the jet of hot air. It may be best to pin down larger components by first soldering two diagonally opposed corners, and then work along the edges, moving the jet on as soon as a joint can be seen to have filled with molten solder.

Bench-mounted equipment

A variety of bench hot-air or hot-gas soldering equipment of varying degrees of sophistication, automation and complexity is on the market. These machines are used for the soldering of single multilead components such as PLCCs, SOICs, QFPs, etc. to boards already populated with the bulk of their components.

The stream of hot air or gas is guided to the array of joints through an interchangeable nozzle, shaped to fit the footprint pattern. It is important that the nozzle should have a low heat capacity so as to heat up quickly and not chill the airblast unduly as soldering begins (Figure 5.24).

During soldering, and until the solder has solidified afterwards, the component must be held down against the board under gentle pressure. This ensures that all component legs sit firmly on their respective pads, even if their coplanarity is not ideal (see Section 7.1), and that the component does not move during soldering.

It is advisable to plan the soldering strategy so as to create the solder depots for this operation during the preceding soldering stage, where the bulk of the components are soldered. Hot-air levelling or wavesoldering will leave enough solder on the pads for subsequent hot-air or hot-gas soldering. The same applies to the solderpaste deposit on unoccupied footprints; the solderpaste will have melted during a preceding normal reflow operation. In most cases the pretinned pads will have to be fluxed again before the component is placed in position.

This consideration applies generally to the area of ultrafine-pitch technology (pitch ≤ 0.3 mm/12 mil), where solderpaste reaches the limits of its printability and its freedom from bridging.

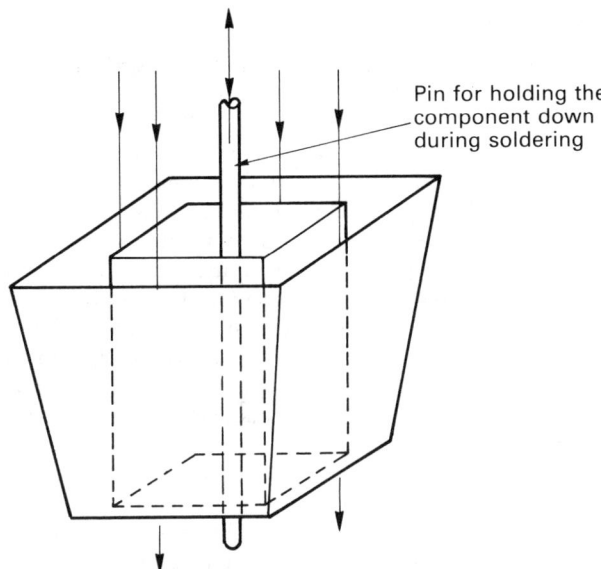

Pin for holding the
component down
during soldering

Figure 5.24 *Hot-air nozzle for soldering a multilead component*

Hot-nitrogen reflow techniques for the soldering of the outer leads of TABs to their pretinned footprints (outer-lead bonding – OLB), which use a so-called hot-air thermode (HAT), have been developed by several companies (e.g. IBM, Fuji and SRT).[33]

With many bench-mounted hot-air soldering machines, visual or video aids are provided to assist with the precise placement of fine-pitch multilead components. For quantity production, machines are on the market where a placement head and the hot-air nozzle operate sequentially, with the board remaining stationary during the operation, its correct position assured by fiduciary holes or markings. With this type of equipment, the temperature, timing and duration of the hot airblast and the hold-down of the component during and after soldering are all programmed.

Preheating

In order to keep the duration of the molten-solder confrontation as short as possible, especially with heavy multilayer boards, many bench-mounted machines provide for locally preheating the circuit board from below, with a gentle flow of hot air or a low-temperature infrared emitter. Preheating the board to 60 °C/140 °F–80 °C/180 °F should suffice. Large boards may warp if the local preheat is too sharp. With manual hot-air soldering, placing the board on a warm hotplate is helpful.

5.7   Laser soldering

Laser soldering presents the ultimate non-equilibrium situation: the wavelengths emitted by the lasers used for soldering lie in the infrared region, but because of the mechanics of laser emission, it makes no sense to talk about the temperature of the radiation source, as will be explained in Section 5.7.1. However, controlling the laser dosage with great precision is essential if the joint is not to be vapourized. In contrast to IR oven-soldering, where the total surface of a board is flooded with infrared radiation, its wavelengths spread over a wide spectrum, laser soldering targets each individual solder joint with a measured pulse of infrared energy, of a single wavelength, and gathered in an extremely narrow beam.

The word 'laser' is formed from the initial letters of the term Light Amplification by Stimulated Emission of Radiation. In 1957, the laser phenomenon as such was recognized as a possible practical application of quantum physics. Th. Maiman (USA) translated it into a practical piece of equipment in 1960. Several features make reflowsoldering with laser beams appear attractive. A beam of heat radiation can be targeted very accurately onto a joint. The high energy density at its point of impact allows for very short soldering times and quick solidification, while the duration of a laser light pulse can be controlled to well within a millisecond. Finally, heating is strictly localized and both the components and the board remain cool.[34]

5.7.1   How a laser works

The working of a laser is based on the fact that certain so-called laser-active substances may, when exposed to a strong source of light (termed 'pumping'), emit quite a special type of light themselves. The effect is caused by the pumping light raising the atoms or molecules of the laser-active substance to a higher quantum level of energy, from which they return instantly to their original state.

Matters can be so arranged that this secondary energy emerges from the laser-active substance in the form of laser light, which has some very special properties. Instead of being spread over a spectrum of wavelengths, it has one single wavelength only, which is specific for the 'lasing' substance. Also, it is composed of long 'coherent' trains of lightwaves, not of the very large number of separate brief energy pulses which make up the light emitted by a hot body or gas, or a gas discharge.

A laser usually takes the form of an elongated cylindrical rod or, if the lasing substance is a gas, a transparent tube. To make it 'lase', its sides must be irradiated by the pumping light, an intense pulse usually produced from a powerful flashlight or gas-discharge lamp with ratings in the kilowatt range. Both ends of a laser rod or cylinder are mirrored, the long wavetrains of laser light bouncing back and forth between them, while being constantly reinforced by the pumping light. Eventually they escape through small apertures in one or both of the end-mirrors in the form of monochromatic, coherent and almost perfectly parallel light beams. Lasing starts as soon as the primary light starts pumping, and stops as soon as pumping stops. This makes it possible to pulse laser light very accurately. The efficiency with which the pumping light is converted into laser light ranges from a few per cent up to 30%, depending on the type of laser. The difference between pumping energy and emitted laser energy takes the form of heat, which must be disposed of by efficient cooling, otherwise the laser can destroy itself in a few seconds (Figure 5.25).

5.7.2 Nd:YAG and CO_2 lasers

Two types of lasers can be used for soldering purposes: the Nd:YAG laser and the CO_2 laser.

The Nd:YAG laser

This is a solid laser. It uses a synthetic semiprecious stone, yttrium aluminium garnet, which is dosed with neodymium (a rare earth element) as a lasing substance. Its emitted light has a wavelength of $1.06\,\mu m$, which is located in the near infrared range. It lases with an efficiency of a few per cent, which means it has to be pumped with light of near 1 kw energy to emit a laser beam of 10 watt. The rest of the energy must be removed by an efficient and reliable cooling system.

The light beam from a Nd:YAG laser can be gathered into a bundle of $10–20\,\mu m/0.4–0.8$ mil diameter. For soldering, beam energies of 10–20 watt are normally used.

Solder absorbs thermal radiation of a wavelength of $1\,\mu m$ well (see Figure 5.16), which means that the light from a Nd:YAG laser has a high heating efficiency. Though according to Wien's law the energy maximum of light emitted by a body with a surface temperature of $2600\,°C/4700\,°F$ is located at the $1\,\mu m$ wavelength, it would be misleading to ascribe that temperature to the lasing substance. As has

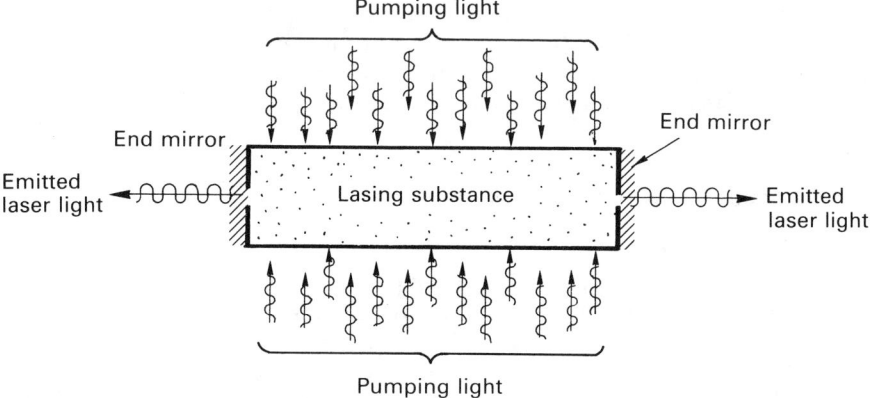

Figure 5.25 *Working principle of a laser with emission at both ends*

been said already, lasing is not a matter of heat and the chaotic oscillations of atoms and molecules associated with it, but of quantum jumps; that is why it is monochromatic.

Therefore, the intense local heat created at the point where a laser beam meets a surface is due not to a high temperature of the lasing substance but to the high energy density at the point of impact: a laser beam of 20 watt, with a diameter of 50 μm, produces an energy density of 10 kw/mm^2 in the spot where it impinges on, for instance, a footprint. It will burn a hole through it in milliseconds, an extreme case of a non–equilibrium situation, which calls for high-precision timing and targeting.

The Nd: YAG beam passes through glass and polymers, including the flux portion of solderpastes, with little absorption. The transparency of glass towards the beam means that it can be manipulated and transmitted by normal optical means, such as ordinary glass lenses, mirrors and glass fibres. It also means that safety goggles do not protect the human eye against it. In turn, this makes it mandatory to place all Nd: YAG laser soldering equipment in completely lightproof enclosures, so that no stray, direct or reflected, laser beam can escape. The beams, being coherent and parallel to a high degree, can traverse great distances without loss of intensity. The Nd: YAG laser beam passes through the cornea, lens and eyefluid of the human eye with little attenuation, and will destroy the retina in milliseconds at the point of impact. Therefore, this type of laser soldering can only be monitored through an indirect video display.

The CO_2 laser

The CO_2 gas laser is another type of laser used for soldering. Its laser–active substance is CO_2 gas, which lases at a wavelength of 10.6 μm, i.e. in the far infrared. CO_2 lasers have an efficiency of about 30%, and they are cooled by pumping the gas through a heat exchanger. The beam from a CO_2 laser can be bundled to a diameter of 50–100 μm/2–4 mil. It is strongly absorbed in polymers, and especially normal glass. On the other hand, solder absorbs only about 20% of its energy (see Figure 5.16). If the beam has to be manipulated, the optics must be made from expensive special glasses. For these reasons, the development potential of the CO_2 laser for soldering applications is limited.

5.7.3 Laser soldering in practice

Points in favour

Laser soldering offers a number of tempting features. The confrontation period is measured in milliseconds instead of seconds or half-minutes, and solidification after soldering is equally rapid. Therefore, the intermetallic layer in a laser-soldered joint is hardly visible under the microscope, and the microstructure of the solder is extremely fine. In consequence, the mechanical properties and the life expectancy of laser-soldered joints are optimal.

Also, the heat input during soldering is so short and so strictly localized that both the board and the components stay at room temperature. This avoid the stresses locked in the joints arising from the different contraction behaviour of substrate and components, as the soldered assembly cools down from the soldering temperature. In fairness, it must be said that laser soldering shares this feature with impulse soldering (Section 5.8).

The narrowly localized soldering spot invites the attempt to direct a fine jet of nitrogen on it and to solder in a controlled atmosphere. The possibility of using this method for fluxless soldering, and thus to avoid cleaning, has been mentioned in some of the literature on the subject.

Points against

Several aspects take some of the shine off laser soldering. One is the one-by-one nature of soldering with a single beam, though double-beam laser machines (see below) reduce this handicap by half. Another problem arises from the extreme non-equilibrium nature of laser soldering. For a given laser-soldering task, every joint with its individual thermal mass and reflectivity demands a precisely defined laser impulse. A slight deviation, such as a bent lead, or a slight change in the amount of solderpaste, may mean a joint left open, or destroyed. To cope with this problem, the temperature of the joint can and should be monitored during soldering. A PbSe(Te) pyrometer focused on the joint is claimed to be able to control the irradiation time to within 0.001 sec. This is the basis of the so-called 'smart lasers'.[36]

5.7.4 Laser-soldering equipment

A double-beam Nd:YAG laser-soldering machine has been commercially available for several years (Baasel Lasertechnik, Germany) (Figure 5.26).

This machine operates with two jointly controlled 20 watt beams of $50 \mu m$ diameter emitted by a double-ended Nd:YAG laser, which provides an energy density of about $10 \, kw/mm^2$ in each soldering spot. The laser beams are manipulated by electromagnetically actuated galvanometer mirrors in such a manner that, with every SMD, two symmetrically opposed joints are irradiated. This symmetry prevents the tombstoning of chips. The duration of a given soldering impulse depends on the mass of the joint members, amounting for example to 0.02 sec for an SOIC gullwing leg. Soldering an SOIC with 28 joints takes approximately one second, half of which is needed to move the beams from joint to joint. During the soldering pulse, the beams oscillate across the joint area in circular or elliptical patterns in order to distribute the heat and prevent local overheating (Figure 5.27). Most good-quality solderpastes have been found to give good, sputter-free laser-soldered joints without the need for predrying.

The movement of the laser spots from joint to joint on a given SMD, and their oscillation during soldering, is taken care of by the galvanometer mirrors. Individual SMDs are brought into soldering position by a servo-operated xy table,

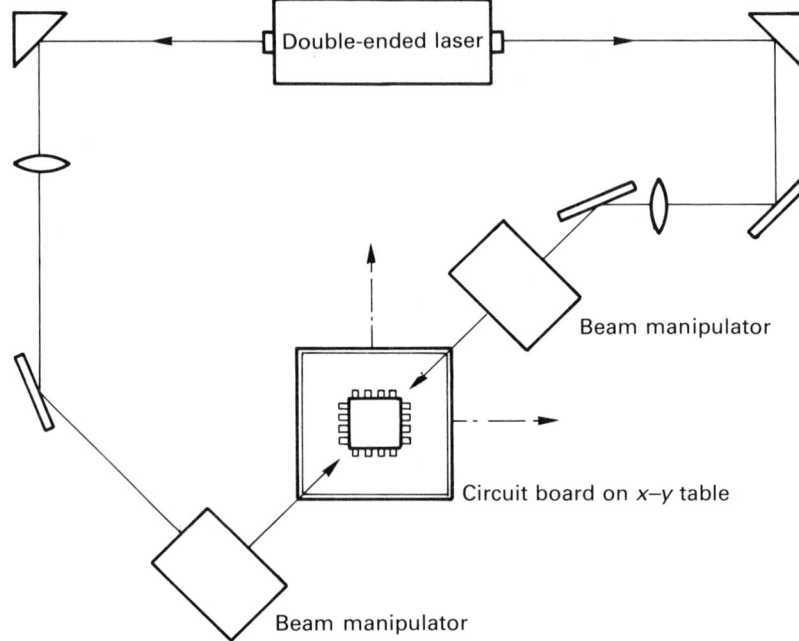

Figure 5.26 *A twin-beam Nd: YAG lasersoldering machine (Baasel, Germany)*

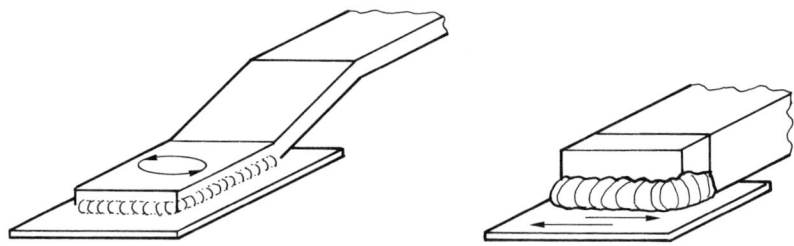

Figure 5.27 *Oscillating laser spot*

on which the board is mounted. The coordinates of the individual soldering points can be derived from the software which creates the solderpaste printing stencil. The optimum exposure times for the individual joints must be established by a teach-in procedure.

Laser soldering is an ideal heating system for soldering individual ICs or TABs with their small footprints, with ultrafine–pitch spacing. The introduction of the twin-beam concept for the soldering of a complete population of ordinary joints with standard pitch spacing was probably premature. It has to use expensive and time-wasting expedients, which may be one of the reasons for its lack of market penetration.

With the spread of fine-pitch and ultrafine-pitch technology, the niche which soldering with smart lasers has found in electronic assembly technology has begun to grow. Four makers of commercial OLB laser bonding equipment have recently been identified in the US and Japan (Panasonic/Matsushita, Hughes Aircraft, Mitsubishi and Sony).[37] Some problems still remain, however. While the ND:YAG laser provides accurately localized heating for densely spaced leads, uniform lead hold-down remains one of the challenges for this particular technique.

5.8 Impulse soldering

5.8.1 Operating principle

The concept of impulse soldering, also known as thermode or hot-bar soldering, derives from the ordinary soldering iron. This time-honoured tool brings a hot, suitably shaped piece of copper into close contact with one or both members of the intended joint, to which flux and solder are either supplied at the same time, unless one or both have been pre-deposited in the joint ('sweatsoldering'). Thermode soldering is a sophisticated derivative of sweatsoldering with a soldering iron.

Prior to soldering, the pretinned footprints are fluxed, the component is placed in position and the heated soldering tool (thermode) is brought down on all the joints simultaneously. It is held down under a slight, controllable pressure until the solder has melted and filled the joints. Heating is then discontinued, and aircooling is sometimes applied to speed up the solidification of the solder. Once it is solid, the thermode is lifted clear of the assembly.

The rate of heat-transfer between the thermode and the joint against which it is pressed is the arithmetical product of the following factors:

* the area of contact between thermode and joint
* the thermal conductivity of the thermode material and the joint
* the time of contact
* the time-integral (the sum total) of the temperature difference between thermode and joint during the soldering process.

Impulse soldering is basically an equilibrium process, because the temperature of the joint equals that of the thermode at the end of the heating-up period, which usually lasts less than one second.

Impulse soldering is used more and more frequently for attaching individual multilead, fine-pitch components like SOICs, QFPs, flatpacks, PLCCs and, particularly, the outer leads of TABs to boards which have already been populated with the bulk of their components. The fact that the thermode holds down all the legs of these components – and their lead-count may rise up to a hundred or more – against their respective footprints with a controllable pressure before, during and after soldering is the principal virtue of impulse soldering.

Because of this, coplanarity of these legs (Section 1.4) is no longer the problem it is with other soldering methods. For obvious reasons, impulse soldering is equally useful for desoldering multilead components, should their removal become necessary for repair or corrective work, and of course equally for soldering the replacement component into place (Sections 10.2 and 10.3).

There may be various reasons for choosing impulse soldering: the components concerned may not be suitable for overall heating because of their construction; equipment for their automatic placement may not be available or may be uneconomic to procure; or the component may be tailor-made for impulse soldering and unsuitable for any other reflowsoldering method, for example TABs. Most frequently its pitch may be too fine to permit wavesoldering or a reliable printdown of solderpaste; heat sensitivity of the component may be an added factor (see the popcorn effect: Section 2.5, Figure 2.5).

5.8.2   The solder depot

Impulse soldering being a reflow method, a solder depot must be provided on one set of the joint members, usually the footprints. On ultrafine-pitch components the tin or solder coating on the component legs may be thick enough to provide enough solder to make the joints.[38]

Principally, a solder-depot thickness of 10–30 μm (0.5–1.5 mil) is sufficient to give well filled joints.[39] The closer the pitch of the footprints, the thinner should be the depot: solder squeezed out from closely spaced footprints can cause bridging.

The impulse soldering of a multilead IC is normally preceded by the wave-soldering or reflowsoldering of all the other SMDs to the board. Wavesoldering will have left a domed solder depot of up to 100 μm/4 mil thickness on the empty pads, as will have hot-air levelling. Reflowing a printdown of solderpaste leaves less solder on the unoccupied pads; with fine-pitch layouts, the thickness of the melted down solder will be around 50 μm/2 mil (Section 5.3.2), when doming is less pronounced.

If a still thinner solder depot is called for, a galvanic solder depot can be used. Unless the preceding reflowsoldering of the other components on the board has fused this electrodeposited solder depot, it ought to be fused in a separate operation (most conveniently in a vapourphase soldering operation). This is advisable because an electrodeposited solder layer is not a coherent alloy coating, but a porous agglomeration of discrete tin and lead particles. After a certain time, oxygen and possibly water vapour penetrate down to the copper substrate and affect its solderability, while the tin and lead particles themselves also oxidize, and are thus less likely to fuse together (Section 3.6.3).

If a printdown of solderpaste is used prior to impulse soldering, it must not be placed on that part of the footprints where the thermode will come down on them, even if the pitch would allow it, because the non-metallic portion of the paste (the previously mentioned mixture of flux, thickeners and thixotropic additives (Section 5.2.4)) would carbonize and adhere to the hot surface of the soldering tool. This impairs its efficiency, and means frequent cleaning, which is costly and can damage the tool. Instead, the paste is printed near the ends of the footprints, away from the component legs. The heat from the thermode will flow along the component legs and the footprints, melt the solder and draw it into the joints (Figure 5.28).

The domed profile of a thick solder depot may cause a thin lead to slip sideways under the pressure of the thermode. Such depots can be flattened by bringing the thermode, whose temperature must be below the melting point of the solder, say 120 °C/250 °F to 150 °C/300 °F, down on the footprints with sufficient pressure ('coining' the footprints).

Before a component is placed, the pads are fluxed with an electronic grade of flux (Section 3.4.3), preferably with a low solids content. Its choice will depend on whatever circumstances govern the choice of flux for the whole board.

The flux is normally applied as a narrow jet of fine spray, either manually or with automated equipment, in a pattern which matches that of the footprints. It is good practice to put down the component while the flux is still somewhat moist, so that the legs of the component receive some of it.

The flux itself is of course of a type suitable for use on electronic assemblies, such as RMA or OA (MIL-F 14256, USA) or UK-BS 5625, Class 4 or 5b (see Section 3.4.3). The exact choice, e.g. whether low-solids or no-clean, will depend on any postsoldering treatment which the finished assembly will have to receive. It would seem that the development of specific fluxes for impulse soldering could be worthwhile, bearing in mind the growth potential of this form of soldering. The same consideration applies to the means for providing an oxygen-free local environment in the immediate vicinity of the thermode during the few seconds of the soldering process itself. Because soldering takes place in a closely defined space, over a time span of only a few seconds, gas consumption could be quite modest.

5.8.3 *The thermode and its heating cycle*

Thermodes are made from a strong, non-tinnable, heat resistant metal, such as titanium or, less frequently, molybdenum, or one of their alloys. A thermode is normally heated by passing a controlled pulse of current through it – hence the term 'impulse soldering' – less frequently by attached resistance heating elements. A thermocouple embedded in the thermode controls its time–temperature profile.

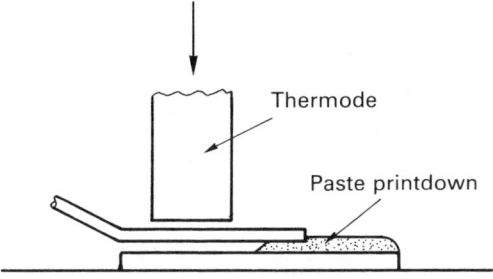

Figure 5.28 *Paste printdown for thermode soldering*

In order to keep the heat pulse short – within the range of a few seconds – the heat capacity of the thermode is kept low by slimming its cross section as far as the mechanical demands on it will allow.

Its outline and dimensions must fit the configuration of the component leads, which implies that every type of component demands its specific thermode (Figure 5.29). PLCCs pose a special problem, because their solder joints are not directly accessible to the thermode: the soldering heat is transmitted to the vertical shanks of the J-leads, while the thermode tip heats the solderpads next to the joints ('collet soldering', Figure 5.30). During the heating cycle and until the solder has solidified, a pin holds the PLCC against the circuit board under slight pressure. With many thermode soldering machines, a hold-down pin is used with all types of component during soldering and postcooling.

The heating current flowing through the thermode necessarily creates a gradient of electric potential along its horizontal bars. Where the ICs to be soldered are of a type likely to suffer damage by such a potential difference between its leads, the underside of the thermode can be given a thin, electrically insulating but thermally conductive coating, e.g. a vitreous enamel or PTFE.

Thermode soldering makes some specific demands on the board layout. Since each solderpad receives the same amount of heat, it is important that all pads have the same heat capacity, which means the same geometry. Potential heat sinks must not be integral with the pad areas, but should be separated from them by narrow bridges, which minimize parasitic heat flow away from a pad (Figure 5.31).

The time/temperature profile of an impulse soldering operation must meet some specific demands:

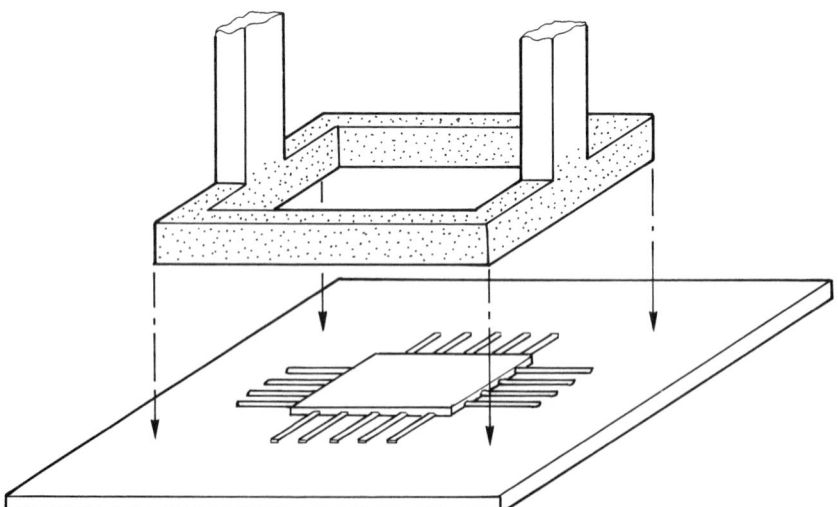

Figure 5.29   *Thermode for ICs with flat leads*

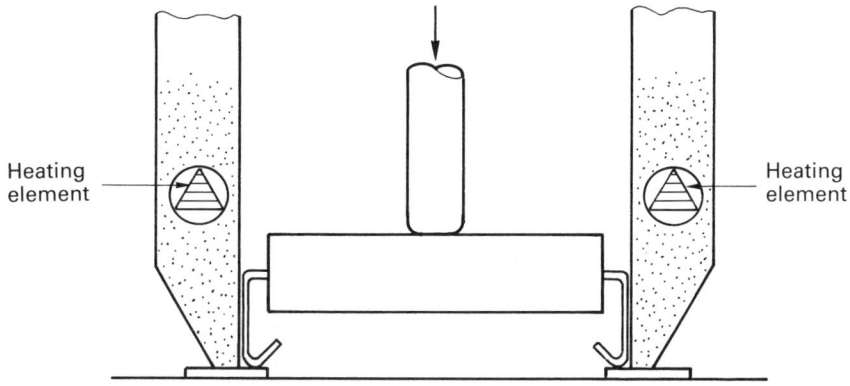

Figure 5.30 *Thermode for PLCCs*

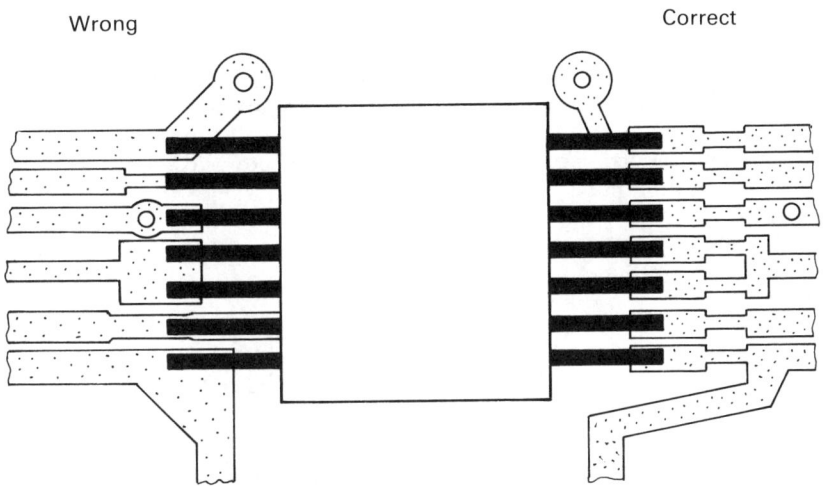

Figure 5.31 *Footprint layout for impulse soldering. After* Soldering in SMD Technology, *Siemens, Munich, Germany*

1. The heat pulse must be kept short for several reasons. The joint gap between a footprint and a component leg is only 0.01 mm/0.4 mil to 0.1 mm/4 mil across, but the joint area may be as much as $1 \text{ mm}^2/1600 \text{ mil}^2$. With such a joint geometry, there is real danger of filling much of the joint gap with brittle intermetallic compounds if the temperature is too high or heating unduly prolonged.[40]
2. Another reason for a short heating pulse is the need to avoid weakening the bond between the footprints and the board laminate. In order to keep the heating pulse short, heavy multilayer boards may need local preheating of the board from underneath, in order to relieve the thermode of the task of heating

a heavy substrate with massive internal copper conductors which form effective heatsinks. This presupposes of course that the underside of the board does not carry its own population of SMDs.

3.   The temperature gradients at the beginning and end of the heating cycle can be quite steep, up to 500 °C/900 °F per second, because the component itself stays cool during soldering and is not in danger from a sudden heat shock.

5.8.4 Impulse-soldering equipment

With most of the commercially available equipment, the thermodes follow a similar design (Figure 5.32). Four horizontal bars, each with separate leads, follow the pattern of the footprints. The pulse of heating current is supplied from a common transformer. The temperature rise at the beginning of the pulse is steep, at 400 °C/700 °F to 500 °C/900 °F per second, and so is the temperature descent at the end of the 3–5 sec heating period. Fast cooling is normally assisted by a blast of cold air.

The design of commercially available impulse-soldering equipment spans a wide range of sophistication. At one end of the scale, single components are placed by hand, their manual alignment being assisted by a simple magnifier. This can be very effective, and a reasonably skilled operator can manoeuvre an IC to within <0.1 mm/4 mil lateral accuracy with a vacuum pipette. Position having been effected, the heating pulse of pre-set intensity and duration is triggered by pedal pressure.

At the other end of the scale, automated production equipment may employ opto–electronic sighting on fiduciary markings or footprint patterns to position

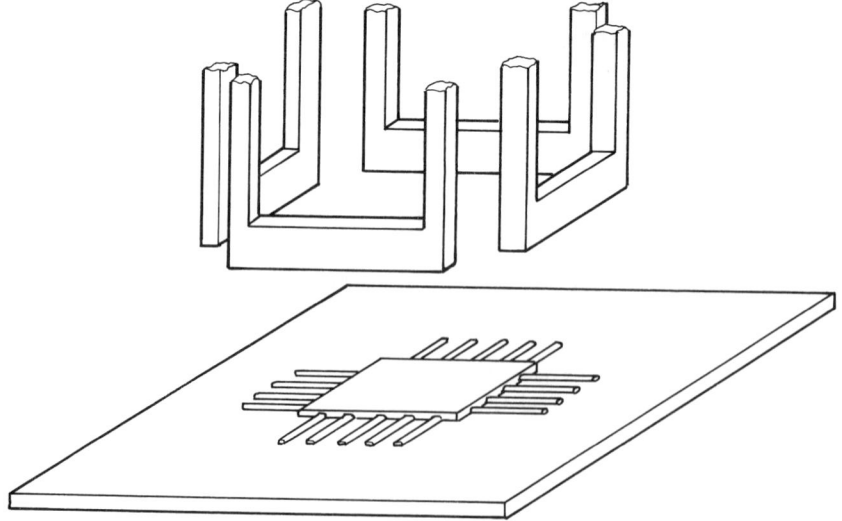

Figure 5.32 *Four-bar thermode*

the component. Often the soldering equipment is integrated with a pick–and–place facility.

If a variety of components is to be handled, thermode heads can be automatically changed. Heating periods, peak–temperature values and dwell periods are numerically controlled. If a number of components are soldered to one board, the board position is normally fixed, with the thermode head mounted on a numerically controlled *xy* gantry. Before a component is placed, the correct pattern of flux can be sprayed automatically on to its footprints.

The steadily rising use of high–pincount ICs, with their pincount showing no sign of slowing their own increase, continues to drive the development of impulse soldering further forward.

5.9 SMD soldering methods – A survey

See Figure 5.33 for temperature profiles of the different soldering methods.

Wavesoldering (Double waves, Combination waves, Jetwaves)

Characteristics: A thermal equilibrium situation – all joints attain the same temperature as the heat source. Flux, preheat and hot solder supplied to all joints sequentially in an in-line process. Capable of being carried out in an oxygen-free atmosphere.

Soldering temperature: As a rule 250 °C/450 °F.

Rate of production: Medium to high.

Duration of confrontation between molten solder and substrate: 2–5 seconds.

Solidification speed of joints: High.

Field of application: Mandatory for boards with a mixed population of components, where SMDs share a board surface with wired-through joints. The SMDs must be glued to that board surface before wavesoldering.

Limits of applicability in normal atmosphere: Suitable for passive components, SOTs, SOs, and all multilead components with a pitch down to 1.27 mm/50 mil. Possible, but more difficult and requiring special layout provisions, at finer pitches, down to 0.75 mm/30 mil, and PLCCs. Unsuitable for ultrafine-pitch components and closely spaced SMDs.

Limits of applicability under nitrogen: Suitable for components down to 0.5 mm/20 mil pitch.

Reflowsoldering in general

Characteristics: Creation of depot of solderpaste [= solder + flux], or of solid solder, then flux, followed by placement of the SMDs, and subsequently the application of heat. The lapsed time between creating the depots, placing the components and applying the heat may be hours or days. Creation of depots footprint-by-footprint or simultaneously on all of them. Subsequent application of heat either joint-by-joint, component-by-component, or simultaneously to the whole board.

Field of application: Suitable for all types of SMD, except bare chips. Unsuitable for all wired-through joints. Reflow methods differ in the way the joints are heated. The different methods of creating the solder/flux depots are applicable to all of them, except to thermode soldering where the use of solderpaste is not always advisable.

Vapourphase soldering

Characteristics: A thermal equilibrium situation: all joints reach the temperature of the heat source, i.e. the working vapour, which is as a rule 215 °C/419 °F. All joints are heated simultaneously. Soldering takes place in an oxygen-free atmosphere. Soldering is carried out on a batch or an in-line basis.

Rate of production: Low to medium.

Duration of molten solder/substrate confrontation: 20–40 sec.

Solidification speed of joints: Medium.

Infrared soldering

Characteristics: A thermal non-equilibrium situation: the temperature of the heat source is higher than the soldering temperature reached by the joints, sometimes by 400 °C/750 °F. Soldering is carried out mostly on an in-line, more rarely on a batch, basis. Capable of being carried out in an oxygen-free atmosphere.

Rate of production: Medium to high.

Usual maximum soldering temperature: 250 °C/480 °F to 300 °C/540 °F.

Duration of molten solder and substrate confrontation: 15–30 sec.

Solidification speed of joints: Medium.

Laser soldering

Characteristics: A non-equilibrium situation: the high energy-density of the laser beam can destroy the joint unless precisely timed. Soldering on a joint-by-joint basis. Capable of being carried out in an oxygen-free atmosphere.

Rate of production: Slow.

Soldering temperature: 300 °C/540 °F–400 °C/750 °F.

Duration of molten solder and substrate confrontation:
0.02–0.04 sec.

Solidification speed of joints: Very high.

Convection-oven soldering

Characteristics: A thermal equilibrium situation: in the final stage, all joints attain the same temperature as the heatsource. An in-line process. Capable of being carried out in an oxygen-free atmosphere.

Rate of production: medium to high.

Maximum soldering temperature: 250 °C/480 °F–350 °C/660 °F.

Duration of molten solder and substrate confrontation: 20–40 sec.

Solidification speed of joints: Medium.

Convection soldering of single components

Characteristics: thermal equilibrium or near-equilibrium situation. Capable of being carried out in an oxygen-free atmosphere.

Rate of production: slow.

Soldering temperature: 250 °C/480 °F–350 °C/660 °F.

Duration of molten solder and substrate confrontation: 10–20 sec.

Solidification speed of joints: Medium.

Impulse soldering of single components

Characteristics: A thermal equilibrium situation. Oxygen-free option possible but rarely required.

Rate of production: Slow to medium.

Soldering temperature: 250 °C/480 °F–300 °C/570 °F.

Duration of molten solder and substrate confrontation: 0.5–3 sec.

Solidification speed: High.

Simultaneous soldering of all joints

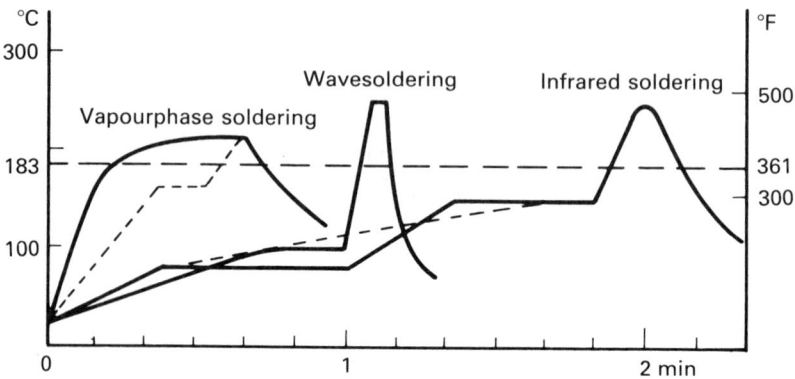

Soldering of individual component or joints

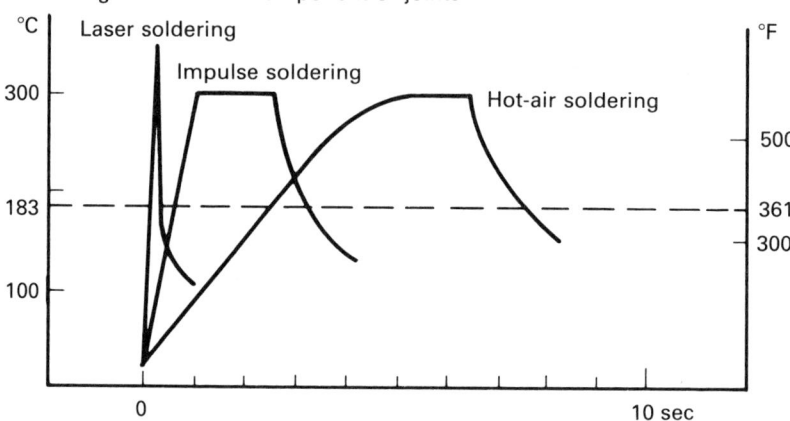

Figure 5.33 *Temperature profiles of the different soldering methods*

5.10 References

1. Strauss, R. (1987) *BABS 5th Intern. Conf. Brighton, UK.* Paper 28.
2. For a detailed discussion of the definition of particle size distribution, see Ruthardt, R. (1990). Die Verfahren zur Lotpulvererzeugung . . . *SMT/ASIC/Hybrid Int. Conf., Nuremberg,* pp. 93–105 (in German).
3. Thwaites, C. J. (1986) Some Metallurgical Aspects of SMD Technology. *Brazing & Soldering, (UK),* Spring 1986.
4. Short, R. H. and Lee, N. C. (1989) Fine Pitch Technology: Optimizing the Role of Solder Paste. *EXPO SMT '89, Nashville, Techn. Proc.,* pp. 83–85.
5. Hobby, A. (1990) Practical Aspects of Printing Solderpaste. *Surface Mount Intern.,* Vol. 4, Issues 4 and 5.

6. Chilton, A. C. and Gaugler, K. W. (1990) Fine Pitch Solder Creams. *Soldering & Surface Mount Technology*, Oct. 1990.
7. Scheuer, I. (1993) The Importance of Viscometry of Solderpastes for Quality Assurance. *6th Academic Soldering Colloquium, Munich, DVS Report 153, Duesseldorf, Germany*, pp. 46–55.
8. Burke, J. (1992) Fine Pitch Screen Printing. *Electron. Manuf. & Test*, Nov./Dec., pp. 33–34.
9. Short, R. H. and Lee, N. C. (1989) *loc. cit.*
10. Burke, J. (1992) *loc. cit.*
11. Burke, J. (1992) *loc. cit.*
12. Short, R. H. and Lee, N. C. (1989) *loc. cit.*
13. Friedrich, D. (1990) Dip pretinning of printed circuit boards for reflow soldering. *Interconnecting Technology for Electronics, DVS Berichte 129, Duesseldorf, Germany*, pp. 164–168 (in German).
14. Maiwald, W. J. (1992) SMD Placement without Solderpaste. *EPP, Leinfelden, Germany*, March, 1992, pp. 56–62 (in German).
15. Chu, T. Y., Mollendorf, J. C. and Pfahl, R. C. (1974) Soldering using Condensation Heat Transfer. *Proc. Nepcon West 74, Anaheim, Cal.*, pp. 101–104.
16. Lea, C. (1988) A Scientific Guide to Surface Mount Technology. *Electrochem. Publ., Ayr, Scotland*, p. 237.
17. Lea, C., Scient. Guide., p. 254.
18. Turbini, L. and Zado, F. (1980) Chemical and environmental aspects of condensation reflow soldering. *Electr. Packaging & Prodctn.*, **20**, No. 1, pp. 49–59.
19. Grikitis, K. (1991) A Renaissance for Vapourphase Soldering. *Electr. Manuf. & Test*, Feb. 1991, pp. 11–14.
20. Bjoerkloef, A. (1990) VPS with preheat, the key to troublefree production. *Nepcon West, Anaheim, Cal.*, 26 Feb.–1 Mar.
21. Reithinger, M. (1991) Reflow Soldering Technology for Surface Mount Assembly. *Electr. Prodctn.*, March 1991, pp. 27–37.
22. Flattery, D. K. (1986) IR Reflow for the solder attachment of surface mounted devices. *Hybrid Circuits*, No. 9.
23. Kolsters, J. W. M. and Beelen-Hendrikx, C. C. M. (1991) Process Requirements for Infrared Reflow Soldering. *Philips Report 52/9OEN*, Jan. 1991.
24. Reithinger, M. (1991) *loc. cit.*
25. Volk, H. (1990) SMT soldering without changing the profile. *Productronic*, **10**, pp. 72–74. (in German)
26. Arslancan, A. N. (1990) IR Solder Reflow in Controlled Atmosphere of Air and Nitrogen. *Proc. Nepcon West (90)*, Anaheim, Cal., pp. 170–178.
27. Ivankovits, J. C. and Jacobs, S. W. (1990) Atmosphere Effects on IR Reflow Soldering. *Proc. SMTCON, Atlantic City, NJ*, pp. 283–300.
28. Gruss, A. and Flattery, D. (1985) Improved reflow oven for SMD technology. *Productronic*, **5**, pp. 30–34 (in German).
29. Bandyopadhyay, N., Kirschner, M. and Marczi, M. (1990) Development of

a fluxless soldering process for surface mount technology. *Soldering and Surface Mount Technology*, No. 4, pp. 23–26. See also USP 4.960.236.

30. Nylen, M. and Norgren, S. (1990) Temperature Variations in Soldering and their Influence on Microstructure and Strength of Solder Joints. *Soldering and Surface Mount Technology*, No. 5, pp. 15–20.

31. Volk, H. (1990) *loc. cit* (in German).

32. Lenz, E. (1985) *Automated Soldering of Electronic Assemblies*, Siemens, Munich, Germany, pp. 113–118. ISBN 3-8009-1449-2 (in German).

33. Palmer, M.J. *et al.* (1991) HAT Tool for Fluxless OLB and TAB. *Proc. Electronic Comps. and Technology Conf.*, pp. 507–510.

34. Lea, C. (1989) Lasersoldering – Production and Microstructural Benefits for SMT. *Soldering and Surface Mounting Technology*, No. 2, pp. 13–21.

35. Lea, C. (1988) *A Scientific Guide to Surface Mount Technology*. Electrochemical Publications, Ayr, Scotland, p. 302.

36. Moeller, W. and Knoedler, D. (1992) Fine Pitch Laserbeam Soldering of TABs. *Verbindungstechnik in der Elektronik*, March 1992, pp. 14–18 (in German).

37. Vardaman, E.J. (1993) International Developments in Fine Pitch TABs. *6th Academic Colloquium on Soft Soldering, Research and Practice; Munich, March, 1993*. DVS Report 153, Duesseldorf, Germany.

38. Zimmer, G. (1993) Soldering Ultrafine Pitch components, esp. TAB Outer Leads. *6th Soft-Soldering Colloquium, Munich*. DVS Report 153, Duesseldorf, Germany, pp. 166–176 (in German).

39. Klein Wassink, R.J. (1989) *Soldering in Electronics, 2nd ed.*, Electrochemical Publications, Ayr, Scotland, p. 587.

40. Strauss, R. (1988) *loc. cit* (in German).

6 The circuit board

6.1 The beginnings

The printed circuit board has come a long way since Paul Eisler filed his pioneering patent in 1943.[1] The patent describes how to create a network of metallic conductors on a flat insulating board by printing its pattern in acid–resisting ink on to a metal foil, which is bonded to the board, and etching away the unprotected area. The idea had to wait till the end of the war in Europe before it was taken up by the America Bureau of Standards and used in anti–aircraft proximity fuses in the Far East. In the early fifties the printed circuit board concept filtered back into Europe, and quickly spread throughout the industrial world. In 1987, the annual production of circuit boards worldwide was estimated as 140 sq. km/55 sq. miles, which is about twice the area of Manhattan Island.[2]

6.2 SMD-specific demands on a circuit board

Mounting SMDs on a circuit board does not place any principally new demands on it, but there are some differences of a qualitative nature:

1. The board must be more rigid. With melfs, chips and SOs, the soldered link between component and board is rigid itself and cannot absorb stresses which arise if the board should warp after soldering. This was no problem with wired joints, and is no great problem with the more flexible leads of PLCCs and multilead devices.
2. The demands on the flatness of the board are higher. Before the advent of SMDs, the permissible bowing of a board across its length was 1%. Over a length of 25 cm/10 in, this would mean a difference in level between edge and middle of 2.5 mm/100 mil, which is unacceptable for the printing down of solderpaste by stencil or screen.
3. Fine-pitch technology demands a circuit pattern of very high dimensional precision and reproducibility. The same is true for the soldermask and its register with the circuit pattern. To meet these requirements, the board

manufacturer must invest in more expensive production equipment, and use base material of a higher specification. This in turn means more expensive boards.

4. For wavesoldering, SMDs must be anchored to the circuit board by adhesive joints which are located between their footprints. If any conductor tracks which run between these footprints, as they often do, are coated with solder, this coating may melt as the board passes through the wave. If the soldermask is a thin lacquer, it will warp (orangepeel effect) and the adhesive joint will lose its footing. For this reason, the conductor tracks of most boards for SMD technology are bare copper (often oxidized to improve the bond between soldermask and track) and only the footprints and the lands of the throughplated holes are tinned. This technique is known as 'soldermask on bare copper' (SMOBC).

5. Boards made from paper-reinforced phenolic resin (FR2) or epoxy resin without glassfibre reinforcing (FR3) are no longer used for SMD technology, except for cheap and undemanding products, because of their low mechanical properties.

6.3   Thermal management

As long as circuit boards carried only wired components, the heat which they might generate during operation posed no great problem: there was sufficient space between components and the board to prevent the board from getting hot, and the lead wires took care of any dimensional mismatch between a warm component and the cool board. With SMDs, the situation is different: the rigid soldered joints between melfs, chips and LCCCs (if used) come under stress when these components get warmer than the board; even if there is no temperature difference between components and the board, there is a mismatch of the thermal expansion coefficient between the board and components with a ceramic body, such as melfs and chips. The relatively stiff legs of PLCCs and SOs can also pose a problem, only the gullwing legs of QFPs and similar components are sufficiently pliable to shield the joints from such stresses (Figure 6.1).

With densely populated boards, and ICs with a large number of switching functions, the heat generated during operation may rise to 0.5 W/sq. cm (1.25 W/sq. in) of board area. This heat must be disposed of, and the consequences of the temperature rise of components and board must be dealt with.

6.3.1   Thermal expansion mismatch

With chips, melfs and LCCCs, the thermal expansion mismatch between component and board was a problem, before the passive components started to become ever smaller and LCCCs began to disappear. Until then, considerable efforts were made to devise circuit board substrates whose thermal expansion

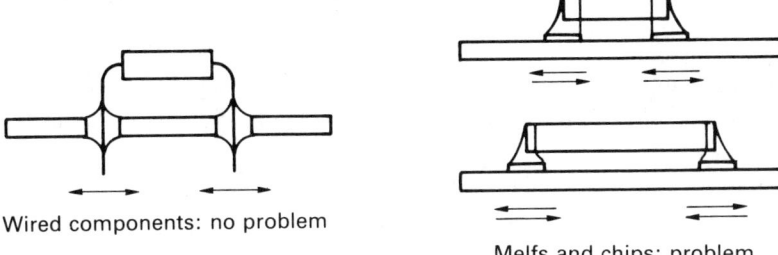

Wired components: no problem

Melfs and chips: problem

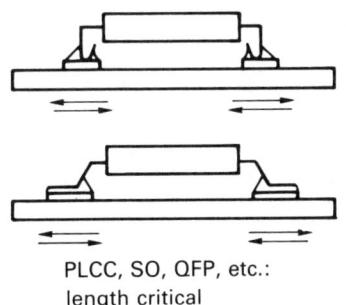

PLCC, SO, QFP, etc.:
length critical

Figure 6.1   *Thermal expansion: the behaviour of SMDs and wired components*

(TCE) matched that of such ceramic bodies better than the polymer-based FR4 (Table 6.1).

Most of these substrates are more expensive than FR4, and they are used only if there is a compelling reason for matching thermal expansion between component and substrate. Metal-cored substrates are sometimes used because of their high thermal conductivity, in situations where a copper layer inside a multilayer board is not enough to conduct the heat created on the board to its edges or the nearest heat sink.

6.3.2   Effects of temperature differences between components and board

Matching the TCE of the component and the board does not solve everything: after an assembly is switched on, its components warm up more quickly than the substrate. After switching off, the board stays warm longer than the components. This temperature difference causes strains, which the joints are expected to accommodate and which are proportional to the size of the component. At a temperature difference between component and board of 50 °C/90 °F, the dimensional difference between the distance between the footprints on an FR4 board and between the ends of the body of the component are as follows:

Ceramic bodies	Dimension	Dislocation
Chip or melf	5 mm/0.2 in long	1.3 μm/0.05 mil
Chip or melf	2 mm/0.08 in long	0.5 μm/0.02 mil
LCCC	25 mm/1 in wide	6.7 μm/0.27 mil
Plastic bodies		
QFP	25 mm/1 in wide	15 μm/0.6 mil

These figures illustrate the stresses which joints can experience under a given operating regime, and the importance of creating thermal links between components and heat sinks on or in a board for certain types of circuit board assemblies, in order to reduce the temperature differences between components under load and the board. The figures also show that as passive components, especially melfs, get ever smaller, the thermally caused stresses on their soldered joints become negligible.

The thermal expansion coefficient of plastics is high, and components housed in plastic like microprocessors are getting larger. But then, their thin leads are sufficiently compliant to take care of the resultant mutual movements.

Depending on the soldering method used, the joints on a freshly soldered board may become stressed after soldering is completed. If the complete assembly is heated to soldering temperature, the board and the components are equally hot. As the assembly cools after soldering, all joints solidify as the temperature passes through 183 °C/361 °F. From then on, board and components cool down to room temperature, each contracting according to its own coefficient of contraction (i.e. expansion). This leaves every joint with a locked-in stress as it starts its life. With wavesoldering, and vapourphase, infrared and convection reflowsoldering, board

Table 6.1   *Substrates and their thermal expansion coefficients (TCE)*

Material	TCE $\times 10^{-6}/°C$	Thermal conductivity $W/cm/°C$	Relative cost
Alumina ceramic	5.5	0.26	—
FR4 (epoxy/glassfibre)	16	0.0016	1
Fibre-reinforced polymer laminates			
Epoxy/Kevlarfibre*	4–8	0.0012	8
Epoxy/Quartzfibre	4–8	0.0013	5
Polyimide/Kevlarfibre*	4–8	0.0013	11
Metal composites			
Cu-clad Invar** coated with epoxy	6	164	1.5
Porcelain-enamelled steel	9–12	60	0.6

*Heat-resistant polymer: © DuPont.
**Invar: Ni–Fe alloy (35–59% Ni) with closely controllable CTE.
Data supplied by Dr S. Smernos, Alcatel, Stuttgart, Germany.

and components get equally hot, and therefore this consideration applies. With laser soldering and thermode reflowsoldering, both board and component remain more or less at room temperature.

With multilayer boards, a continuous copper layer below the board surface, with provision of course to allow for the passage of vias leading to lower layers or the other side of the board, can serve to conduct the heat generated by the components to the edges of the board, where it can be suitably disposed of. Metal-filled, heat-conductive adhesive is often used as the thermal link between the underside of such components and the copper layer. The latter is made accessible to the heat-conductive medium by blind holes leading to it.

6.4 Solderable surfaces

A consistently good solderability of lands and footprints is an essential precondition for soldering with a low reject rate. As a rule, the solderability of a circuit board is enhanced and preserved by applying a coating of tin, or preferably solder to its copper surfaces (Section 3.6.3).

Recently, several vendors of equipment for wavesoldering or reflowsoldering under nitrogen claim that pretinning of footprints is no longer necessary, and plain copper footprints, since they do not oxidize at soldering heat, will suffice. This presupposes that the copper is perfectly free from oxide to begin with, and has not deteriorated during storage or handling. Experience will have to show whether this practice can prevail.

6.4.1 Galvanic coatings

Several manufacturing methods of circuit boards include an electroplating process whereby the pattern of the conductors and footprints is created by the so-called 'patternplating' of a tin–lead deposit onto the copper laminate. This plated pattern serves both as an etch resist in the next manufacturing stage and as a solderability-enhancing and preserving coating afterwards. Because a galvanic tin–lead coating is a somewhat porous layer of discrete tin-rich and lead-rich particles, it must be reflowed in order to be able to preserve the solderability of the substrate beneath (Section 3.6.3). This reflow process serves at the same time as a quality check: if the solderability of the copper underneath the plating is unsatisfactory, the reflowed coating will show conspicuous signs of dewetting. A minimum coating thickness of $5\,\mu m/0.2\,mil$ is considered necessary to ensure long-term solderability.

6.4.2 Hot tinning

Hot-air levelling (HAL)

This process was initially developed for enhancing the solderability of boards with through-plated holes. It involves vertical immersion of a fluxed board in a bath of molten solder, or flooding it with molten solder by passing it horizontally through a solderwave. Afterwards, the molten solder must be blown out of the holes by an array of high-pressure jets of hot air, before it solidifies, hence the name of the process.

HAL leaves the footprints with a domed coating of solder, which may reach a thickness of 25 μm/1 mil in the middle of a footprint or land, but which tapers down to as little as 1 μm/0.04 mil towards its circumference. With through–plated holes, this happens at the shoulder between the barrel of the hole and its land (Figure 6.2). This phenomenon, which is called a 'weak shoulder', can be avoided by careful control of the operating parameters of the tinning machine. It can create a solderability problem. After a time, the steady growth of the intermetallic layer between the copper and the solder will cause the $Cu_6Sn_5(\eta)$ to reach the surface. Once exposed, it soon oxidizes and becomes almost insolderable, even with strong fluxes (Section 10.4.2).

The domed profile of HAL-tinned footprints, on the other hand, may cause problems when multilead components are to be impulse soldered: the thin, flexible leads may slip sideways under the pressure of the thermode and give a defective joint (Figure 6.3). For this reason, HAL tinning is only appropriate for boards with through–plated holes, i.e. for mixed SMD/leaded component technology. For pure SMD boards, other pretinning methods are more appropriate.

Rollertinning

Rollertinning is a cheaper alternative to HAL tinning, suitable mainly for the tinning of single-sided boards. It involves passing the circuit board between two horizontal rotating rollers, of which the lower one is made of steel, and half immersed in a bath of molten solder (Figure 6.4). Rollertinning gives a coating

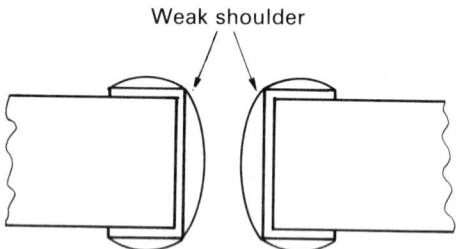

Figure 6.2   *'Weak shoulder' of HAL-tinned through-hole*

Figure 6.3   *Domed HAL-solder coating on a footprint*

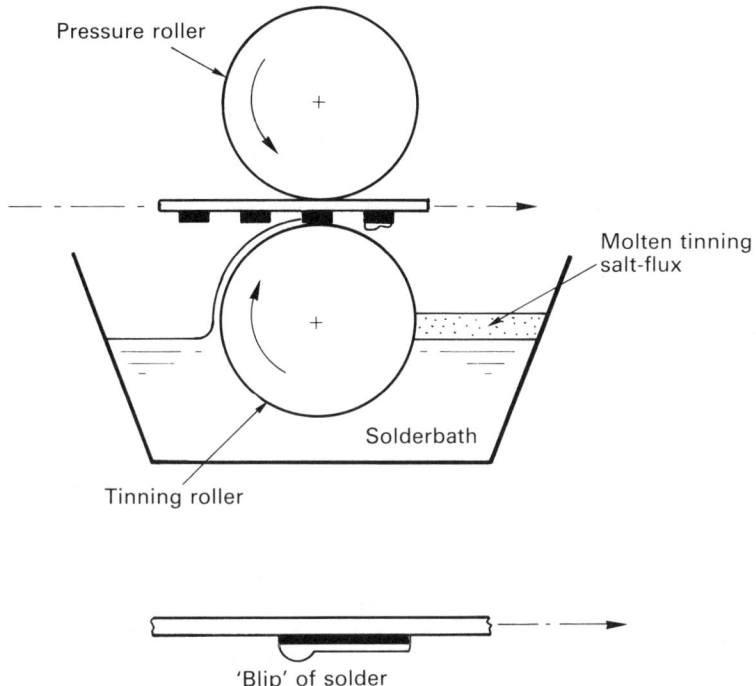

Figure 6.4 *Rollertinning*

thickness of between 2 μm/0.08 mil and 5 μm/0.2 mil, which is thick enough to protect footprints for about one year's storage before the η layer comes to the surface.

On the other hand, every footprint carries a small 'blip' of solder, which can be up to 100 μm/4 mil high at its trailing edge, where the footprint emerged from the nip between the two rollers. It is caused by part of the peelback between the tinning roller and the board hanging on to the footprint. Some board users object to these blips on cosmetic grounds, while others report problems with component placement.

6.4.3 Flat solder depots

In recent years, methods for creating a flat solder depot which is fused to the copper laminate on every footprint have been proposed. The depots are in the form of a flat pad of solder, 0.1 mm/40 mil to 0.2 mm/80 mil thick, which is fused to and fully covers every footprint. The thickness of the solderpads corresponds normally, but not necessarily, to the thickness of the surrounding soldermask (Figure 6.5).

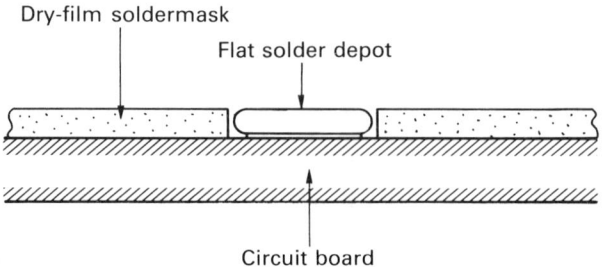

Figure 6.5   *Flat solder depot*

Providing these solder depots is conceived as a part of board manufacture. This means that the board vendor will supply circuit boards with the solderpads already in position. The user applies flux selectively to the pads only, or to the whole board, before the SMDs are placed in position. Preferably, the flux also acts as an adhesive, in order to fix the components temporarily before they are reflowsoldered to the board by any of the conventional methods.

Thus, the board manufacturer finds himself saddled with a process which involves the handling of molten solder or the handling and fusing of solderpaste, all of them technologies new to his trade. Naturally, he will have to be paid for this.

For the board user, the proposition is an attractive one: he is relieved of the increasingly demanding task of printing a special grade of solderpaste onto ever narrower and ever more closely spaced footprints, with the associated problems of process and quality control. In brief, it lets him escape from a technology which is being pushed to the limits of its capabilities. Naturally, he will have to pay for this.

Two ways have been proposed to create these pads. In one of them, molten solder is applied to the circuit board in such a way that it tins the prefluxed footprints and covers them, while the surrounding film of solder resist acts as a shim which defines the thickness of the solderpad. Its flat surface is defined by a plate pressed against the circuit board.[3]

With the other system, solderpaste (which need not be of fine-pitch quality) is squeegeed over the footprints, with the surrounding soldermask acting as a stencil. Alternatively, a stencil is placed over the dry–film resist, which allows a thicker paste printdown. Subsequently, the solderpaste is reflowed, so that it forms a domed solderdepot on every footprint. Finally, these domed depots are flattened by a heated platen, with the soldermask acting as a shim which defines their thickness (Figure 6.6).[4]

Soldermasks which have the properties required by the SIPAD (© Siemens AG) process, and fluxes which can act as adhesives, are commercially available.

Technologically, the idea of a depot of solid solder is attractive: placing components on sticky, solid solderpads is simpler and less risky than putting them down on squashy paste deposits. There seems to be little, if any, risk of bridges and solderballs. Quality inspection is simplified, and the technology appears to open the way to still closer leadspacing without serious technical problems. The future

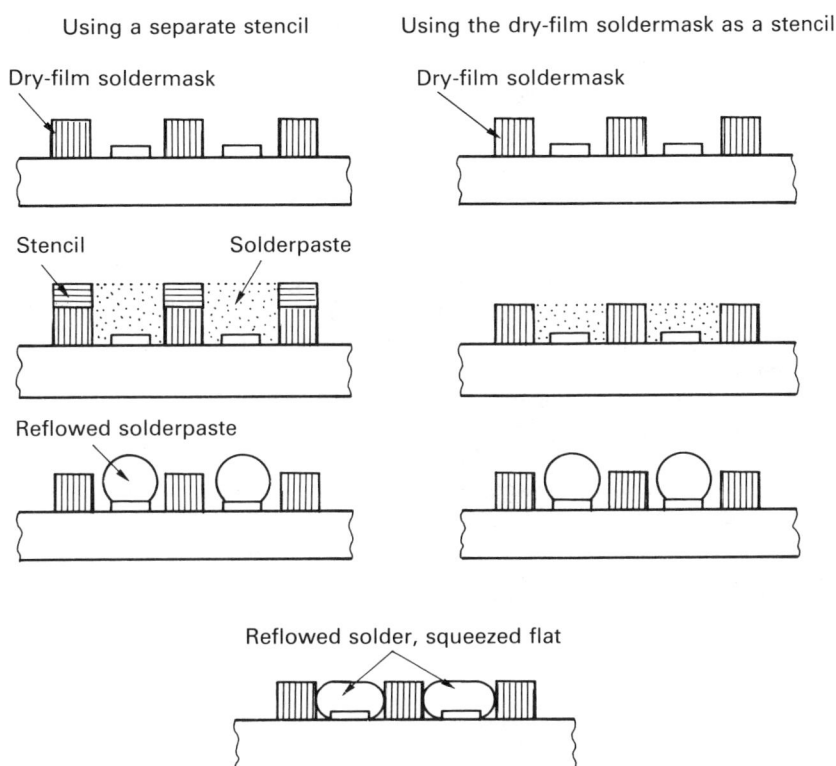

Figure 6.6 *Two options for creating flat solder depots (the Sipad©process)*

will have to show whether one or the other of the proposed processes is commercially sufficiently attractive to both board makers and board users to get it off the ground.

6.5 The soldermask

The soldermask was originally introduced in the fifties, primarily to reduce solder consumption by preventing the unnecessary tinning of every conductor line, and also in order to reduce the copper contamination of the solder in the wavesoldering machine. Its other advantages of improving joint quality and facilitating inspection by confining the solder to the lands soon became apparent. By the time SMDs appeared, the use of the soldermask was universal.

SMD technology does not place any specific requirements on a soldermask. With fine-pitch boards, creating the very narrow strips of mask between adjacent footprints is becoming increasingly difficult, and places high demands on process control. The necessity of separating the footprints by these strips of soldermask is beginning to be queried by several board makers and users, and no adverse effects of their omission have been reported.

6.6   Layout

A good board layout must be user-friendly, in the sense that the initial user of a finished circuit board is the person who must solder the components to it. The more difficult this becomes, the greater the likelihood of something going wrong. Soldering faults mean rework and rework means increased costs, while reworked joints are likely to be less reliable.

Naturally, the first task of the layout designer is to create a circuit which functions electronically and which fulfils the tasks for which it is intended. But if the layout places excessive demands on its manufacture, the commercial success of the end product may be at risk.

To make a layout user-friendly in the above sense, the circuit designer must know which soldering method is to be used. Some layout rules are specific for wavesoldering, others for reflow.

6.6.1   Layout for wavesoldering

Avoiding skipped joints

The main problem with wavesoldering SMDs is the 'skipped joint', the joint which the solderwave could not reach because of the 'shadow effect' (Section 4.4.4), which even the best double wave cannot always overcome. A good wavesoldering layout helps the molten solder to find its way to joints which are hidden behind or partly underneath a component (such as the legs of PLCCs).

An extended footprint, or a short length of conduit which is not covered by the soldermask, helps to lead the solder to a joint close to a high component housing; footprints for melfs and chips should extend far enough to provide an aspect angle of about 60° (Figure 2.4). This allows for a slight misplacement of the component, so that in no circumstances does the angle get steeper than 45° (Figure 6.7).

Setting difficult components such as PLCCs at an angle of 45° against the direction of travel of the board is another way of helping the solder to reach the joints on the shadow side (Figure 6.8). Of course, this measure wastes valuable real estate on the board, and is therefore often unacceptable. As an alternative, the whole board, if it is small enough, can be set obliquely on the transport carriage. Some machine vendors provide a third alternative by offering wavesoldering machines where the soldering wave is set at an angle towards the line of the conveyor belt. This makes it possible to solder full-size boards (see also 4.4.3 and 4.4.4 and Figure 4.17).

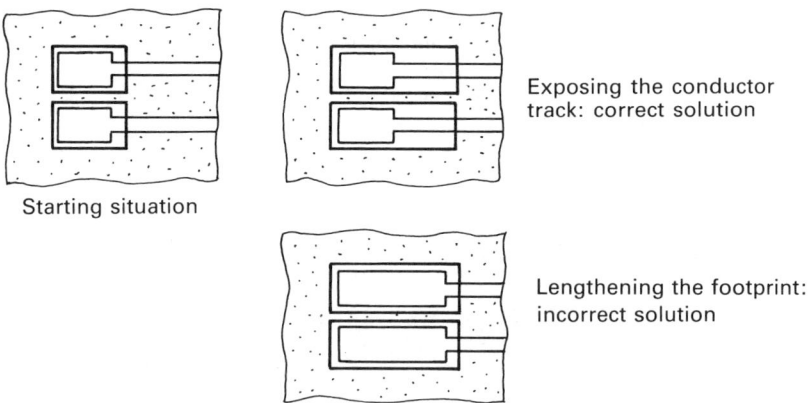

Figure 6.7 *Letting the conductor track lead the solder to the joint*

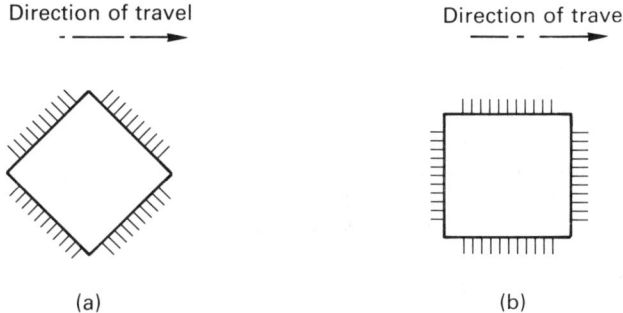

Figure 6.8 *Mitigating the shadow effect. (a) Favourable alignment; (b) unfavourable alignment*

Avoiding solderbridges

Solderbridges are the result of some idiosyncrasies of the peelback behind the solderwave (Section 4.4.3). With a row of footprints, provided the pitch is not below 1.25 mm/50 mil, it is easier to avoid bridges when they emerge from the wave in single file rather than all of them together in a broad front (Figure 6.9). Wavesoldering in a nitrogen atmosphere extends bridge-free soldering below that limit (Section 4.4.6).

As a row of footprints leaves the wave in single file, the peelback seems to jump from print to print, until the last two emerge, when a bridge tends to form between them. Placing a somewhat larger dummy footprint, called a 'solderthief' (see Figure 4.15, Section 4.4.3), at the end of the row draws the bridge to a place where it does no harm. With the footprints for square multilead components, a solderthief is placed into each corner so as not to restrict the choice of the direction of travel. This expedient is often written into the software of the layout program.

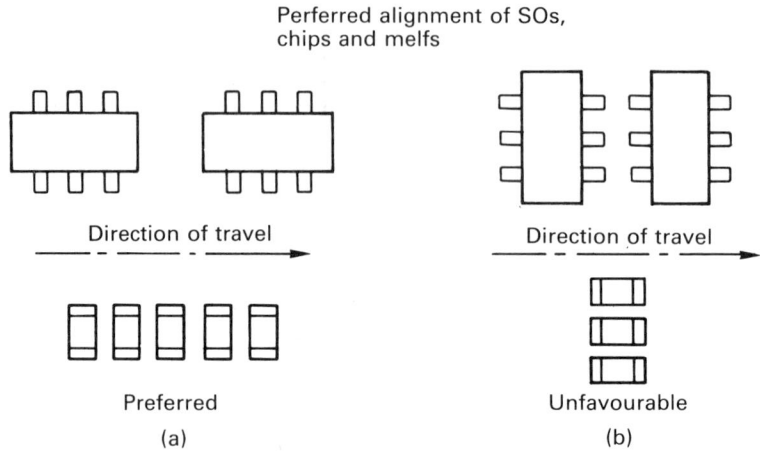

Perferred alignment of SOs,
chips and melfs

Direction of travel

Direction of travel

Preferred

Unfavourable

(a)

(b)

Figure 6.9   *Preferred alignment relative to direction of travel. (a) Favourable alignment; (b) alignment resulting in the formation of bridges*

Soldering in a nitrogen atmosphere: its effect on bridging

Under the terms of surface tension, a solder bridge should not really exist, and it could be called a 'metastable' state: because it does not represent the smallest possible surface area of the molten solder, it ought to pull back onto the footprints, and the surplus should break up into drops of solder. This does not happen, either because there is no time for it before the solder solidifies, or more likely because a thin skin of surface oxide reduces the mobility of the molten solder. Experience with soldering equipment working in a nitrogen atmosphere with roughly <20 ppm of oxygen, be it a wavesoldering machine or a reflow installation, confirms that in the absence of oxygen, bridging is very much reduced. Some of the solderballs which are found on boards which are wavesoldered under nitrogen (Section 4.6), and which represent the trade-off for this advantage, are the debris of broken-up solder bridges.

Layout for boards with a mixed SMD/leaded component population

With such mixed technology boards, the following contingencies must be borne in mind:

1. Should the SMDs have to be placed and glued to the board as the first step, they must be spaced so as to allow enough room for the crimping tool of the machine which inserts the leaded components afterwards. If this is not done, some SMDs might get knocked off the board. It is advisable to know which type of insertion machine is to be used when the layout is drafted.
2. If the SMDs are placed after the leaded components, the space requirements of the placement head of the pick-and-place machine must be taken into account when the spacing of the SMDs is being decided.

6.6.2 Layout for reflowsoldering

The layout rules for reflowsoldering are governed by three considerations:

1. The components must stay firmly in their allocated positions.

    Leaded components and wavesoldered SMDs are fixed in their positions before soldering starts. In reflowsoldering, the SMDs float on molten solder during a crucial period. They are then at the mercy of the forces of surface tension and of capillarity, which act between the footprints. These forces also act between the metallized endfaces in the case of melfs and chips and the component legs in the case of all the others. Depending on the geometry of the footprints, this can be a help or a cause of trouble.

    Impulse soldering and laser reflowsoldering are exceptions: with the former, the soldering tool firmly presses the legs down onto the footprints during soldering; with laser soldering the joints are soldered one-by-one, so the component is never free to float.
2. All of the available solder depot must contribute to forming the joint; none of it must be tempted to flow away from it.
3. With many reflowsoldering methods, both the amount of heat supplied to the joints, and the soldering time are limited. Because it is important that all joints of a component reach roughly the same temperature at the same time, the heat capacity of footprints should be uniform.

Surface tension and capillarity

Surface tension tries to make the free surface of a liquid as small as possible; capillarity pulls two flat, wetted surfaces towards one another as closely as possible (Section 3.6.2). As long as all the footprints of a component have the same shape and size, surface tension and capillarity tend to pull every solderable surface towards the middle of its footprint (see Figure 3.17). This 'self-alignment' is effective with chips, melfs and SOs. Multilead components are heavier and less mobile: with these, the self-alignment effect is hardly effective, if at all.

If there is a lack of symmetry between the footprints of a component, the footprint which offers more capillary force than the other, or the footprint on which the solder melts earlier, tries to pull the wetted surfaces of the component towards itself, while the component is free to move (Figure 6.10).

Lack of symmetry is also behind the phenomenon of 'tombstoning' of chips (see Figure 3.17). There may be several reasons for this asymmetry. The solder may melt and wet the metallized face of the chip earlier or better at one end than at the other, because of a difference in heat capacity or because one end is more solderable than the other. A faulty paste printdown may also be to blame: there may be less solderpaste on one footprint than on the other, or maybe none at all. Finally, yet another layout fault may be to blame: if there is a possibility of capillary action robbing a footprint of some of its solder, for example through the proximity of a via (Figure 6.11), the molten solder on the affected footprint will exert less pull on the component than the solder at the other end.

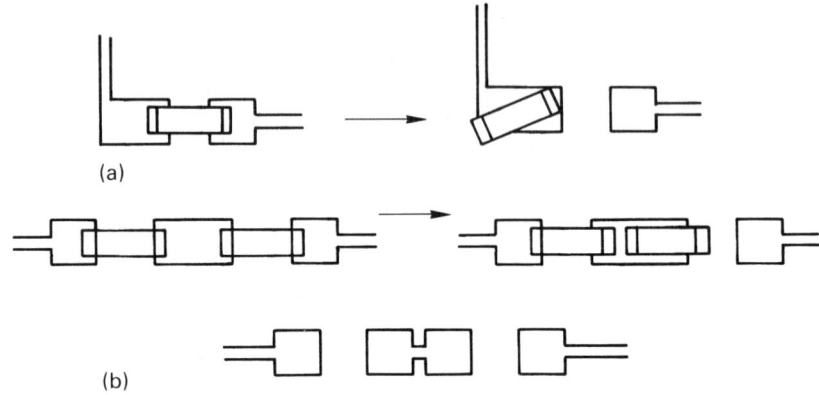

Figure 6.10   *The dangers of asymmetry. (a) Floating caused by lack of symmetry between left and right footprint areas; (b) remedy*

Figure 6.11   *Footprint starved of solder by neighbouring via. Remedy: a bridge of solder resist between the via and the component footprint*

The better end wins, because its surface forces are stronger or start to act earlier, and so they pull the weaker end off its pad. As a consequence the affected chip finishes by standing upright on its better end. Tombstoning affects only bipolar SMDs, and among those mainly chips. It is rarely observed with melfs.

Layout for impulse soldering

With impulse soldering, equal heat capacity of all footprints of a component is important. Large footprints heat up more slowly than small ones, and since with impulse soldering the heat pulse is the same for all joints and must be as short as possible, the consequences of asymmetry are either badly soldered large joints or overheated small ones (see Figure 5.31).

6.7   References

1.   British Patent 639,178,02; 02 (1943) Strong and Eisler, *Manufacture of Electric Circuits and Circuit Components.*

2. Klein Wassink, R. J. (1989) *Soldering in Electronics, 2nd ed.*, Electrochemical Publications, Ayr, (Scotland), p. 376.

3. Friedrich, D. (1990) A Circuit board Surface which Optimises the Metallurgy and Processing Technology of Reflowsoldering. *Proc. SMT/ASIC/Hybrid Internat. Conf., Nuremberg, Germany*, June 1990 (in German).

4. Maiwald, W. J. (1993) Can SIPAD© Technology Provide Better and More Reliable Board Assemblies? *DVS Report 153*, Duesseldorf, Germany, pp. 124–130. (© Copyright, Siemens AG.) (In German.)

7 Component placement

7.1 The task

'Pick–and–place', the common name for equipment which puts SMDs on circuit boards, is an apt description of its tasks:

1. The correct component must be selected from a magazine or feeder.
2. It must be placed on the correct footprints, in the correct orientation, and with the required precision.

These two straightforward tasks are getting ever more demanding, as the diversity of components grows and as footprints continue to get smaller and more closely spaced. Some components, like melfs, are getting smaller, and others like QFPs and TABs get larger, with over a hundred closely spaced leads. Minimelfs 0204 (3.6 mm/142 mil × 1.4 mm/56 mil dia.) have become common, and micromelfs 0102 (2.0 mm/80 mil × 1.1 mm/44 mil dia.) have begun to appear in vendors' catalogues. PLCCs and QFPs reach dimensions of 29.4 mm/1.16 in or above. These demands are the driving force for the increasing sophistication and specialization of the pick-and-place equipment which is on offer.

Depending on whether the boards are to be wavesoldered or reflowed, the placement operation is either preceded or followed by an additional process. For wavesoldering, placement of the components is preceded by putting down a metered amount of adhesive between the footprints. The components having been placed, the assembly is heated to cure the adhesive before soldering (Section 4.9). Placement for reflowsoldering is preceded by covering the footprints with a metered amount of solderpaste.

The placement of SMDs as such is simpler, and pick-and-place machines can handle a greater variety of components than insertion machines for leaded components, because insertion demands different equipment for axial, radial and DIL components (Table 7.1).

The various forms of packaging in which SMDs reach the user are given in Section 1.4. Normally, a user has a choice of packaging systems to suit his placement methods:

Table 7.1   *The difference between the tasks of placing SMDs and of inserting wired components*

| Inserting wired components | Placing SMDs | |
	For wavesoldering	For reflow
Placing components on tape★	Placing components in blistertape or magazines★	
Sequencing components	Placing the adhesive	Printing the solderpaste
Preform the leadwires	Collect a component	
Insert a component	Place a component	
Crimp leadwires	Cure the adhesive	
Solder on the wave	Solder on the wave	Reflow

★Normally the task of the vendor

Choice of packaging systems

Chips	Paper or plastic blistertape; bulk package
Melfs	Paper or plastic blistertape; bulk package
SOs	Plastic blistertape; linear magazine
PLCCs	Plastic blistertape; linear magazine
QFPs	Waffle trays
TABs	Film

Automated placement equipment places specific demands on paper and plastic blistertape: the coverfoil must peel away smoothly and without jerking from the tape, otherwise components are liable to jump out of the tape and jam the machine. Paper tape must not form fluff which might interfere with the placement mechanism.

7.2   Reliability of placement

The automatic placement of SMDs introduces five potential defects:

Missing components
Incorrect components
Incorrect polarity
Misregistration with the footprints
Mounting in the wrong place

The first three, which used to be the major sources of defect, are largely avoided by the now universally available 'smart' placement machines which check the presence, identity, polarity, and often the functional integrity of every component between picking it up and putting it down (see below). The elimination of the last two is the aim of the continuing refinement of today's placement equipment.

Positional accuracy

With fine-pitch lead spacing, the width of the footprints is half their distance. Fine-pitch below 0.5 mm/20 mil and ultrafine-pitch with 0.25 mm/10 mil spacing have become real demands. This means footprints 0.125 mm/5 mil wide, and lateral placement accuracies of ±0.06 mm/2.5 mil. Pick-and-place equipment with a placement accuracy of ±0.05 mm/2 mil, and a repeatability of ±0.02 mm/ 0.8 mil is already commercially available.

Angular accuracy of placement also matters. With a QFP of dimensions 25 mm/1 in × 25 mm/1 in, an angular twist of 1° means a lateral displacement of 0.22 mm/8 mil at every corner. As a result, with a fine-pitch layout, about half the legs would sit on the wrong footprint (Figure 7.1). Therefore, with large fine-pitch components, angular placement accuracy has to move into the ±0.1° bracket.

Component identity and functionality checks

The number of placement errors is a measure of the reliability of a placement system. Until not so long ago, wrong-value chips and melfs in blistertape or bulk packages contributed significantly to manufacturing reject rates. Smart placement systems, which detect and correct such errors during placement 'on-the-fly' brought a drastic improvement.

7.3   Placement options

A user must decide between two strategies of component placement: on the one side are the manual and semi-automatic manual methods, on the other the fully automatic ones, which also fall into two categories, the sequential and the simultaneous systems. The choice between them depends on several factors:

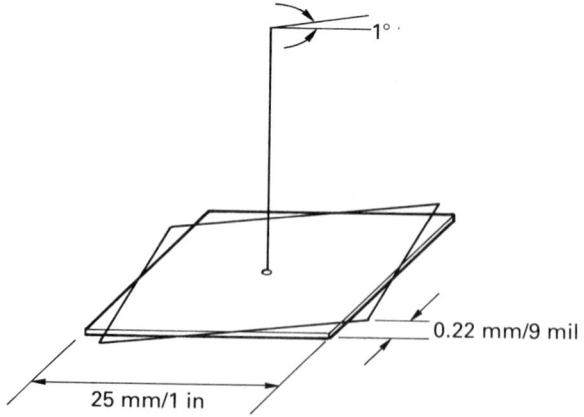

Figure 7.1   *Angular accuracy of placement*

1. A newcomer to SMD technology who operates on a small-to-medium scale will tend to opt for a manual or semi-automatic system, unless he is part of an organization where in-house know-how and technical assistance with fully automatic systems are available.
2. The type of product and the volume of production are crucial factors. If there are no more than about 50, at most 100 components, on a board, however complex their function and the layout, and if the number of boards does not exceed a few hundred per working day, a manual, probably semi-automatic placement system may well be the best choice.

As the number of boards to be processed per day rises, the cost effectiveness of purely manual placement soon drops. Semi-automatic placement reaches its maximum cost efficiency in the middle range of production volume, particularly where full-time working is not always guaranteed.

A further factor which affects the choice of system is the product mix: if the boards are all customer-specific boards, each with a short or unpredictable length of run, and if production must be flexible and capable of coping with frequent changes, semi-automatic manual placement may be best. It is worth noting, though, that recent years have seen the arrival of several fully automatic pick-and-place machines of great flexibility, with facilities for a rapid change-over from one working program to another.

In the last resort, the size of the necessary investment must be decisive. Naturally, manual and semi-automatic equipment is cheaper, and writing it off is less of a problem when faced with a fluctuating and highly differentiated demand. On the other hand, where one or a number of soldering lines must be fed with assembled boards, without the risk of costly interruptions, one or more fully automatic pick-and-place installations are the best, if not the only choice.

7.3.1 Fully manual placement

Like every placement system, placing SMDs by hand involves two steps: finding and fetching the component, and then putting it down on its footprints, having rotated it into its correct orientation (Figure 7.2). With boards to be wavesoldered, placement is preceded by putting down the spots of adhesive with a handheld syringe or by hand stencilling. The footprints of boards to be reflowsoldered are provided with solderpaste, again from a syringe, a metering dispenser, or by manual stencilling.

As with all manually operated processes, good ergonomic design of the work place is the precondition to get the best possible results with a minimum of errors. Even so, manual placement without any additional aids like assigning to each type of component its proper set of footprints, needs good housekeeping and unremitting concentration. It should be practised only for assembling small numbers of prototype boards, or very simple assemblies with few types of components.

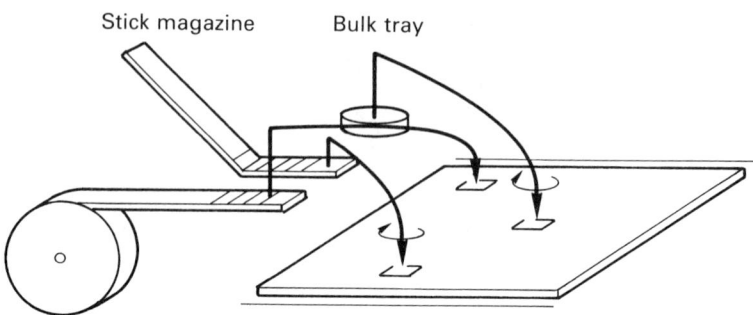

Stick magazine   Bulk tray

Tape dispenser

Figure 7.2   *The basic tasks of manual placement*

Even with entirely manual placement, no component should ever be touched with bare fingers. As has been said several times already, however clean fingers are, they will transfer fatty acids and salts to the components and their soldering surfaces. This affects their solderability, especially if adhesive joints must be cured before soldering. The curing heat greatly intensifies the damaging effect of any surface contamination on a soldering surface.

Tweezers can be used for handling the components, but a vacuum pipette with a finger-actuated rotatable head is much more convenient. The pipette may be handheld, or mounted on a gantry which is operated by a 'joystick'. With both, the accuracy of placement depends on the normally very high degree of coordination between the human eye and the human hand. Most operators, with some training, have no problem in putting down components on fine-pitch footprints without smudging the solderpaste. It may be worth remarking here that, so far, no vendor seems to have found it worth his while to provide manual placement equipment which can be converted for left-handed operators.

With manual placement, there are three kinds of possible errors: picking the wrong component, putting it in the wrong place and, with active components, placing it the wrong way round. Every manual placement system, however well conceived, which uses a bulk feed system of loose melfs or chips contains one further, and dangerous, source of placement error: one or more stray components may have found their way into the wrong compartment of a carousel or feeder box. If such an error is noticed, or maybe only suspected, it may be simpler and cheaper to discard the contents of the whole compartment than to try to find the rogue components amongst several hundred correct ones.

It is important to allow the operator regular intervals of rest, at least five to ten minutes every hour. It has been found that, depending on the complexity of the board and the number of component types, the error rate rises rapidly with less than that amount of rest time. With ten to twenty components per board and not more than ten different types of components, placement rates of up to 500 or 600 components per hour can be realized with purely manual placing. This can drop to 300/hour with more complex boards.

Components may be picked from a turntable (carousel) which is subdivided into a number of compartments holding loose components, from a row of horizontal feeders, which present the components from the open ends of blistertapes, from stick magazines or waffle trays.

With purely manual placement, the error rate may drop below 0.1%, i.e. <1000 ppm. But errors there will be, and in order to avoid expensive rework it is strongly advisable to inspect every board for correct placement before soldering it. Corrections are made by lifting off the offending component. Melfs and chips, unless valuable, are best discarded. SOs and gullwing-legged components can be re-used, after the legs have been cleaned with isopropanol. The footprints are wiped clean of solderpaste with a small piece of cotton or linen soaked with isopropanol. Dots of adhesive are best left alone, lest they be smeared over a footprint, which will become unsolderable. Adhesive vendors may be able to recommend or supply a suitable solvent to clear off an adhesive spot. The place having been cleaned up, fresh solderpaste (or adhesive) is put down and the component is replaced.

7.3.2 Semi-automatic placement

Semi-automatic placement machines take over some of the tasks of manual placement. These are principally those where human error could creep in, such as picking the right component and putting it in its correct place. Moving the components from their feeders to their footprints, and putting them down with the necessary precision, is still left to the sensory and muscular feedback system between the human eye and hand, a system which can be replicated by electromechanical and opto-electronic means only at great expense.

All semi-automatic manual placement machines are linked with a computer system, which can be either integral to the machine or operated by a separate PC. Such a system can be programmed to indicate by an LED or a spot of light the feeder from which the next component must be fetched. At the same time, the place where it has to be put down is illuminated by a beam of light, or indicated in some other way.

These functions can be refined and added to. Some semi-automatic placement machines make only the correct feeder accessible, while the others are covered. Feeders are mechanized to automatically present the next component after the preceding one has been collected. The vacuum pipette which handles the component can be mounted on a traversing mechanism which first guides it to the correct feeder, and then to the correct location above its placement position. The operator then lowers the pipette and guides it to its exact position. As the component touches down, the vacuum in the pipette may be released automatically. Finally, the controlling computer can be programmed to work out the most economical placement sequence to save time-wasting movements. Depending on the complexity of the board and the mix of components, the capacity of semi-automatic placement systems with full computer support can rise to about 900 components her hour.

7.3.3 Fully automatic sequential systems

General features

The term 'sequential' means that, as with the manual systems, the machine always places one component at a time (Figure 7.3). Apart from that, a fully automatic sequential placement system is like a semi-automatic manual system, with the human element replaced by opto-electronic and electromechanical means. Taking out the human element has several consequences:

1. Fully automatic equipment is more expensive, by one or more orders of magnitude.
2. Fully automated sequential placement is faster, with a capacity of normally about 2000 to 2500 components per hour.
3. The high accuracy of placement demands a robust and stable platform on which the moving parts are mounted. This in turn means a design based on the technology of a precision machine tool rather than of a mechanical plotter.

Flexibility

With most users of pick-and-place equipment, flexibility of the system is a crucial requirement in order to reduce down-time of the expensive equipment to a minimum. In this context, flexibility means the ease and speed with which a machine can be switched from one type of board to another. Such a change-over involves changing the array of feeders and the pick-and-place sequence. The latter needs little time if the operating software already exists. If it does not, it can often be created while the machine is still busy placing components on another board.

Assembling the array of feeders and loading them with the required tapes and magazines takes more time. Many state-of-the-art machines have mobile feeder arrays which can be assembled away from the machine. An array can be fully loaded with the tapes and magazines for the next run while the machine is still busy on the preceding one. For the change-over, feeder arrays can be quickly exchanged against one another. Working out the best sequence of components in a feeder

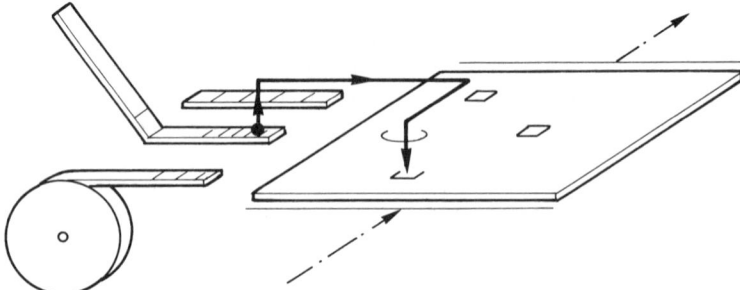

Figure 7.3 *Sequential placement*

array, together with the most economical pick-and-place sequence, is a matter of software and programming.

Additional functions

Apart from picking and placing components, most automatic pick-and-place machines can put down single drops of adhesive for SMDs which are to be wavesoldered.

'Smart' machines

Smart machines observe and react to circumstances, and detect errors and take the appropriate action. In the context of component placement, this includes the following:

- Feeder units identify the tapes or magazines loaded into them, and in the case of an error give a signal or prevent operation until the mistake is rectified.
- Placement heads identify the components they have picked up, and check some of their basic functions ('in-flight testing'). Multilead components are checked for the coplanarity of the lead ends.
- Vision centring systems align fine-pitch components with their footprints before setting them down.

Grouping of placement units

Placement lines can be assembled from individual automatic placement units, which are linked together and operated as a CIM system. This makes it possible to multiply the capacity of the placement section while still maintaining its flexibility to respond quickly to changing production requirements.

7.3.4 Simultaneous placement systems

Some types of electronic product like domestic audio and video equipment involve long runs of similar boards which are not particularly complex but which are produced in large volume. These boards are soldered on high-capacity lines, which may be wave machines or reflow ovens. They in turn must be fed by placement equipment of equally high output.

For this task, simultaneous placement systems are better suited. As the name implies, a number of components are picked and placed at the same time at every working stroke of the machine (Figure 7.4). A number of placement arms, each of which can choose between several feeders, pick up components simultaneously. The length of the stroke of each arm determines where the component comes down. If the boards are to be wavesoldered, a dot of adhesive is placed on the underside of each component as it moves forward to be placed. A machine with 16 placement arms, for example, each of which can choose from six feeders, can place 96 different components on a board at a rate of 60 000 SMDs per hour.

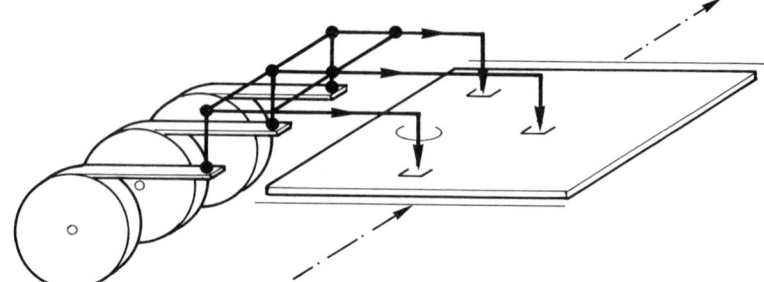

Figure 7.4   *Simultaneous placement*

7.4   The practice of automatic component placement

7.4.1   The range of choice

An unremitting driving force behind the development of automatic placement equipment is the evolution of the SMDs themselves. In some respects this seems to have reached a plateau, at least a temporary one: the miniaturization of melfs and chips ('birdseed' in the language of component users) has probably stopped with components 1 mm/40 mil wide and 2 mm/80 mil long. The size of multilead components may have reached a temporary limit with 55 mm/2.2 in square, which with a 0.5 mm/20 mil pitch gives a leadcount of about 400. The makers of automatic placement equipment have responded, many of them in collaboration with their customers.

A recent survey[1] lists 24 vendors in the USA, Europe, and Japan. The functional capabilities of the equipment which they offer is classified in Table 7.2.

Choosing a placement system poses formidable problems. In terms of size of investment, it may well equal if not exceed the cost of the soldering equipment. There is no single machine which is simultaneously the fastest, the most flexible,

Table 7.2   *Functional capabilities of commercially available automatic SMD placement equipment (1993)*

Capability	*Percentage of available equipment offering the capability*
Capable of in-line integration	100%
Placing >10 000 SMDs/hour	70%
Machine vision to assist placement	95%
Dispensing metered adhesive drops	95%
Handling SMDs with <0.4 mm/16 mil pitch	75%
Handling 0402 chips and 0102 melfs	85%
Handling TABs	60%
Soldering them	25%

the most accurate and the cheapest. All attributes must be traded off against one another.

However, most of today's automatic placement machines are conceived as modules which can be added to one another into an integrated line, and this makes the choice easier: a newcomer to automatic placement can start with a unit of comparatively modest capability and cost, without pre-empting his later options for expansion.

7.4.2 Classes of placement machines

Entry-level and mid-range machines

At what is sometimes termed the 'entry-level' and the 'mid-range' of placement machines, up to 100 feeder stations are provided with a changeable mix of bulk-feeders, feeder tape, and stick-tray and waffle-tray magazines. With many machines, sets of feeder stations are mounted on individual, interchangeable trolleys which can be quickly detached from, or linked to, the machine to increase flexibility and speed up the change-over from one type of board to another. The feeder stations on each trolley are assembled according to a computer-generated sequence to suit a given type of board. This enables the machine to switch from one type of board to another very quickly.

The single placement heads collect, centre, rotate and put down single components in sequence. Most of these single-head machines have a maximum theoretical placement rate of 4000 components per hour. In practice, the necessary allowance for travel time and stop-and-start movements reduces the practically achievable rate of placement to about 2400 components per hour.

This type of operation is suitable for placing melfs, chips and SO components with standard pitches from 1.25 mm/50 mil down to 0.65 mm/26 mil. For placing components with finer pitch, machines are either equipped, or can be retrofitted, with opto-electronic placement aids. Facilities for in-flight verification of component identity and functional integrity can also be provided.

With single-head placers, putting down drops of adhesive on boards destined for wavesoldering would require a second placement station. For this reason, it is economical to feed this type of machine with boards to which the adhesive has been pre-applied by one of the methods described in Section 4.8.3.

Fine-pitch placement machines

Next in line are fine-pitch placement machines, which can cope with placement accuracies up to 0.4 mm/15 mil and component sizes up to 55 mm/2 in square, which means maximum xy and rotational accuracies. Such machines are fully equipped with opto-electronic sighting, verification of the coplanarity of component legs, and automatic adjustment of placement pressure so as not to smudge or squeeze out the paste printdown on the narrow footprints. With a single head, the maximum output is as above.

High-speed 'chip-shooters'

The next step in sophistication are the high-speed 'chip-shooters'. Different equipment makers have chosen different paths to this end. With one system, for example, the placement head takes the form of a rotating disc (revolver head) with twelve circumferential stations, each of which grips one component (Siemens). With a two-head machine, one disc loads up while the second one puts down its load of components on the board. Instead of the second revolver disc, a single placement head for very large multilead components can be substituted. A further head for pre-placing metered amounts of adhesive can also be fitted. The placement rate of such machines is quoted at 13 000 components per hour.

Very high speed placement machines

Boards which are populated with comparatively simple components, capable of being handled by tape feeders, permit the use of very high speed placement machines. One example is a simultaneous system which operates with sixteen placement heads moving only in the x direction, while the board is indexed past them in the y axis. Placement speeds of up to 60 000 components per hour are quoted (Philips). Naturally, this type of equipment is designed for running downstream of a high-volume soldering line which is tuned to long production runs of one, or at most a few, types of board.

7.5   Reference

1.   *Vendors of Automatic SMD Placement Equipment.* Productronic (Germany), June 1993, p. 18 (in German).

8 Cleaning after soldering

8.1 Basic considerations

If cleaning must be considered, its problems can be reduced to three questions:

1. What has to be removed?
2. Why?
3. How?

The first question has a simple answer: principally the residue from soldering fluxes, and sometimes from wavesoldering oils. This is why cleaning is often called 'defluxing'. Sometimes, contaminants from manufacturing steps which precede soldering can also be present in amounts which affect the need for cleaning and the choice of cleaning strategy, as well as the result of subsequent tests for cleanliness. These contaminants are listed alongside the principal flux residues in Table 8.2.

The answer to the 'why?' is less simple. It clarifies the issue to some extent to rephrase the question like this: 'How clean has your circuit board got to be, considering who wants to use it, where, and for how long?' Even then, at the present state of knowledge, it is not always possible to give a reasoned and quantifiable answer to this question.

Generally it is true that any contamination on a soldered assembly can reduce its reliability. Impaired reliability means that the affected assembly is likely to malfunction, or to stop functioning altogether before its designed, expected or guaranteed lifespan. 'Contamination' in this context means the presence of any foreign substance – either too much of it, or in the wrong place.

Flux residues are the most common example of such contamination. The problems they cause can be electrical and/or chemical ones; if they are visible, they may also affect the marketability of the product. All these aspects will be fully dealt with in the next chapter.

An important point must now be made before we go any further: unless there is a compelling and unanswerable reason for cleaning a soldered board, do not clean. Cleaning is expensive, and a badly cleaned board is worse than it was before cleaning: inefficient cleaning is liable to spread flux residue to places where there was none before.

Moreover, the halogenated solvents, principally the CFCs and CHCs, which were until very recently the principal and most convenient cleaning solvents for the electronic industry, have now been recognized as posing severe environmental threats, and they will have to be phased out in the very near future (see Section 8.3.5).

Therefore, the whole technology of cleaning has been forced into different cleaning media and strategies. Some of them, like cleaning with water, are already well established; others are opening up new approaches. Anyone who is still cleaning with a halogenated solvent must sooner rather than later change over to an alternative cleaning method. Now is the time to reconsider whether cleaning is really unavoidable.

If with your given market or customer situation, and your present soldering strategy, cleaning is or seems to be obligatory or unavoidable, examine carefully whether you cannot solder in such a way that cleaning is no longer necessary. Possible solutions might be the use of low-solids or no-clean fluxes or solderpastes (Section 3.4.3) or soldering in an oxygen-free atmosphere (Sections 4.6 and 5.4.5). A word of caution is necessary at this point: having changed your flux and/or your procedure so that your product looks and is clean to your own satisfaction, check with your existing customers and test your market to find out whether the product still meets their expectation of reliability and life expectancy before sending it out into the world.

Once the need for cleaning has been recognized as definitely unavoidable, the 'how?', that is the strategy of cleaning, depends entirely on the choice of flux. Here, a flux with a fully watersoluble residue is well worth considering: with full watersolubility, there is no need to replace a soon inadmissible halogenated hydrocarbon with an alternative organic solvent, with potential problems of flammability, regeneration or disposal, or a saponified waterwash. A plain waterwash and some semi-aqueous systems have far fewer environmental problems, and zero-effluent cleaning systems are already available. Sections 8.3, 8.4 and 8.5 deal with these various cleaning options.

8.1.1 Reasons for cleaning

Cleaning can become necessary for one or several of the following reasons:

Electrical problems

Flux residues left on a soldered assembly can cause leakage currents, ionic migration between neighbouring conductors (formation of dendrites), or insulating deposits on contact surfaces such as pin-connectors, relays, or trimmer-tracks, but above all on the contact pads for testbed pins.

Here, the insulating effect of the flux residue falsifies the test results. If it is sticky it contaminates the contact points of the test needles, which then need expensive and time-consuming maintenance. This problem was the initial driving force behind the development of modern low-solids fluxes. Though needle-adaptor testing may begin to reach its limits with the growing complexity and population density of modern circuitry, and other testing strategies are beginning to take its place, as long as needle testing is being practised any flux residue on test pads must be such as not to interfere with it.

Impaired surface insulation resistance is a serious defect with boards carrying ICs with closely spaced leads and impedances of 10^6 to 10^{12} ohm. Leakage currents of 10^{-12} amp may be sufficient to cause malfunctioning of binary gates.[1]

Chemical problems

Corrosion

In hostile operating environments, flux residues can cause corrosion or promote fungus growth. Electronic assemblies which have to operate in humid tropical climates are especially at risk from the latter (Figure 8.1).

Polar, watersoluble components of residual flux (e.g. activator residues, or indeed residues from any watersoluble flux) can, in the presence of moisture, cause electrolytic corrosion between adjacent surfaces of dissimilar metals. For example, moist ionic flux residue sitting on the dividing line between solder and the silver metallization on a chip can form an electrolytic cell with a potential of 2.5 V (see Figure 8.2).[2]

Formation of dendrites

All organic films, including protective lacquers and conformal coatings, are permeable to water vapour. Residual activator, trapped under a coating of lacquer between two adjacent metallic conductors with an electric potential between them, causes not only leakage currents, but can initiate the growth of dendrites between the conductors, which may lead to short circuits (see Figure 8.3).

Under the conditions described above, the formation of dendrites is most frequently observed between surfaces of Ag, but it can also occur between Pb/Sn, Cu, AuPd and AuPt. Their rate of growth can be up to 0.1 mm/min.[3]

Problems arising with follow-on processes

Flux residue can be responsible for bad adhesion of conformal coatings, lacquers or potting compounds. The vendors of several low-solids or 'no-clean' fluxes claim that the residues from their fluxes do not cause such problems.

Reasons related to the nature of the product

Some classes of electronic product require the highest possible degree of reliability, and hence cleanliness, which can only be achieved by a cleaning procedure after soldering, irrespective of the flux or soldering method used. These classes include professional electronic equipment which falls under categories I, II and sometimes III, set out in Chapter 9. Professional electronic equipment in categories IV to VI

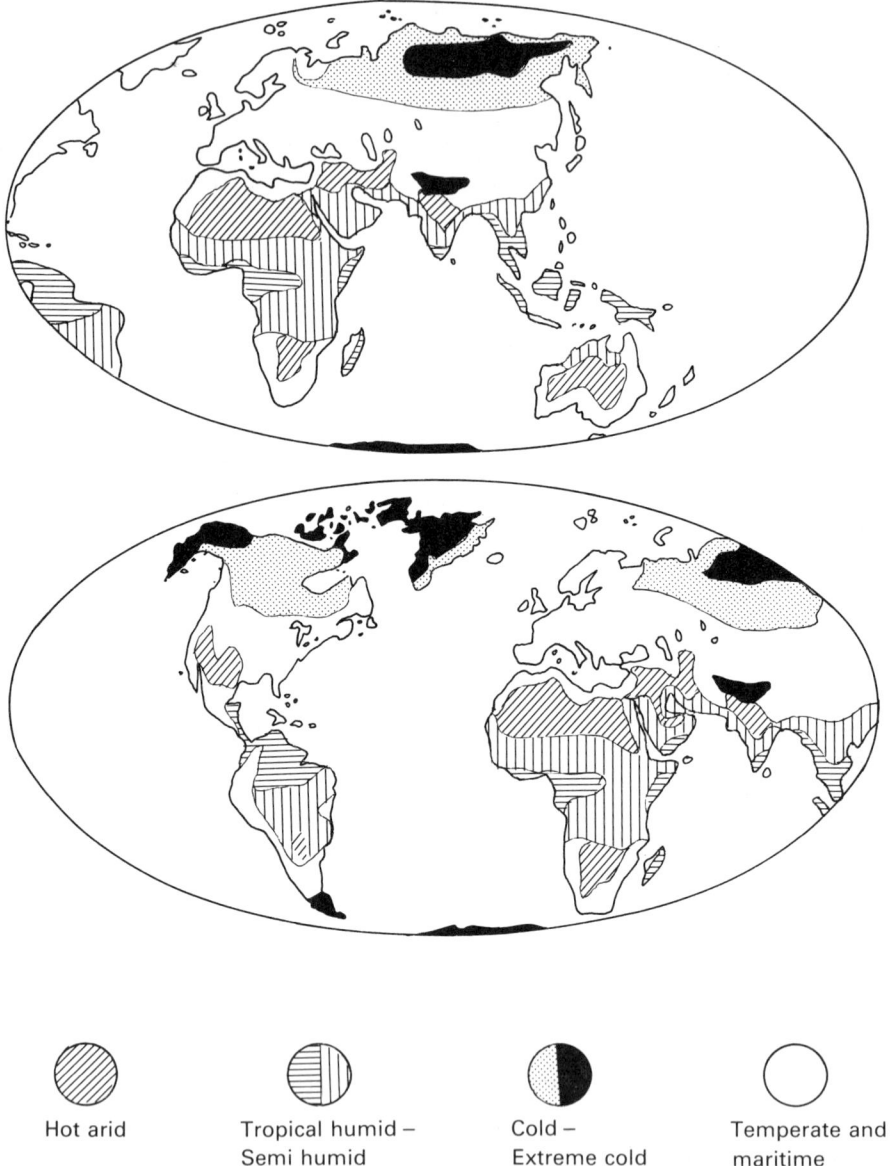

Figure 8.1   *Map of world climates. After Britten, R. and Matthews, G. W. (1988) Plessey Assessment Services; Electronic Production*

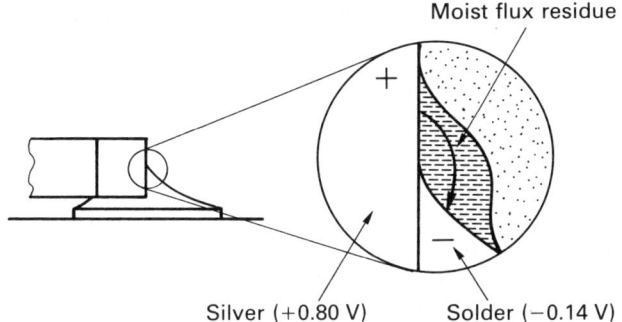

Figure 8.2 *Electrolytic corrosion*

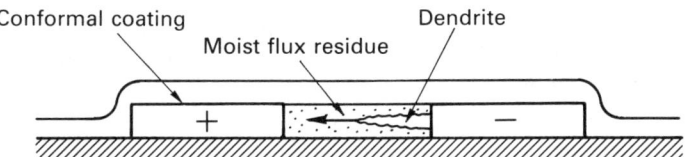

Figure 8.3 *Formation of dendrites*

requires cleaning only in certain circumstances, e.g. for service in damp tropical climates or hostile environments. Products of the class 'Consumer Electronics' in category V rarely need cleaning.

Reasons related to marketing considerations

A number of non-technical considerations fall under this heading, such as:

- the customer wants (and pays for) cleaning
- the competitors clean
- cleaning improves the appearance of the product and makes it more saleable in a competitive market.

8.1.2 Designing for cleanability

Once cleaning has been accepted as unavoidable, designing the board and specifying the components in such a manner that successful cleaning does not become unnecessarily difficult, expensive or even damaging becomes an important consideration. To be readily cleanable, high-profile components should be spaced as far apart from one another as economic and functional considerations will allow. Braided wire should not be used for jumper connections, because flux and cleaning solvents are likely to get trapped in them. Plastic sleeving on such wires must be tested for its compatibility with the cleaning process to be used. Shrinkable sleeves should not be used on boards which must be cleaned.[4]

Components should be tested for their compatibility with the cleaning method chosen. Military Standard MIL-STD-202 demands the testing of component markings for their resistance to the cleaning solvent which is going to be used. British Standard BS 9003 provides also for the testing of component bodies, sealing materials, varnishes and encapsulating compounds. It has been proposed that components ought to be marked with a cleanability code in the same manner in which garments carry a code of appropriate cleaning and washing procedures.[5] Detailed cleanability testing schedules for complete electronic assemblies are outlined by Lea.[6]

The clearance between the underside of a soldered SMD and the surface of the board (its 'stand-off height') has a critical influence on the ease with which flux residue trapped in this gap can be removed. The effectiveness of removing such residues increases with the fourth power of the stand-off height. Multilead components present a serious problem here: with their stand-off height approaching zero, cleanability is often totally neglected in their design. This aspect is dealt with fully in Section 8.2.2.

8.1.3 What must be removed?

Flux types and their residues

The nature of a flux residue and its response to cleaning depend obviously and principally on the type of flux used. The various fluxes which are available to the industry are all subject to national and international standard specifications. As far as flux composition is concerned, a standard specification does not say what a given flux actually contains, it only states what kind of substance might be found in it. On the other hand, it states precisely what a given flux or class of fluxes must not contain. Generally speaking, standard specifications classify fluxes into various groups, though these groups do not fully overlap between different countries.

All countries distinguish between rosin-based or rosin-containing fluxes, and rosin-free fluxes. In the US, MIL-F-14256 gives all rosin-containing fluxes the letter 'R'. ANSI/IPC-SF-818 is an exception because it does not classify fluxes according to their chemical composition, but by the behaviour of their residues. This whole, rather complex, subject is dealt with in Section 3.4.3.

What matters in the present context is whether a flux residue is watersoluble or not. Residues from fluxes containing rosin or a synthetic resin are not watersoluble, but require an organic solvent, or water with an added saponifier. Rosin-free fluxes generally leave a watersoluble residue.

If for technical or marketing reasons 'defluxing' becomes mandatory, the non-corrosive nature of the residue from rosin-based fluxes, which is their principal virtue, becomes irrelevant. Instead, they turn into a liability: their removal involves the use of solvents with all their problems, or the addition of saponifiers to the washing water.

In the context of cleaning, rosin-free fluxes are nowadays gathered under the collective description of 'watersoluble fluxes'. Solderpastes which are based on such fluxes are also called 'watersoluble'. This classification relates to the residues they leave behind. With watersoluble fluxes, cleaning after soldering used to be advisable, even if the soldered product as such did not demand cleaning, and if corrosion tests as specified in some flux standards (e.g. British Standard 5625 or Federal Specification QQ-S-571) did not disclose any corrosive action.[7]

In recent years, low-solids fluxes and 'no-clean' fluxes, where the vendor specifically guarantees that cleaning is not required, have become available. They leave a very thin layer of a dry, sometimes powdery residue (see Section 3.4.3) which is of obvious advantage when an assembly is subsequently tested on a pin adaptor or 'bed of nails'.

With 'no-clean' fluxes, especially when they are used on assemblies of a critical nature, the buyer may be well advised to verify that the vendor's claims apply to his specific product and its market. It must be remembered that the action of all fluxes is based on their ability to react with metal oxides and that therefore all flux residues contain traces of metallic compounds. For that reason, flux residues ought to be removed whenever the service requirements of the finished assembly are at all critical (which applies for example to the Product Classes I, II and III as set out at the beginning of Chapter 9).

The flux residue from rosin-based soldercreams contains not only rosin, but also the remnants of various additives which such creams contain, such as stabilizers, thickening agents, etc., which control their flow-properties during and after printing or dispensing. These substances are normally no more corrosive than the rosin itself, but they must be considered in the cleaning process.

The effect of the soldering method and its parameters

Wavesoldering

With wavesoldering, the whole underside of the circuit board is covered with flux, not all of which is washed off when the board passes through the wave. On the other hand, with wavesoldering, SMDs have to be glued to the underside of the circuit board, and the adhesive takes up a considerable part of the gap between the component and the board. This can mean that there is less flux residue to be removed from under wavesoldered SMDs, and such residue as is present is near the edge of the component and therefore more accessible (Figure 8.4).

On the other hand, if cracks have opened in the adhesive joint during the preheat or the passage through the solderwave, they may harbour some flux residue which can leak out during later service, even if the MIL test for cleanliness has given the board a clean bill of health.

Exposure to heat causes rosin to polymerize and to become less soluble. Therefore, overheating the board in the pre-drying stage before it passes through

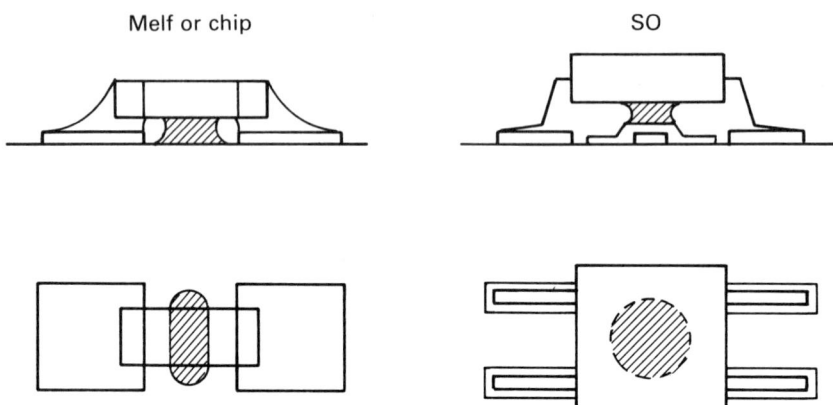

Figure 8.4   *Effect of the adhesive joint on the accessibility of the flux residue*

the solderwave, and above all undue delay between soldering and cleaning, will make the cleaning process more difficult.

Wavesoldering in a protective and practically oxygen-free nitrogen atmosphere was introduced in the early eighties (see Section 4.4). This makes it possible to use a gaseous flux, or a liquid flux which leaves extremely little residue. For many applications, boards soldered in nitrogen need not be cleaned. Whether boards used in electronic products falling under the Classes I, II and III mentioned above can be left in an uncleaned state even when soldered under nitrogen is still subject to discussion. With products of this type, it is advisable to clear the position with the purchaser or the end-user of the equipment.

Reflowsoldering

With all reflowsoldering methods, soldercream or flux is applied only to the solderpads. Thus, the flux-residue is confined to the neighbourhood of the footprints, and consequently near to the outer edge of the components. It is therefore more accessible and easier to remove. Thus, needle-adaptor testing is no problem with reflowsoldered boards, since testpads are, or ought to be, free from flux residue.

Vapourphase soldering and infrared soldering in a nitrogen atmosphere are claimed to leave rosin flux residues in a more soluble condition, because they do not oxidize during soldering and should therefore be easier to remove.

Recently, several solderpastes have become available which are claimed to leave no residue at all. One of them, containing a flux based on a synthetic resin, requires infrared reflowing in a special reactive atmosphere.[8] Another low-residue paste demands reflowing in a low-oxygen atmosphere such as nitrogen, or in a vapourphase system.

Other residues

Wavesoldering methods where oil is injected into the wave (Hollis) or applied to the wave surface (Kirsten, Soltec/Smartwave) can leave a certain amount of oil residue on the board. After prolonged use, such oils can polymerize and become less soluble, so that routine cleaning does not remove them fully. For this reason, it is important with such machines to follow the manufacturer's operating instructions very carefully. In any case, wavesoldering in the presence of oil places an additional load on any subsequent cleaning process, and this factor has to be considered in the choice and the operation of a cleaning installation.

Apart from flux residues and perhaps soldering oil, a well conducted assembly process should not leave any other substances on the soldered boards. It has been pointed out, though, that markings made with certain felt-tipped pens during inspection may be difficult to remove by normal cleaning methods, the marking ink having been specifically formulated to resist removal. Lea recommends the use of peelable stickers instead of marking with a pen.[9]

Chemicals or surface contamination arising from preceding manufacturing processes of the board, or from handling the boards during assembly, can seriously affect the performance of a board, its cleanliness as determined by the MIL test, or its surface insulation resistance after it has been soldered.[10] It can therefore become advisable to write cleanliness requirements into purchasing contracts, or to subject incoming boards and sometimes components to cleanliness tests before assembling and soldering them.

8.2 The theory of cleaning

8.2.1 The physics of cleaning

The removal of a flux residue consists of two separate tasks:

1. The deposit of solidified flux residue must be dissolved or at least made mobile in the cleaning medium or solvent.
2. The resultant solution or suspension of flux residue must be flushed away and replaced by fresh, clean solvent which soaks up or dislodges more flux and which must be flushed away in turn.

Thus, every solvent–cleaning process is in principle a progressive dilution, and absolute cleanliness can only be approached asymptotically. Therefore, a rational and economic cleaning process demands a sensible definition of the required degree of cleanliness, and often a precise measurement of the degree of cleanliness which has been achieved. SMD technology, with its complex geometry of densely populated boards, with joints hidden in narrow gaps between neighbouring components and with flux residue trapped between SMDs and circuit board, has made cleaning more difficult. It makes more stringent demands on the efficiency of cleaning equipment and solvents than did boards with simple, inserted components.

The 'stand-off' height

The distance between the underside of an SMD and the board surface – the 'stand-off height or gap' – is a critical parameter as far as cleaning is concerned. With stand-off heights below 75 μm/3 mil, cleaning with a liquid medium becomes practically impossible. This poses a problem with many modern multilead components, which sit close to the board, and still awaits a practicable solution. With greater stand-off heights, procedures which are supported by an input of hydrodynamic energy, such as ultrasonic vibrations or high-pressure jets, are generally successful.

For simple bath-type cleaning systems, a stand-off height of at least 500 μm/20 mil has been postulated.[11]

With wavesoldered boards, the stand-off height is less critical than with reflowsoldered boards, because the adhesive which secures the SMDs to the board during wavesoldering occupies between fifty and seventy-five per cent of the gap-volume. This means that the flux residue is located mainly near the outer edge of wavesoldered components (see Figure 8.4). However, should the adhesive joint crack during or after hardening because of faulty formulation or processing, matters get worse instead of better: flux residue is bound to enter the cracks, from where they may exude later and reach the soldered joints.

The stand-off height itself is governed by a number of factors, including the structure of the board surface and the shape of the component legs (Figure 8.5).

The dimensions given in Figure 8.5 must be understood as averages, and they can vary from one supplier to another. Further factors are the thicknesses of the copper laminate and of the solder resist on the circuit board (Figures 8.6 and 8.7).

As Figures 8.6 and 8.7 show, the thickness of the solderpads on which the components sit raises the stand-off height, while solder resist and conductor tracks, which run underneath a component, reduce it. Before deciding on the cleaning strategy for a given board, it is sometimes advisable to check the actual stand-off height of critical components with a feeler gauge. Obviously, components with a high profile and dense packing make cleaning more difficult. A further factor is the 'picket-fence' effect of the legs of components like PLCCs or QFPs, which obstruct the flow of the cleaning medium into the gap between component and circuit board.

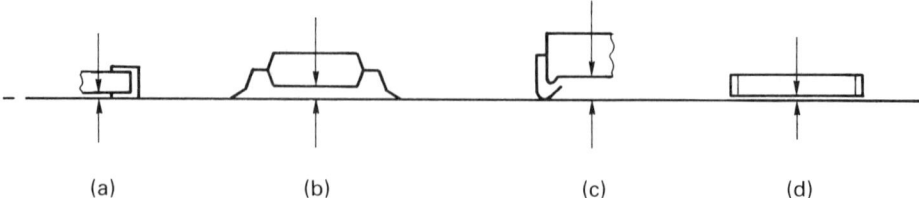

(a) (b) (c) (d)

Figure 8.5 *Stand-off heights of various SMDs. (a) Chip resistors and ceramic condensers 25 μm/1 mil; TA condensers 100 μm/4 mil; (b) SOTs 125 μm/5 mil; SOICs, QFPs < 250 μm/10 mil; (c) PLCCs 250 μm/10 mil; (d) LCCCs < 25 μm/1 mil*

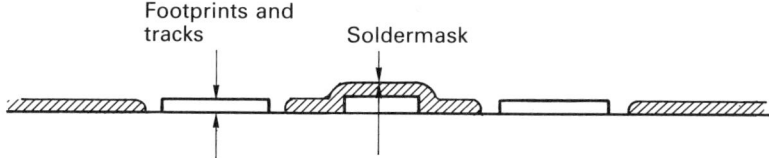

Figure 8.6 *Height of various surface features on a circuit board. Footprints and tracks 35–135 µm/1.5–5.5 mil; soldermask lacquer 10–20 µm/0.4–0.8 mil; soldermask dry-film 50–100 µm/2–4 mil*

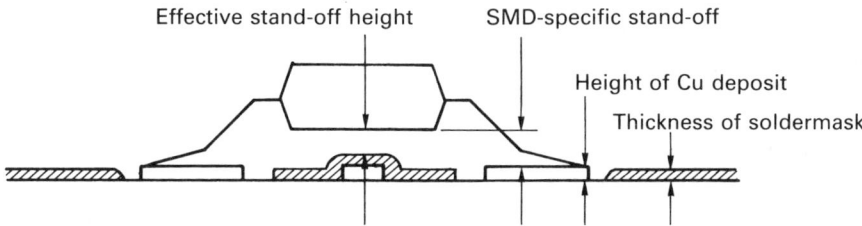

Figure 8.7 *Effective stand-off height = SMD-specific stand-off − thickness of soldermask*

The hydrodynamics of cleaning

Clearly, flux residue can only be removed from a crevice or from under a component if the cleaning medium penetrates into the gap first. This penetration depends on the wetting behaviour of the medium towards the surfaces involved, which include the board laminate, the solder resist and the solder. It also depends on its viscosity (Figure 8.8; Table 8.1).

Compared with organic solvents, water penetrates faster into a given gap. Unfortunately, the same capillary force which pulls the medium quickly into a gap holds it there firmly once it has entered. Effective cleaning demands that the

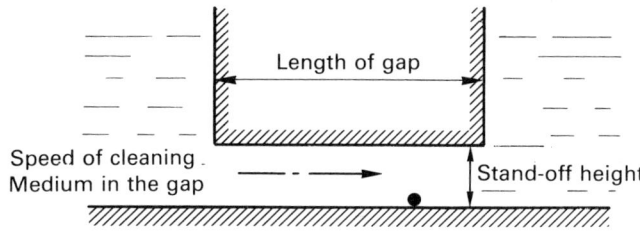

Figure 8.8 *The hydrodynamics of cleaning*

Table 8.1 *Penetrating speeds of different solvents*[13]

Solvent	Temperature	Penetrating speed	
		cm/sec	in/sec
CFC/alcohol azeotrope	40 °C/104 °F	2830	1130
CHC/alcohol azeotrope	73 °C/163 °F	4056	1620
Water	70 °C/158 °F	11 380	4550

solvent, having dissolved or loosened the contaminant, is expelled from the gap by a pressure differential between its opposing ends and replaced by fresh solvent. At a given pressure differential, the force which acts on a body lodged in a gap is proportional to:

- the density of the cleaning medium
- the fourth power of the stand-off height

and is inversely proportional to:

- the square of the viscosity of the cleaning medium.[12]

These relationships underline the decisive effect of the stand-off height: with a stand-off height of 75 μm/3 mil, the washing power of a fluid medium is 81 times higher than with a stand-off of 25 μm/1 mil, all other circumstances being the same.

Practical measures to improve the kinetics of washing

Ultrasonics

Ultrasonic energy introduced into a liquid creates a pattern of closely spaced wavefronts of high and low pressure which travel across its volume. If the pressure in the low-pressure regions drops below the vapour pressure of the liquid, small bubbles of near-vacuum form there. They collapse immediately with the arrival of the next high-pressure front. Where the wavefronts impinge upon a solid surface the collapsing bubbles act like innumerable small hammerblows against that surface (cavitation), loosening adherent contamination and thus assisting cleaning.

If the liquid is boiling, or near-boiling, the mechanism is a different one. Because the vacuum would immediately fill with vapour, no bubbles can form. Instead, the ultrasonic pressure waves exert a strong pumping action on gaps and crevices by creating local pressure differentials between opposite ends of a gap.

The ultrasonic pressure applied to the circuit boards immersed in a cleaning bath is normally held at 2–4 bar, with the total power introduced into a cleaning bath amounting normally to about 10 watt per litre/38 watt per US gallon. With ultrasonically assisted cleaning, circuit boards must not be packed too tightly in fixtures or cleaning baskets, and they should be oriented in such a way that the ultrasonic energy has unimpeded access to all surfaces where it is needed.

The ultrasonic frequencies are normally in the range 35–40 kHz. There has been much argument in the past about the danger of damaging wirebonds or circuitry inside ICs by setting up a resonance between the bonding wires and the ultrasonic oscillations or their harmonics, thereby causing fatigue damage. Practical experience has shown that there is very little danger of this happening provided the frequency is kept within the above proven range.

Safe ultrasonic cleaning regimes, in terms of ultrasonic power density and exposure times, have recently been established.[14] On the other hand, cracks in outside gullwing legs, which are made from Fe-containing copper alloy, have sometimes been blamed on fatigue damage caused by vibrations induced during ultrasonic cleaning. However, no clear conclusions can be drawn so far, and research is continuing.

High-pressure jets

Ultrasonic cleaning requires full immersion in a solvent bath and it is therefore more suited for batch processes. With jet-cleaning, which is mostly used in conveyorized in-line installations, the soldered assemblies pass between arrays of high-pressure jets. The boards travel on a horizontal open–mesh conveyor, and the jets impinge on them from above and below, at a carefully balanced intensity so as not to make them flip over.

When working with a hot organic solvent, the jets have diameters in the range of 1–2 mm (0.04–0.08 in). Their operating pressure is governed by the lowest stand-off height of the various SMDs which populate a given board. Experiments with a CFC solvent have shown that with stand-offs below 25 μm (1 mil), jet pressures of 15–17 bar are required. With stand-offs above that figure, jet pressures of 2–5 bar suffice to give a cleanliness to MIL standard within 10 seconds.

With the new water cleaning systems, the hydrodynamics of the water jets have received a great deal of attention. These matters are discussed in Section 8.4.4.

The glass plate test for checking cleaning efficiency

Two thin glass plates (e.g. microscope slides), separated from one another by metal shims and pressed together by clamps at both ends (e.g. 'Bulldog' paperclips) form a convenient and readily evaluated model of the stand-off gap underneath an SMD (Figure 8.9). The thickness of the shims is chosen to represent the minimum stand-off of the board's population of SMDs.

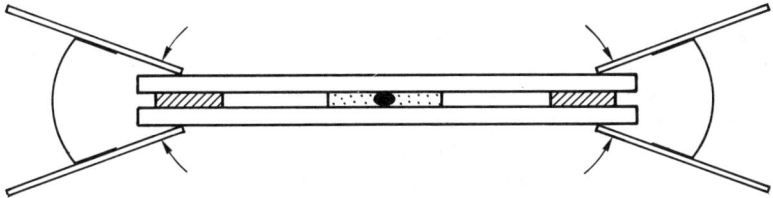

Figure 8.9   *The glass plate model for testing cleaning efficiency*

For the test, a small amount of the same flux or soldercream which is used on the boards to be cleaned is placed in the centre of the gap between the two glass plates. The model is then put through the same temperature cycle which the assembly will have to undergo before it reaches the cleaning stage. Afterwards, the model is put through the cleaning procedure which is to be assessed. The depth to which the flux residue has been flushed from the gap between the glass slides is readily visible and provides a measure of the efficiency of the cleaning treatment. The addition of a small amount of fluorescent dye to the flux or cream, and subsequent examination under UV light, makes the result even easier to observe.

This method is very useful for evaluation tests when comparing different cleaning solvents or different cleaning procedures. It is also useful for comparing the response of different fluxes or soldercreams to a given cleaning procedure, in situations where a change of product or supplier is contemplated. However, for a safe prediction of cleaning results in an existing situation it must be regarded with some caution: the dimensions of the glassplates differ too much from the actual components, and in addition the gap has a simple geometry and is not obstructed by a picket fence of component leads or by neighbouring components.

A concluding thought for this section: in practice, the problem of removing flux residue from under large, flat components is not as daunting as it seems: for one thing, such components with their closely spaced leads are normally not wavesoldered, but are reflowsoldered by one of the various methods. This means that the flux residue is located near the perimeter of the component, close to the 'picket fence' which is formed by the component leads, so that it need not be fetched from far inside the stand-off gap. On the other hand, the capillary action of a low stand-off gap, which pulls the flux from a solderpaste underneath the component body, must not be forgotten.

8.2.2 The chemistry of cleaning

The choice of the most suitable cleaning medium for a given cleaning task depends on the chemical nature of the substance which must be removed. Polar solvents are best for polar contaminants. Non-polar contaminants can be removed with non-polar solvents, though there are polar solvents, for instance alcohols, which can remove most flux residues and contaminants encountered in practice. Tables 8.2 and 8.3 list and classify the commonly encountered contaminants and cleaning media.

It follows that residues from activated rosin-based or rosin-containing fluxes can be removed with the (now doomed) CFC- or CHC-azeotropes (Section 8.3.1) and with their environmentally acceptable polar alternatives, such as plain and modified alcohols and esters (Section 8.3.7).

Water with added saponifiers (Section 8.4) is another viable alternative. The cleaning effect of a water/saponifier mixture on these residues is based not on their solubility in these media, but on a chemical reaction which turns them into water-soluble soaps.

Table 8.2   *The chemical nature of various contaminants*

Soldering residues	
Natural rosin	non-polar when solid; slightly polar when molten or in solution
Synthetic resin	non-polar
Flux activators	polar
Cover-oil for wavesoldering	
Regular	non-polar
Watersoluble	polar
Solderpaste additives (stabilizers and fillers)	slightly polar
Hot-air levelling flux	polar
Pre-soldering residues	
Plating and etching residues	polar
Swarf from drilling, sawing or scoring	non-polar solids
Fingerprints	polar/non-polar mixture
Protective handcream	polar/non-polar mixture

Table 8.3   *The chemical nature of some cleaning media*

Water	polar
Saponifier	polar
Wetting agents	polar
Chloro-fluoro-carbons (CFCs)	non-polar
Chloro-hydrocarbons (CHCs)	non-polar
Alcohols	polar
CFC- and CHC-azeotropes (see Section 8.3.1)	polar
Modified alcohols, esters	polar
Terpenes, ketones and hydrocarbon mixtures	non-polar

Terpenes and ketones are excellent solvents for rosin, resin and oils, but not for the polar activators. Cleaning systems based on them need an added waterwash if residues from rosin-based or resin-based fluxes have to be removed (semi-aqueous systems).

Residues from rosin-free fluxes are soluble in plain water, as are the residues from watersoluble wavesoldering cover oils. Because all cover oils oxidize or polymerize after prolonged use, and so their residues become less soluble, suppliers' instructions as to their conditions of use should be strictly followed.

Temperature has a decisive influence on every cleaning process: as has been mentioned when fluxes were discussed (Section 3.4), a temperature rise of $10\,°C/18\,°F$ roughly doubles the speed of a chemical reaction. The same is true for the speed with which a contaminant dissolves in a given solvent. For this reason, almost all cleaning methods operate with warm solvents. It must be remembered, however, that for reasons explained above ultrasonic energy loses its principal cleaning mechanism if it acts in a boiling or near-boiling solvent.

Time-interval between soldering and cleaning

Cleaning should follow as soon as possible after soldering. The longer the time gap between the two, the greater is the effort needed to remove the flux residue. Resinous deposits, like any varnish, oxidize in air and become less soluble. Watersoluble residues can become dry and hard, and their solubility decreases. Experiments have shown that immediately after wavesoldering, rosin-residues could be removed completely by spraying with a CFC for 20 seconds. Thirty minutes later, even spraying for 50 seconds removed only 75% of the residues.

For this reason, with in-line soldering installations, soldering and cleaning machines are placed as close together as possible. However, two provisos must be made:

1.   If a board is immersed in a cleaning medium while it is still hot, the medium may be sucked into the interior of the components as they cool, thus damaging them.

2.   Care must be taken that vapours from any solvent-cleaning equipment which is still operated with CHCs or CFCs cannot drift back into the hot zones of wavesoldering machines or IR heating ovens. Vapours of these solvents can generate highly toxic decomposition products at elevated temperatures, particularly in the presence of hot solder-dross (see Section 8.3.5).

8.3   The practice of cleaning

The cleaning of a soldered electronic assembly involves, almost without exception, a chemical and physical interaction between a liquid medium – the solvent – and the flux residue left on that assembly. In general practice, these liquid cleaning media are either organic solvents or water, with suitable additives if necessary. Mechanical cleaning with dry rotary brushes is only used for removing solderballs (Section 4.4.5). Cleaning in a plasma atmosphere has been proposed and is being investigated; at the time of writing (1993) a pilot plant, based on a combination between plasma treatment and water vapour absorption, is in operation. Wavesoldering machines incorporating a plasma pretreatment stage are commercially available.[15] The industrial use of cleaning by placing the assembly in a pressure vessel filled with supercritical carbon dioxide, at a pressure of 140 bar/2000 psi–540 bar/8000 psi, and at a temperature of 40 °C/104 °F–50 °C/122 °F has been reported.[16]

8.3.1   Organic solvents

Ever since rosin-based fluxes were introduced for wavesoldering in the fifties, organic solvents have been used for removing their residues whenever that was necessary. The lower alcohols are excellent organic solvents for both rosin and its usual activators. Isopropyl alcohol (isopropanol) is in fact the standard solvent base in fluxes containing rosin or synthetic rosin substitutes. The flammability of these alcohols has in the past been to some extent a deterrent against their industrial use for cleaning soldered assemblies, but recent developments have helped to overcome this resistance against them (Section 8.3.7). It must also be remembered that large quantities of alcoholic flux solutions are in daily use worldwide in soldering machines, in close proximity to the electric preheaters and to molten metal. Such mishaps as do occasionally occur are mainly due to some operational misjudgement.

In contrast to the alcohols, halogenated hydrocarbons are only slightly flammable or non-flammable. They are therefore widely used as industrial solvents for organic substances such as rosins, oils, paints or rubbers. Halogenated hydrocarbon solvents are for the most part simple hydrocarbons such as ethane, methane or propane, with some or all of their hydrogen atoms replaced by a halogen, mostly chlorine and/or fluorine. A CHC contains chlorine and hydrogen apart from carbon; CFCs contain only chlorine, fluorine and carbon, but no hydrogen. Chlorinated hydrocarbons have been in use for a long time, while fluorinated ones are relative newcomers. The attraction of the low or non-existent flammability is outweighed by their pronounced health risk, especially of the chlorinated hydrocarbons. Most of them are either poisonous or narcotics, or both.

The rise and fall of the CFCs

About twenty years ago, the non-polar hydrogen-free chloro-fluoro-carbons (CFCs) and in particular CFC 113* ($C_2F_3Cl_3$), trichloro-trifluoro-ethane, which is known by a number of tradenames such as Freon or Arklone, became industrially available. CFC 113 in particular quickly became very popular for defluxing, for a number of reasons. It is non-flammable, chemically very stable and not toxic, nor does it smell. It has a conveniently low boiling point, and the flux-loaded solvent can easily be recovered on site by distillation, and recirculated. It does not attack the circuit board, the solder resist, the printing on the board nor the components. Its relatively high density improves the dynamics of ultrasonic or pressure-jet assisted cleaning. Above all, it is, or was, freely available at a reasonable cost.

By itself, CFC 113 is not an especially good solvent for rosin, but adding a small amount of ethyl alcohol or methyl alcohol makes it into an efficient cleaning solvent for removing the residue from activated rosin fluxes. Its worldwide consumption in 1989 was estimated at 200 000 tonnes, fifty per cent of which was used for defluxing soldered electronic assemblies.

Unfortunately, the chemical stability of CFC 113 and its relations turned out to be their undoing. They sooner or later evaporate, and their vapour eventually reaches the upper layers of the atmosphere. There it contributes to the global

---

*This shorthand classification was originated by the DuPont Chemical Company in the thirties for proprietary reasons, and it is now used worldwide. If the substance contains chlorine, the first capital letter is C. The second is F, if fluorine is present; the third is H, if any hydrogen is left in the molecule. The last C stands for carbon. Of the numerals, the first is the number of carbon atoms minus one, the second the number of hydrogen atoms plus one, the last the number of fluorine atoms.

warming of our planet, and causes the depletion of the ozone layer, both of them effects of the utmost environmental menace (Section 8.3.5). For that reason, the use of CFCs and related solvents, not only for cleaning but for refrigeration and as propellants in aerosol packages, has come under severe attack. Their use is already restricted and will be phased out completely in the near future (Section 8.3.6).

Nevertheless, the CFCs and their less virulent, but still in the long run unacceptable, cousins have to be considered here, together with the various types of cleaning equipment in which they are used and the operating precautions which must be observed with them. The reason for their inclusion in this book is the large investment which these installations represent, and the possibility of using this plant with less objectionable, even if in the long run equally inadmissible, alternative fluorine and/or chlorine containing solvents. Industry will have to live with them and use them for the time being, while taking every possible precaution to prevent or at least minimize the escape of these solvents into the environment by observing certain rules and practices of handling CFCs, which are outlined later.

Table 8.4 lists some important properties of the two main halogenated hydrocarbons which were, and for the present still are, important cleaning solvents in the electronics industry, together with the properties of ethyl alcohol and isopropyl alcohol, which are both flammable but are very good solvents for flux residues. Water is added to the list for comparison, since it forms the basis for a growing range of cleaning processes. There are many other halogenated hydrocarbon solvents, some of them with very interesting properties and excellent solvent power, but most of them are disqualified from industrial use in the electronics industry. Examples of these are methylene chloride, which attacks

Table 8.4 *Some solvents and their main properties*

Solvent	Density	Boiling point	Flash point*	MAC ppm**	Polarity
113-Trichloro-Trifluoro-Ethane (CFC 113, 'Freon', 'Arklone')	1.56	47.6 °C 117.7 °F	n.a.	1000	non-polar
111-Trichloro-Ethane (CHC 111, Methyl-Chloroform, 'Genklene', 'Inhibisol')	1.46 165.4 °F	74.1 °C	n.a.	350	non-polar
Water	1.0	100 °C 212 °F	n.a.	n.a.	polar
Methyl alcohol (methanol)	0.791	65.1 °C 149.2 °F	12 °C 54 °F	200	polar
Ethyl alcohol (ethanol)	0.789	78.5 °C 173.3 °F	14 °C 57 °F	1000	polar
Isopropyl alcohol (isopropanol)	0.786	82.4 °C 180.3 °F	13 °C 55 °F	400	polar

*Tag closed cup flashpoint
**MAC = Maximum allowable concentration (see Section 8.3.4)

many polymers and marking inks, and trichlorethylene, which poses severe health risks.

Flammability

Flammability of a given substance implies that a fire risk is involved in using or storing it. Once a substance catches fire, it is not the solid or the liquid itself which burns but its vapour. Combustion needs oxygen to start and to proceed. Hence there are so-called flammability limits for any vapour: with too much vapour in a given space there is not enough oxygen to start and sustain combustion; with too little vapour there is not enough of it to start and maintain combustion.

The 'flashpoint' of a substance gives a measure of the fire risk involved in using it. It is the lowest temperature at which the combustible liquid (or solid) gives off enough vapour to enable an air–vapour mixture within the combustibility limits to ignite on the application of a flame. The 'Tag Closed Cup' test is designed to determine that temperature. A sample of the substance is placed in a closed container with an aperture in its lid. The temperature is raised slowly until a small flame which is applied to the aperture at regular intervals manages to ignite the vapour which issues from that aperture.

Liquids with a flashpoint below 37.8 °C/200 °F are commonly classed as 'flammable'; with a flashpoint above that limit they are classed as 'combustible'. Flammable liquids are known as class I liquids, combustible ones fall into classes such as II, IIIA and IIIB. Regulations concerning handling and working with flammable or combustible liquids vary from country to country. In the US they fall within the competence of the US National Fire Protection Association.

If a flammable or combustible liquid is present in the form of a spray or mist, i.e. an aerosol, there is the risk of an explosion, that is of a very rapidly spreading combustion, if that aerosol is ignited locally, even if its bulk is below the flashpoint of the liquid involved. The temperature of each individual droplet can rise very rapidly to above the flashpoint, and the flame, once started, travels quickly through the aerosol. The working of the internal combustion engine, aptly called 'explosions-motor' in the German language, is based on this phenomenon.

In the context of solvent cleaning, jets or sprays of non-flammable but combustible liquids with flashpoints above 37.8 °C/200 °F, such as terpenes, pose an explosion risk. Equipment using such liquids in the form of jets or sprays must be made explosion-proof, or the jet–cleaning section of the machine must be filled with nitrogen (see Section 8.5).

Solvent power

CFC 113 and CHC 111, both of them non–polar, are useful solvents for rosin and synthetic resin, but not for the polar additives which all activated rosin fluxes contain. However, unless these activators are removed together with the rosin, they or a portion of them will remain on the cleaned circuit board, usually in the form of a whitish deposit. This not only makes the board look unattractive and sometimes unsaleable, but the polar activators, now no longer safely encapsulated in the neutral rosin, can cause corrosion and leakage currents in the presence of moisture. Thus, a board cleaned in an unsuitable CFC or CHC is worse off than it was before cleaning.

Alcohols, being polar, are excellent solvents for all activators (see Section 8.2.1) and one or another of the alcohols given in the list is therefore added to the above non-polar solvents to enable them to remove all of the flux residue. The ratio in which the alcohol is added to the non-polar solvent must be such that during boiling the composition of the mixture does not change, and this means that the vapour must have the same composition as the boiling liquid. For a given solvent and a given alcohol there is just one such ratio, and mixtures of this type are called 'azeotropes'. In all solvent-cleaning processes which involve heating or boiling, only azeotropes should be used. Otherwise, the composition of the solvent would have to be constantly monitored and corrected. Table 8.5 lists some commonly used azeotropes together with their solvent power for rosin.

The kauri–butanol value

The kauri–butanol (K_b) value given in Table 8.5 is a commonly used indicator for the solvent power of a given solvent for resinous substances. It is an arbitrary unit, measured with the aid of a standard solution of kauri-gum in butyl alcohol, hence its name. The higher the K_b value of a given solvent, the greater is its solvent power against not only rosin as used in fluxes, but also other resinous substances. For this reason, solvents with high K_b values like CHC 111 must be tested before being used on a given assembly, to make sure that they do not damage the circuit board, its printed markings, or any component housings.

The relevance of the K_b value as a yardstick for the defluxing efficiency of a given solvent is open to doubt. It was originally conceived as a measure of the

Table 8.5   *Some commonly used azeotropes and their solvent power for rosin*

Solvent	Boiling point	Solvent power K_b value
CFC 113	47.6 °C/117.7 °F	27
CHC 111★★	74.1 °C/165.4 °F	124
CFC 113/Methanol azeotrope	39.6 °C/103.3 °F	45
CFC 113/Isopropanol azeotrope	46.4 °C/115.5 °F	47
CFC 113/Methanol/Dichloroethylene azeotrope★	38.4 °C/101.1 °F	not reported
CFC 113/Ethanol/Dichloroethane azeotrope★	45.0 °C/113.0 °F	not reported
CHC 111/Isopropanol azeotrope★★	72.0 °C/161.6 °F	126
Isopropanol	82.4 °C/180.3 °F	infinite

★Reduced ozone depletion potential
★★Very low ozone depletion potential
See Section 8.3.5

solvent power of hydrocarbon solvents for gums and rosins in comparison with that of butyl alcohol (butanol) for kauri-gum. This gum is a hard, fossil rosin totally different from the rosins or their synthetic substitutes as used in soldering fluxes. More relevant for defluxing efficiency are various solubility parameters, which take, amongst other things, the speed of flux removal into consideration.[17,18] Nevertheless, the K_b value continues to be used as the main indicator of the defluxing efficiency of a given solvent.

Environmental acceptability

When judging the environmental acceptability of a cleaning solvent for industrial use, its potential to cause damage can be quantified by four principal parameters. Its maximum allowable concentration (MAC) quantifies the harm which it might cause to the persons working with it. This aspect will be discussed in Section 8.3.4. Its ozone depletion potential (ODP) and its global warming potential (GWP), or in popular parlance its 'greenhouse effect', quantify the threat which the substance poses to life on our planet if it escapes into the atmosphere. Its photochemical ozone creation potential (POCP), a relative newcomer to the list of environmental villains, relates to the threat to the quality of life in our cities and industrial areas from the phenomenon of man–made 'photochemical smog'. The ODP, the GWP and the POCP form the subject of Section 8.3.5. Table 8.6 lists the environmental risk parameters of the main solvents discussed so far.

The MAC is defined as the peak concentration, in parts per million, of the potentially noxious vapour in the atmosphere at the workplace to which a person may be exposed, based on an eight-hour day and a five-day week. High levels of MAC indicate a low level of health risk.

The ODP relates the ozone depletion potential of the solvent to that of trichlorofluoromethane (CFC 11), an until-now widely used refrigerant and foam-blowing agent, which is taken as 1.00. The GWP compares the global warming potential of the solvent against that of carbon dioxide, the universally known 'greenhouse gas', which is taken as 1.0. The POCP relates the photochemi-

Table 8.6   *Environmental risk parameters of the principal CFCs*

Solvent	MAC (in ppm)	ODP	GWP	POCP
Methanol	300	zero	zero	12
Ethanol	900	zero	zero	27
Isopropanol	800	zero	zero	15
CFC 113	1000	0.85	2100	zero
CFC 113/Methanol azeotrope	300	0.75	not reported	not reported
CFC 113/Ethanol azeotrope	900	not reported	not reported	not reported
CHC 111	350	0.13	37	zero

cal ozone creation potential of a given solvent to that of isoprene, a naturally occurring volatile organic compound (VOC), which is arbitrarily taken as 100.

Solvents based on CHC 111 have the great attraction of having a much lower ODP than CFC 113. The GWP of CHC 111 and its azeotropes, which is a measure of their contribution to the 'greenhouse effect', is by more than one order of magnitude lower than that of CFC 113. Nevertheless, neither of them is zero, and therefore they are under notice of restriction or phase-out.

The POCP of the CFCs, as well of the other non-flammable halogenated solvents, none of which contain oxygen, is zero. On the other hand, the otherwise environmentally safe alternative solvents, like the terpenes and the modified alcohols which replace them, do have non-zero POCPs. Whether these are at all significant in practical terms is however far from certain. These aspects will be dealt with more fully in Section 8.3.5.

8.3.2 Solvent-cleaning installations

Manual cleaning

Before describing the various mechanized cleaning processes with their more or less elaborate capital equipment, it may be useful to mention a simple manual cleaning method which is based on an old-establish practice of watch- and clockmakers. This method has its legitimate uses in small establishments, for short or laboratory runs, or in touching-up work, whenever not more than 6–12 boards have to be cleaned at a time. Its main virtue lies in the important fact that it uses none of the halogenated hydrocarbons, but plain isopropyl alcohol (isopropanol).

Isopropanol is inflammable, and sensible precautions must therefore be taken, such as the avoidance of open flames, and especially a strict prohibition of smoking (which should not be allowed in any case wherever a cleaning process is carried out). It is worth remembering that most rosin-based fluxes contain isopropanol as a solvent and are therefore just as inflammable as the plain solvent. Handled sensibly, they do not represent a dangerous fire risk, as normal wavesoldering practice proves.

In order to minimize the risk, a mixture of 75% vol isopropanol and 25% vol demineralized water can be used, which is a very good solvent for rosin-flux residues at room temperature. It is used for that reason in the MIL test for checking the cleanliness of circuit boards after a given cleaning treatment (Section 8.6.2). As with all manual cleaning procedures, polythene gloves should be worn.

The equipment required consists of three flat dishes, large enough to take 1–3 boards at a time (photographers' developing dishes will serve very well), and three brushes with long, fairly stiff bristles. For small boards, a toothbrush is quite suitable; bigger industrial brushes for larger boards are obtainable from most equipment suppliers. The board to be cleaned is allowed to soak for a while in dish no. 1. It is then scrubbed with the brush, using a circular motion. The procedure is repeated in dish no. 2, with a second brush, and the final wash is given in dish no. 3, with brush no. 3. If after a while dish no. 1 becomes visibly contaminated, the solvent in it is discarded (which means it is emptied into the appropriate drum provided for the disposal of contaminated solvents by an accredited disposal company, which should be obligatory in every cleaning department).The dish is cleaned, filled with fresh solvent and becomes dish no. 3. The former no. 3 is demoted to no. 2, and no. 2 becomes no. 1. This procedure is repeated whenever it becomes necessary. To assist with the removal of flux residue from underneath large components like PLCCs, the use of a dental jet-cleaner operated with the solvent instead of water has been favourably reported. Again, it must be ascertained beforehand that the solvent does not damage the apparatus. If operated with due care and given the right circumstances, manual cleaning can be very efficient and cost effective.

Mechanized cleaning processes

This cleaning equipment which will be described in this section is designed to operate with non-flammable halogenated hydrocarbons, principally CFC 113 and CHC 111, and their azeotropes. However, the forthcoming demise of these solvents does not mean that these machines are already irrelevant. Large numbers are in daily use worldwide (80 000 in the European Union). As long as they continue to be operated, it is important that their users are not only familiar with their working principles, but that they operate them in such a way that the escape of these solvents into the environment is reduced to an absolute minimum (see Section 8.3.4). Also, the working principles, if not the actual design, of cleaning installations which operate with the alternative solvents discussed in Section 8.3.7 are the same as those on which the machines for halogenated solvents are based.

Rotary-brush cleaning

This is a simple cleaning method, often employed for defluxing the underside of wavesoldered boards which carry no SMDs. The soldered board passes over a number of successive cylindrical brushes, which are partly immersed in a bath of solvent, and which rotate against the direction of travel of the board. This so called 'kiss-cleaning' method is unacceptable for working with halogenated organic solvents because of its high vapour loss to the atmosphere. Apart from that, it is unsuitable for the cleaning of boards which carry SMDs: it does not remove flux residue from spaces close to or underneath SMDs, and in fact it could brush flux residues into dead corners.

Batch cleaning by immersion, twin-tank method

Immersion methods employ a combination of full immersion and vapourphase cleaning. A 'twin-tank cleaner' contains two adjoining solvent tanks, which are surrounded by a high-walled enclosure fitted with condensing coils well below its upper rim (Figures 8.10). The solvent in the first tank, which is heated, is kept boiling. Consequently, the space above this boiling sump is filled with saturated, heavier-than-air solvent vapour up to the level of the cooling coils which surround the freeboard of the twin-tanks at a safe distance below their top rim. Solvent

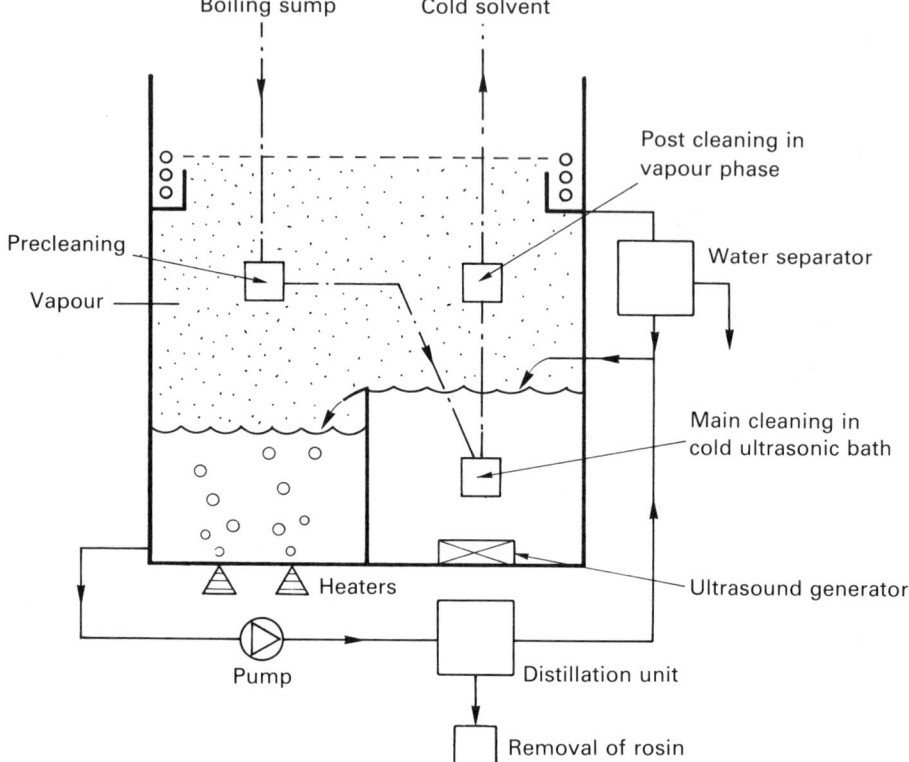

Figure 8.10 *Twin-tank solvent cleaning unit*

which condenses on these coils flows via a water separator into the second tank, which is not heated but is fitted with ultrasonic generators which are normally located on the tank bottom.

The water separator removes moisture which might have condensed on the cooling coils along with the solvent. Though halogenated solvents like CFC 113 or CHC 111 contain stabilizers which reduce decomposition, the presence of water may cause the formation of hydrochloric or hydrofluoric acid, with fatal consequences for the cleaning plant. For this reason, it is best to maintain the temperature of the cooling water in the condensing coils at room temperature. To refrigerate the cooling water is risky: drops of dew, which form on the chilled coils, can collect HCl or HF from the solvent vapour and cause perforation. From the bottom of the boiling sump, contaminated solvent is continuously or periodically withdrawn and returned to the second tank via a distillation unit, which separates out the dissolved flux residues.

The plant is operated as follows. The circuits to be cleaned are loaded into a stainless steel basket and lowered first into the solvent-vapour space above the boiling sump. Clean solvent condenses on the cold circuit boards and drains back into the sump, carrying the bulk of the flux contamination with it. For this to happen efficiently, the boards should be stacked in the basket vertically and not too close together. Once the boards have reached the vapour temperature, the basket is moved across to the second tank and immersed in the clean, cold solvent, where the boards are exposed to the effects of the ultrasonic agitation and cavitation. After a suitable interval, normally a minute or two, the basket is raised out of the solvent and allowed to remain in the vapour space for a further short period to allow dragged-out solvent to drain back into the tank and to evaporate into the surrounding vapour, thus reducing solvent drag-out. This type of twin-tank installation has proved its worth over many years, while wavesoldered boards were populated with inserted components only. It is still sufficiently effective with SMD-populated boards of simple design, with components not too closely spaced, and above all wavesoldered boards, where the stand-off gap is to a large extent filled with adhesive. For closely populated and specifically reflowsoldered boards, a three-tank installation like the one shown in Figure 8.11 is a better choice.

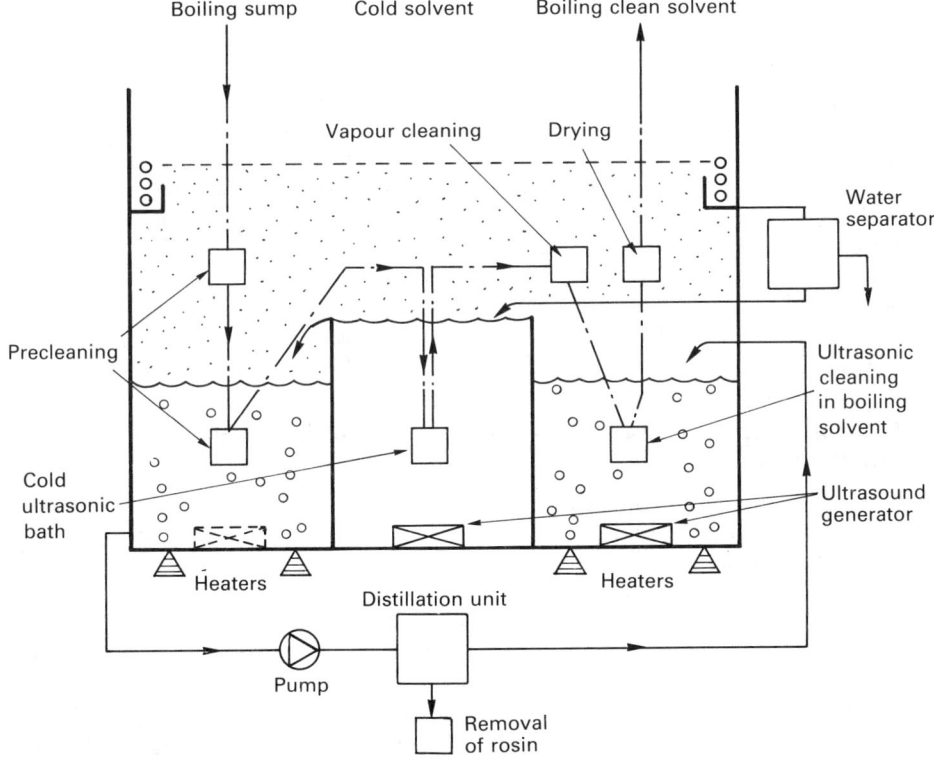

Figure 8.11 *Three-tank solvent cleaning unit*

Three-tank method

After the initial condensation-cleaning stage above the boiling sump, the circuits are lowered into the boiling sump of tank I, which is optionally fitted with ultrasonic agitation. As has been explained in Section 8.2.2, in a boiling medium ultrasonic impulses produce a strong rinsing effect in the space between SMD and the board surface. From the boiling sump, the boards move to the second tank, which is filled with clean, cold solvent. This tank too can be fitted with ultrasonic agitation, in order to remove flux adhesions from the free board surfaces between the components. It is fed with clean solvent condensate from the cooling coils, again via a water separator. After a short dwelling period, the boards move to a third tank, which contains clean, boiling solvent and is fitted with ultrasonic agitation for a final cleaning stage. From here, the circuits are lifted into the vapour space where any adhering solvent is allowed to steam off in order to reduce solvent loss.

Typical dwelling times in the various stages are reported as one minute for the first stop in the vapour volume, two minutes in the boiling sump, one minute each in tanks II and III, and a final half-minute dwell in the vapour volume before lifting the basket clear of the cleaning plant.

Energy requirements

To turn, for example, CFC 113 azeotrope into vapour, starting from room temperature, requires approximately 0.08 kW/hr per litre (0.3 kW/hr per US gallon). In actual cleaning practice, this means an installed heating power of 0.1 kW for each distilled litre of solvent per hour (0.4 kW for each US gallon distilled per hour). Depending on the size and the type of a given installation, the complete solvent content of a machine may be turned over as distillate between one to ten times every hour. Therefore, with a large plant, fitting the plant with a heat pump which cools the condensing coils and helps to drive the evaporators or to heat one of the tanks may be a sensible investment. With a small plant, it could take years to write it off, longer than the normal lifespan of the machine.[19]

Environmental safety rules for the operation of immersion cleaning installations

1. For both environmental and economic reasons, a number of measures must be taken to prevent solvent or its vapour from escaping into the working environment.
   (i) Circuit boards must be stacked vertically and not too tight in the holding basket, so that they can drain readily. This prevents not only carry-over of contaminated solvent from a dirty bath into a clean one, but also drag-out of solvent from the plant.
   (ii) The freeboard of the tank-surround must be at least three-quarters of the width of the tank, to prevent vapour from 'sloshing' over the edge of the tank when a transport basket is lowered into or raised out of the tank (Figure 8.12). Edge exhausts must on no account be fitted.

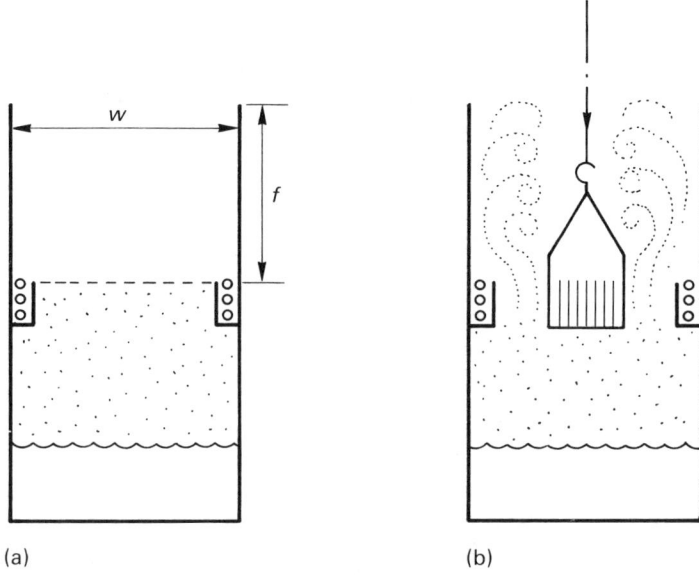

(a)   (b)

Figure 8.12 *Safety measures for immersion-cleaning tanks. (a)* h > 0.75w; (b) *avoiding the 'piston effect' when lowering a working basket*

    (iii)   Baskets must be lowered and raised slowly, no faster than 3 m/min (9 ft/min), in order to minimize the 'piston effect' (Figure 8.12). Programmed mechanical handling of the baskets is a main factor in reducing solvent loss and thus environmental damage.

    (iv)   Draughts of air in the workroom must be rigorously guarded against. When a tank or an installation is not in actual use, it must be covered. Permanently fitted sliding lids are best.

2.   All solvent-cleaning installations must be fitted with sensors which monitor and limit the bath temperature and the height of the vapour blanket above the solvent.

3.   All heated solvent tanks must be fitted with cooling coils, which are turned on as soon as the plant is shut down. Cooling is continued until the solvent has reached room temperature in every tank.

4.   Solvent must never be poured from a bucket (it should not be in a bucket in the first place) or a canister. To move solvent from one container into another, a pump, manual or motorized, must be used at all times.

5.   The formation of acid in the solvent must be monitored, and if necessary corrected. With a small installation this can be done periodically and manually, but larger installations should be fitted with automatic acid-control. In fluorinated or chlorinated solvents, ingress of water causes the formation of HF or HCl (see above) with grave consequences for the plant.

6.   Each solvent tank should be provided with a well fitting stainless steel mesh, held in a frame slightly above the bottom of the tank, so that it can be lifted

out. This screen intercepts any metallic or non-metallic items which may occasionally drop into the tank. Though almost all solvents are stabilized, a number of metals like copper, aluminium or brass are attacked when in contact with stainless steel in the presence of solvent. Tramp-metal parts can lead to local attack or perforation of the tank bottom if left there for any length of time.

8.3.3 In-line cleaning plants

An integrated soldering line is best served by an in-line cleaning machine. The relatively long dwell times which a circuit board must spend in a static or even an ultrasonically energized tank filled with solvent would make an in-line cleaning plant too long. Therefore, in-line cleaning must be assisted by more powerful means than ultrasonic energy.

High-pressure solvent jets are used to shorten both the washing and the rinsing stages of the cleaning process. In this context, it is worth remembering that the required jet pressure is determined by the SMD with the lowest stand-off height on a given board, and that doubling the stand-off height reduces the kinetic energy needed to flush a contaminant from under the SMD by a factor of eight (Section 8.2.2). Table 8.7 shows the relationship between stand-off height and solvent jet pressure.

In all in-line cleaning plants, the boards travel on a stainless steel belt of open mesh construction through successive arrays of jets, which impinge on the boards both from above and below. The pressure and volume delivery of the upper and lower jet arrays must be accurately balanced, so as not to dislodge the boards as they travel through the cleaner. The in-line cleaners at present on the market are either pure jet cleaners or combined jet and immersion cleaners.

Pure jet in-line cleaning

An example of such a system is shown in Figure 8.13. The circuit boards travel through the cleaner on an open-mesh endless stainless steel belt. Cooled, inwardly sloping entry and exit tunnels minimize the escape of the heavier-than-air solvent vapour.

The conveyor carries the boards through a cooled, downwards sloping tunnel into the machine. There they pass through three successive arrays of hot,

Table 8.7   *Relationship between stand-off height and solvent jet pressure*

| Stand-off height | | Jet pressure | |
μm	*mil*	*bar*	*lb/sq. in*
< 25	< 1	15–17	220–250
> 25	> 1	2–5	30–75

The data were established with a CFC 113 azeotrope solvent

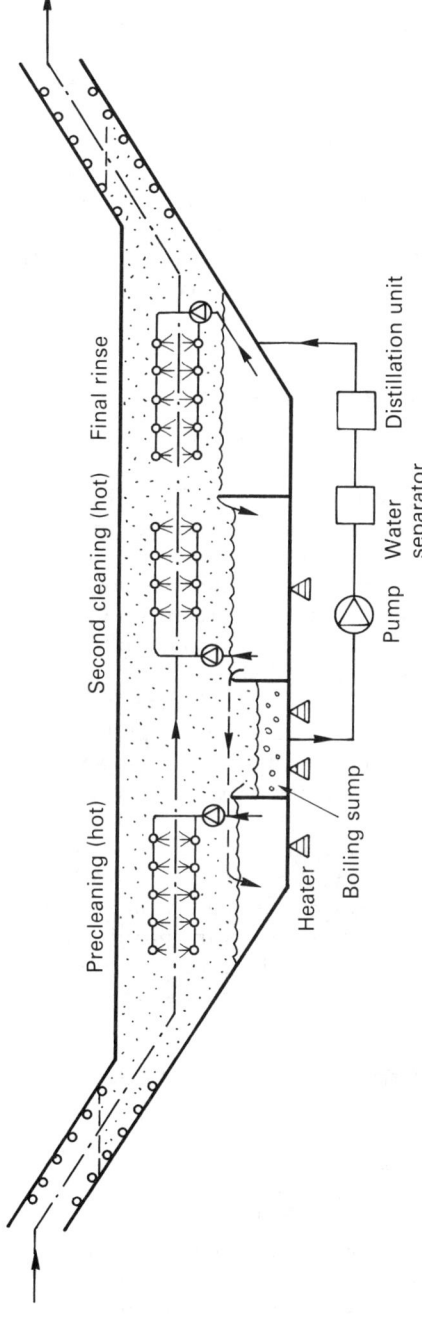

Figure 8.13 An in-line pure jet cleaning plant (OSL)

near-boiling solvent jets, each of which is fed from a tank beneath, with the boards moving in an atmosphere of practically saturated solvent vapour. Tank I is supplied with clean solvent, derived from the cooling coils of the downward sloping entry tunnel. It is followed by a boiling sump which provides the vapour atmosphere for the whole unit. A pump withdraws soiled solvent from this sump and feeds it to the exit tank III and its jets via a water separator and distillation unit. The solvent cascades from the exit tank via tank II back to the boiling sump. The boards leave the machine through a cooled, upwardly sloping tunnel.

If a machine of this type is fitted with a forced draught exhaust in order to prevent any solvent diffusing into the workroom, the exhaust duct must of course be fitted with a cooled heat exchanger to prevent solvent escaping to the outside air.

Combined immersion and jet cleaning

Again, the cooled entry and exit tunnels slope inwardly. On entering the machine, the boards dip into a flat tank filled with warm solvent and fitted with optional ultrasonic agitation. Halfway along the tank, boards pass underneath a vertical bulkhead, which forms an airlock to contain the saturated solvent vapour in the interior of the machine. The central cleaning stage consists of two jet arrays in series, which are fed with hot solvent from two boiling sumps in the bottom of this section. Having passed the jets, the boards enter a flat exit tank, containing hot, clean solvent. They dive underneath a second bulkhead and leave the machine through the upwardly sloping, cooled exit tunnel.

The solvent flows through the machine from tank to tank in a direction opposite to the travel of the boards, which leave the cleaner through a bath of clean, freshly distilled solvent. Solvent from the entry tank is pumped via a water separator into a distillation unit, which removes the dissolved flux residues and returns the clean distillate to the exit tank.

With in-line solvent cleaning machines, the drag-out of solvent by the boards which leave the machine could be a problem with closely spaced SMDs on a board. The combined immersion and jet cleaning machine shows an advantage here: the solvent in the exit tank is hot and so are the boards on leaving it. This gives the boards a chance to dry as they travel through the exit tunnel. Warming the exit spray of the pure jet cleaning machine might bring a similar improvement.

8.3.4 Halogenated solvents: safety and health

Technical aspects

Most chlorinated and fluorinated solvents intended for industrial use contain stabilizing additives, which are designed to protect them against thermal or chemical decomposition. Nevertheless, a number of safety precautions are mandatory when working with these solvents in an industrial environment.

Water contamination of the solvent – for instance through water condensate formed on cooling coils – causes the formation of hydrochloric or hydrofluoric

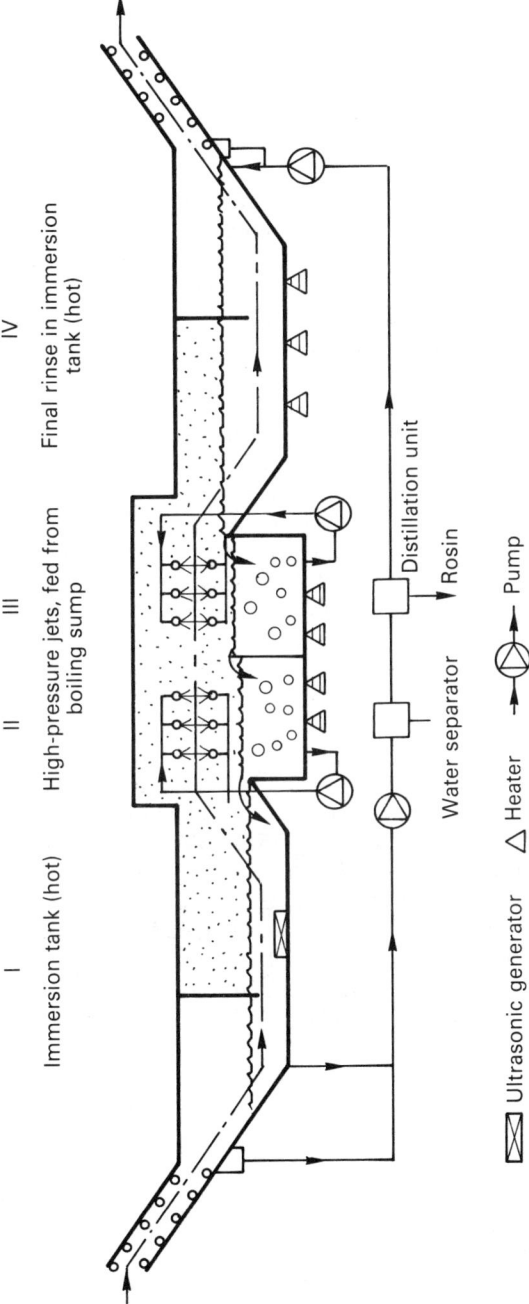

Figure 8.14 Combined jet and immersion cleaning machine (Electrovert)

acid in the system, with dire consequences for the installation. For that reason, the solvent circulation loops of both batch and in-line installations must be fitted with reliable and regularly serviced desiccators of sufficient capacity, which should be located upstream of the distillation unit. Either molecular sieves (zeolite) or silica gel can be used, and manufacturers' instructions must be carefully followed as to their regular regeneration or replacement. Gravity water separators must not be used on installations which work with azeotropic solvents: the alcoholic component of an azeotropic mixture is watersoluble, and would be removed together with the separated water. Most equipment vendors can supply simple test kits for the detection and measurement of water contamination of a CFC or CHC.

Alternatively, a simple test for excessive water content of the solvent works as follows. A test tube is half filled with the solvent to be examined, and a few grains of anhydrous cobalt chloride are added. If, on shaking, the colour of the solvent changes from blue to red, it is time to replace the zeolite or the silica gel in the desiccator.

Any undue accumulation of flux residues in the boiling sump must be avoided. If the rosin content of this sump rises above 3%, the solvent could begin to decompose, forming injurious decomposition products.[20]

All metallic components of a solvent cleaning installation should be made of stainless steel. Metals like aluminium, magnesium, zinc, copper or brass are not resistant to long-time exposure to chlorinated or fluorinated solvents. Therefore, any fittings or components added by the user to a solvent cleaning installation must be made of stainless steel.

Health aspects

The occupational exposure limit (OEL) is a measure of the health risk of a substance to a person using it, and the definition of an OEL varies from country to country and from industry to industry. With volatile substances such as solvents, the normally used OEL is the 'maximum allowable concentration' (MAC) of the solvent vapour in the working atmosphere, expressed in volume parts per million (ppm). The MAC of a volatile substance is the maximum concentration (in ppm) of its vapour in air to which a worker may be regularly exposed, on the basis of a 40 hour working week. Thus, the concentration of the vapour may be at or near its MAC during the whole working week, but the OEL must never be exceeded, for however short a time.

MAC values may vary to some extent from country to country. Generally speaking, the lower the MAC value of a given substance, the higher is its toxicity. Table 8.8 shows MAC values for some common solvents.

As the table shows, CFC 113 has the same low toxicity as ethyl alcohol. Its azeotrope with ethyl alcohol is slightly more toxic than either, while the addition of the toxic methyl alcohol drastically increases the toxicity of the azeotrope. Good ventilation of the working premises is in any case essential, preferably both at floor level, because CFC 113 vapour is heavier than air, and above the normal working height.

Table 8.8 *MAC values of some common solvents*

Solvent	MAC in ppm
Methanol	200
Ethanol	1000
Isopropanol	400
CFC 113	1000
CHC 111	350
CFC 113/Methanol azeotrope	300
CFC 113/Ethanol azeotrope	900
CFC 113/Ethanol/Methanol azeotrope	800

Above all, as has been said already, smoking must be *strictly* prohibited in any room where CFC or CHC solvents are used or handled. Both form extremely poisonous decomposition products at temperatures above 350 °C/662 °F, which means that inhaling their vapour through a burning cigarette can be very dangerous, if not fatal. Equally, neither food nor drink should be consumed in rooms where solvents are employed. Furthermore, Ellis mentions a physiological effect of exposure to the vapour of isopropanol over a full working day.[20] Absorption of this vapour is cumulative and additive to the effect of ethyl alcohol in terms of slowing physiological reaction times. This means that just one aperitif at the end of a full working day which has been spent in an atmosphere containing isopropanol vapour can have a pronounced effect, and he advises drivers to be aware of that. He also recommends that anybody suffering from a drink problem should not take on a job which involves the handling of organic solvents or exposure to their vapours.[21]

8.3.5 The three environmental threats

In recent years, scientists, industrialists and politicians have become critically aware of the effects of our industrial activities on the living environment of our planet. The organic solvents which concern us here are just a few amongst a very large number of environmentally damaging substances. Being volatile, their vapours enter the atmosphere, where they make their effects felt. Some of these effects have recently been recognized as potentially disastrous.

All halogenated solvents are potentially dangerous, the CFCs more so than the rest. The tonnages in which they have been used until recently were considerable. The average annual global consumption of halogenated hydrocarbons in the late eighties amounted to approximately five million tonnes, which is 1 kg/2 lb for every person on this planet. Amongst these, the CFCs accounted for about one million tonnes in 1986, with seventy-five per cent consumed in Europe and the United States. Twenty per cent of this tonnage was used as solvents, while the rest went into refrigerating and airconditioning plant, aerosol cans and foamed plastics, in roughly equal parts.[22] Eventually, all of this tonnage finishes up in the atmosphere.

The ozone layer

Ozone (O_3) is a colourless gas with a distinctive, sharp odour (its name derives from the Greek word for 'smelly'). It is highly toxic, causing damage to mucous membranes when above 10 ppm in air. At ground level, ozone is present in the atmosphere in minute quantities, but its concentration increases at higher levels until in the stratosphere, at about 28 km/92 000 ft above ground level, it reaches a maximum of approximately 8 ppm by volume. At greater heights, the concentration tails off again.[23]

To speak of an ozone 'layer' is therefore not quite accurate, but in the public mind the term has created a powerful image of its concrete existence and of the threat to it; thus the term serves a useful purpose. The ozone layer's crucial importance for life upon earth is its function as a powerful filter for UV radiation below the wavelength of 320 μm. Radiation within that range has many different effects on humans, animals, plants and also man–made materials. Most of them are harmful and some of the damage that can be caused will be discussed presently.

Ozone owes its presence in the stratosphere to a delicate balance between the processes of its formation and of its depletion. Ozone is formed from oxygen present in the stratosphere through a photochemical reaction, under the influence of the intense UV radiation which prevails at those heights. Its depletion is due to a breakdown of two ozone molecules into three oxygen molecules through a series of steps, initiated by the catalytic effect of certain 'free radicals' like Cl or NO, which are naturally present in the stratosphere. These free radicals have diffused upwards as products of naturally occurring processes down at ground level, such as the decomposition of organic matter.

The arrival in the stratosphere of additional amounts of man-made free radicals, like Cl, NO, and also Br, has now tipped the balance towards the decomposition of ozone. Measurable overall depletion has set in. It is particularly severe in the Arctic regions, and especially in the Antarctic where the ozone layer periodically disappears completely (ozone hole). In the latitude of the UK too, the ozone loss since 1969 has been estimated at approximately 7%.[24] By now, public awareness of the ozone problem has risen to a level where ozone figures are given in the daily press.

The situation is aggravated by the catalytic nature of the depletion process, which means that the offending radicals remain in the stratosphere for a long time.

It has been estimated that one Cl radical in the stratosphere can destroy 10^4 to 10^5 ozone molecules before diffusing back to the troposphere, where 'the weather happens'.[25] Once down there, at a height below 12 000 metres/36 000 feet, it eventually comes back to earth as acid rain.

At this point is must be emphasized that the account given here of the mechanisms of ozone depletion is a highly simplified and by no means authoritative summary of very complex chemical and physical processes. A more detailed and authoritative overview of the subject has been written specifically for the solvent–using electronic industry by Dr Colin Lea (see Reference 1). Table 8.9 shows the ozone depletion potential of some solvents.

The damage potential of the more intense UV radiation which will reach the surface of the earth as the ozone filter gets weaker is very great, as can be seen from the following. Primitive life forms are endangered, especially plankton in the surface layer of the oceans. Fewer plankton not only affects the whole marine food chain, but also restricts the role of the oceans as a global sink for CO_2, which the plankton normally assembles into organic matter. The loss of this mopping-up mechanism for CO_2 increases in turn the threat of the 'greenhouse effect' which will be dealt with presently. Meanwhile, further problems due to increased UV exposure include a higher risk of skin cancer for people with light coloured skin, suppression of the body's immune system and increased damage to the eyes.[26]

Global warming

The presence of CFCs in the atmosphere poses a further environmental problem. The radiation received on earth from the sun extends from the UV part of the spectrum over the visible light into the near infrared wavebands. The radiation re-emitted from the surface of the earth lies in the middle and far infrared. Much of this outgoing infrared is absorbed in the atmosphere (see Figure 5.19), which thus forms a warming blanket, maintaining the more or less comfortable temperature to which life on the planet has adjusted itself (except for glacial periods, the last one of which ended about 10 000 years ago).[27]

Like the maintenance of the ozone layer, the thermostatic control of the annual average global temperature is a matter of delicately balancing several parameters, in this case the transmission and retention of infrared, that is warmth, which is reflected back from earth into space. The main naturally occurring infrared absorbing gases, now termed 'greenhouse gases', are water vapour and CO_2. Since the start of the industrial revolution, steadily growing amounts of carbon dioxide have been added to the atmosphere, most but not all of which can still be absorbed by the naturally occurring sinks like the oceans. The rise in the average surface

Table 8.9   *Ozone depletion potential of some solvents*

Solvent	Ozone depletion potential (ODP)	Stratospheric lifetime
Alcohols	nil	nil
CFC 11 (Freon 11)	1.00	281 years
CFC 113	0.85	138 years
CHC 111 (Methyl chloroform)	0.13	6.3 years

CFC 11 is a widely used low-boiling solvent, though not used for cleaning soldered electronic assemblies. Its ODP = 1.00 has been chosen as an arbitrary reference point

temperature of the earth by 0.5 °C/0.9 °F over the last century has been attributed to the rising level of CO_2.

CFCs are a newcomer of deadly efficiency upon this scene. On a tonne-for-tonne basis, CFCs have 6000 times the warming potential of CO_2, because just within the wavelength window between 8 and 13 μm, where the carbon dioxide allows the infrared radiation to pass through, the atomic bonds of C-Cl and C-F absorb it.[28]

All fluorinated and chlorinated solvents have been given ratings to indicate their potential to inflict damage on the environment, based on experimental evidence. The rating figures refer to their respective ozone depletion potential (ODP) and their global warming potential (GWP), both potentials being rated against those of CFC 11 (Trichlorofluoro-Methane) arbitrarily taken as 1.0. CFC 11 is a low-boiling CFC solvent which is not normally used in the electronics industry (Table 8.10).

The projected effects of global warming, like all projections, are still a subject of debate, as is also the question whether the recent apparent changes in the weather pattern are the beginning of a trend or due to statistical scatter. The rise of the mean sea level through thermal expansion of the oceans, as they get warmer, could be an early effect. On the other hand, a melting of the polar ice caps is thought to take a long time. Moreover, according to a recent study, a rise in the average annual temperature leads in fact to thicker polar ice caps. This is explained by the increase in precipitation and a consequently increased snowfall in the polar regions, with the Greenland icecap thickening by 0.45 mm and the Antarctic one by 0.75 mm.[29]

What seems certain is that a shift in climatic behaviour over large land areas will have early and serious consequences for farming and crops on a global scale. All this has strengthened the arguments which motivated the delegates at the London Conference 1990 in hastening the regulatory process concerning the halogenated solvents.

Photochemical smog

The 'photochemical ozone creation potential' (POCP) of the vapour of a volatile organic compound is a further worry of environmentalists, for the following reason. In contrast to the stratospheric ozone, which protects life down below, ozone at street level is a noxious poison. Exposure to more than 0.1 ppm of ozone in the breathing air for longer than one hour is considered dangerous to a healthy adult, while 10 ppm in air are positively poisonous. Ozone concentrations at 10^{-10}

Table 8.10 *GWP values of some solvents*

Solvent	Emission (1985) in kilo tonnes/yr	GWP
CFC 11	280	1.00
CFC 113	138	1.35
CHC 111	8	0.024

to 10^{-1} levels are one of the constituents of so-called photochemical smog, a noxious cocktail of ozone together with organic peroxides, nitrates and other unpleasant chemicals. The smog and its ozone constituent are termed 'photochemical' because they are formed from otherwise less harmful, if not innocuous, vapours of organic compounds by the photochemical action of sunlight upon them.

Photochemical smog is mainly an urban and local industrial problem, affecting cities such as Los Angeles, Mexico City and Athens, or areas of dense industrial conglomerations. The principal man-made culprits are the exhaust gases from cars not fitted with catalysers, the emissions from petrol refining and from the inefficient burning of coal or oil, but there are many other minor ones as well. Recently, even isopropanol, the most commonly used solvent in soldering fluxes, has come under attack (Section 3.4.4).

The compounds whose vapours are liable to be affected in this manner are collectively known as 'volatile organic compounds' (VOCs). Their POCP is measured in arbitrary units, from zero to 100, the latter being the POCP of isoprene, a constituent of natural and synthetic rubber. At the present state of the art, there is no general agreement on the method of determining or computing the POCP of a given volatile organic compound. What is certain is that the POCP of all CFCs and their relatives is zero. Unfortunately, on the other hand, isoprene with its POCP of 100 is a building block of most terpene molecules, the terpenes being one of the classes of organic solvents which at present are among the most favoured substitutes for the CFCs. So far, no POCP values for commercial defluxing terpenes have been published. Alcohols, too, some of which are as promising as the terpenes for the purpose of defluxing, are classed as VOCs, and have their specific POCPs (Section 8.5.3).

In judging the environmental risk posed by a given VOC, its volatility is a decisive factor. In this context, the sparingly volatile defluxing terpenes and modified alcohols, which are such promising alternative defluxing solvents, seem fairly innocuous.

To put the situation further into perspective, the quantities of other VOCs with non-zero POCPs which enter the atmosphere on a global scale ought to be considered. All plants release terpenes into the atmosphere, and their global annual terpene emission amounts to probably gigatons, admittedly much of it far from urban habitations. Furthermore, the vapour emitted by VOCs used for defluxing in electronic manufacture, most of which is carried out in enclosed plant, is infinitesimally small compared with that given off by non-catalysed car exhausts and the inefficient burning of coal and oil refining. Nevertheless, the VOCs have attracted the attention of the various international bodies concerned with the protection of the global environment, and along with them the terpenes and alcohols, as will be discussed in the next chapter.

8.3.6 Restrictions on solvent usage

The nature and size of the environmental damage caused by the emissions from the various halogenated solvents have now been recognized as global concerns. The

United Nations Environmental Program (UNEP) in 1981 set up a cooperative framework for monitoring, research and information exchange on all matters concerning the ozone layer and related global topics.

In September 1987, UNEP called a convention in Montreal, Canada, with the aim of agreeing on regulatory measures concerning the production and use of halogenated solvents. The outcome was the 'Montreal Protocol', signed by the EC countries, the USA, Japan and 21 other countries. The then USSR and six further countries signed later. The Protocol relates to the present and future use of, amongst other substances, five CFCs (some of which are not only used as solvents but also as aerosol propellants, blowing agents for plastics and refrigerants). As far as solvents are concerned, the Protocol and its successors directly affect the cleaning practices of the entire electronic assembling industry.

In June 1990, representatives from 92 countries met in London under the auspices of UNEP for a ten-day conference to agree on a redrafting of the 'Montreal Protocol'. As a result, all fully halogenated CFCs will be phased out by the year 2000, while the consumption level of not fully halogenated solvents such as methyl chloroform (CHC 111) is to be reduced by seventy per cent by that date, before being fully phased out by the year 2005. Not the least problem of the conference was the need to persuade industrially developing countries like China and India, who between them account for forty per cent of the world's population, to agree to these costly environmental measures, and for the developed nations to find ways to internationally fund them.

Within the EC countries, the UNEP restrictions are constantly being updated, the last conference at the time of writing (1993) having taken place in December 1992. The EC phasing-out schedule reads now as follows:

CFC 113   75% phase-out by end 1993
          complete phase-out by 1995

CHC 111   Complete phase-out by end 1995

HCFCs   By 1996 reduced to <3.1% of the 1989 CFCs' plus HCFCs' ODP-weighted tonnage. Complete phase-out by 2030.

Regulatory measures to be taken concerning the VOCs are still being vigorously debated within the UNEP committees. The VOCs have been classified into four classes, graded according to their POCP values, with most of the alternative defluxing solvents within the lower categories. Solvents may come under restriction according to their area of use rather than as specific solvents. Such areas are for instance car painting, printing, industrial painting, wood preservation and dry cleaning. Defluxing is not among them. A full account of these developments is found in Reference 1.

8.3.7   Non-flammable organic solvents with reduced environmental risks

Two features combine to give the CFC molecules their high environmental damage potential: their chlorine content and the high stability of their structure.

The chlorine causes the destruction of the ozone in the stratosphere, and the high stability of the molecule helps the chlorine to get there and to stay around for a very long time. Replacing one or more of the halogens by a hydrogen atom destabilizes the molecule to some extent, so that it decomposes in the troposphere, at a height below 12 000 metres/36 000 feet, before reaching the stratosphere. On the other hand, any halogen set free in the troposphere will eventually come back to earth in the form of acid rain (or something worse, see below).

To avoid confusion among the various alternatives to fully halogenated solvents, Table 8.11 lists the elements present in the different classes of halogenated organic compounds.

Though the CHC 111 molecule has a shorter life in the atmosphere than the CFCs, the behaviour of the Arctic and Antarctic ozone holes has become more alarming and the Cl content of 111 poses a sufficiently serious threat for the 1992 London Conference to reduce its use to zero by the end of 1995.

The 1991 conference of the European Community moved the reference year from 1986 to 1989.

Of the large number of possible non-flammable HFCs and HCFCs, only a few have been evaluated so far for their use in defluxing.[30] These are given in Table 8.12.

The new HCFCs such as HCFC 123 and azeotropes of HCFC 141b have lower ODPs and GWPs than CHC 111, and the parties to the revised Montreal Protocol have agreed to describe them as 'transitional' substances. They called for their 'prudent and responsible use', prior to their phasing out between 2020 and 2040, by which time the CFCs and CHCs will already have disappeared.

Furthermore, some academics are suspicious of the HCFCs. Not only do they fear an intensification of acid rain, but also the potential formation of trifluoroacetic acid (TFA), a possible byproduct of the breakup of the HCFC molecule, which the rain could bring down to earth.[31] The fear is that TFA could enter the food chain through the roots of plants, and thus endanger health.

Pentafluoro-propanol (tradename Pefol) is in a different category: it is a fluorinated, non-flammable propyl-alcohol, which is produced in Japan. At the time of writing (1993) the commercial availability of these solvents is limited, and evaluation is still proceeding.

Table 8.11 *Nomenclature of halogenated organic solvents*

| Class | Elements present in the molecule | | | |
	Hydrogen	Chlorine	Fluorine	Carbon
FC	−	−	+	+
CC	−	+	−	+
CFC	−	+	+	+
HCFC	+	+	+	+
HCC	+	+	−	+
HFC	+	−	+	+

Table 8.12   *Nonflammable alternatives to CFC 113 and CHC 111*

Solvent	Boiling point	K_b value	Atmospheric life (years)	ODP	GWP (CFC 11 = 1.0)
CFC 113	47.6 °C/117.7 °F	27	90	0.85	1.3
CHC 111	74.1 °C/165.4 °F	124	6.3	0.13	0.02
HCFC 123	28.0 °C/82.4 °F	68	1.6	0.02	0.02
HCFC 141b	32.0 °C/89.6 °F	61	12	0.15	0.18
123/141b-Methanol azeotrope	30.0 °C/80.0 °F	85	—	—	—
'Pefol'★	80.7 °C/177.3 °F	36	—	0	—

★'Pefol' is the tradename of pentafluoro-propanol, a propyl alcohol with five of its hydrogen atoms replaced by fluorine

Once these CFC 113 substitutes have proved their worth and are fully commercially available, they could be used as drop-in substitutes for CFC 113 in existing cleaning installations, thus extending the useful life of sometimes large investments for a limited number of years. Meanwhile, CHC 111 can be used in these installations, with all due handling precautions, for a short while, though with a steadily declining availability.

8.3.8 Flammable solvents

Isopropyl alcohol (isopropanol) is one of the best solvents for all flux residues, with a clean environmental bill (except for its POCP, Section 8.3.4). As with all lower alcohols (Table 8.13), its ODP is zero, its vapour is not listed under the greenhouse gases and its health risk is low. However, it is flammable, and therefore it requires specially designed flame-proof and explosion-proof, and consequently more expensive, cleaning equipment.

Amongst the lower alcohols, isopropanol is the normal choice not only as a solvent for liquid fluxes, whether rosin-based or rosin-free, but also for defluxing. Methyl alcohol is toxic and can cause blindness. Ethanol, which is the alcohol contained in wine and liquor, carries a high tax in most countries, unless made unpalatable. There is also the risk of addiction in operators. n-Propanol is more toxic and more expensive than isopropanol; it is mainly used in countries where isopropanol carries a tax.

One way of rendering explosion-proof a defluxing line which runs with isopropanol is to fill its interior with nitrogen. With a conveyorized in-line machine (Figure 8.14) with immersion in a cold, ultrasonically agitated bath, this is a practical possibility. The oxygen level in the nitrogen atmosphere need not be as low as in a nitrogen-filled wavesoldering machine. On the other hand, the

Table 8.13   *The lower alcohols*

	Boiling point	Flash point	MAC (ppm)
Methyl alcohol (Methanol)	64.5 °C/ 148 °F	11 °C/ 52 °F	200
Ethyl alcohol (Ethanol)	78.2 °C/ 172.8 °F	14 °C/ 57 °F	1000
Isopropyl alcohol (Isopropanol, IPA)	82.3 °C/ 180.1 °F	17 °C/ 33 °F	400
n-propyl alcohol (n-propanol)	97.2°C 207.0 °F	22°C 42 °F	220

electrical side of the machine must be fully flame- and explosion-proof, which adds to the cost of the installation.

Another method of rendering boiling isopropanol explosion-proof by adding to it a high-boiling perfluorocarbon (tradename Flutec) has recently been proposed.[32] The name 'perfluorocarbon' implies that the molecule contains only carbon and fluorine, but no hydrogen. The ozone depletion potential of Flutec is zero, though its vapour is a greenhouse gas.

Flutec itself is not miscible with isopropanol, but the vapours of the two fluids are, and the vapour mixture is not flammable. Plant for putting this process into practice is commercially available. Because of the GWP of the perfluorocarbon and its high cost, the design of the plant aims at minimum vapour loss.

The low boiling points of the lower alcohols and their low heat of evaporation make it easy to incorporate a re-distilling facility into any cleaning plant working with alcohol. This avoids disposal problems of spent solvent, and adds to the environmental attraction of alcohol cleaning.

8.4   Cleaning with water

8.4.1   *Chemical and physical aspects*

Water is the universal and most readily available inorganic polar solvent. Using water instead of an organic solvent for defluxing soldered circuit boards solves the problems of ozone destruction, global warming and flammability, but raises several others.

Chemical aspects

Rosin and residues from most wavesoldering oils are practically insoluble in water. Therefore, alkaline additives, so called saponifiers, must be added to the washing water to render these residues soluble. Rosin-free, fully watersoluble fluxes, instead of rosin-containing ones, and watersoluble wavesoldering oils avoid this problem: the washing water needs no saponifiers and, as a further bonus, the dissolved residue of a watersoluble flux is more mobile and flushed away more readily than that from a saponified rosin. This means less rinsing effort, cleaner boards, possibly a smaller washing unit with a lower energy consumption, and fewer effluent problems.

Therefore, there is a strong case for choosing a rosin-free flux with a fully watersoluble residue for soldering boards which for some good and unanswerable reason must be cleaned. Removing such residues with plain water is simpler, and a high degree of cleanliness can be achieved with less trouble and expense. As an added bonus, a more active watersoluble flux can be used than would be admissible with a rosin-containing flux, and this means fewer faulty joints, less expensive rework and a more reliable product.

Physical aspects

Table 8.14 lists the values of some physical constants which are relevant to cleaning for water, isopropanol, CHC 111 and CFC 113. Their significance, as far as cleaning is concerned, is as follows:

1.   Water has a higher surface tension than the organic solvents. Therefore, it is retained more firmly in a given gap, such as under an SMD, and it requires more effort to push it out in order to achieve efficient rinsing.
2.   Chlorinated and fluorinated solvents have a higher density, hence a solvent jet of a given pressure and diameter has a higher kinetic energy than an equivalent water jet. Therefore, water-jet cleaning machines must operate at higher pressures than solvent-jet cleaners.
3.   Water has a much higher heat of evaporation and is much less volatile than organic solvents. This means that drying a circuit board washed with water needs much more energy, and therefore costs more, than drying a solvent-cleaned board. In fact, drying the latter is mostly taken care of by the stray heat present in a solvent cleaning installation, so that normally no special drying stage is needed.

Table 8.14   *Comparison between the physical properties of water, isopropanol, CHC 111 and CFC 113*

Parameter	Water	Isopropanol	HC 111	CFC 113
Boiling point	100 °C/	82.3 °C/	774.1 °C/	447.6 °C/
	212 °F	180.1 °F	165.4 °F	117.7 °F
Density at 20 °C/68 °F	1.0	0.79	1.46	1.56
Surface tension $(N\,m^{-1})$	72	22.6	26	19
Viscosity (mPA sec)	0.40	2.40	0.42	0.52
at	70 °C/	20 °C/	73 °C/	40 °C/
	158 °F	68 °F	163 °F	104 °F
Heat of evaporation (cal/g)	539	163	35	53

For these reasons, an in-line water washing machine consumes more energy and must be longer than a solvent-cleaning one with comparable output. At a given travelling speed, a board must receive a higher amount of kinetic and thermal energy for the same cleaning and drying effect. A large in-line water washing installation may be rated at 25 kW or more. Alternatively, a batch operating water washing machine demands a longer dwelling time per charge than an equivalent solvent plant. Both aspects imply a higher demand on expensive factory space compared with solvent cleaning. All this adds up to the fact that running a water cleaning plant needs more thermal and kinetic energy, and thus costs more money than running a solvent-based one.

A good deal of kinetic energy is dissipated when the high-pressure water jets, and the blasts of hot air issuing from the air-knives in the drying stage hit the circuit boards and the surrounding machinery. This means that a water washing line can be quite noisy unless soundproofing measures, which may be expensive, are taken.

Recently, the makers of water washing equipment have paid closer attention to the hydrodynamics of cleaning with water. Sprays consisting of individual drops of water, however intense, have been recognized as inefficient. The aim now is to devise coherent, fast-moving water jets which, on hitting the board surface, convert their kinetic energy into equally fast-moving sheets of water which flow tangentially across the board surface and penetrate into the spaces underneath and between the components, and around and behind the soldered joints, instead of shattering into a chaotic, multidirectional spray which may leave puddles of stagnant water between closely-set components. Fan-shaped jets which impinge on the boards at an angle are amongst the designs aimed at improving cleaning efficiency.

8.4.2   Water quality

The degree of cleanliness achieved in any water cleaning procedure depends on the quality, that is the purity, of the water employed. 'Hard' water contains salts of calcium and magnesium. A distinction must be made, however, between 'temporary' and 'permanent' hardness. The former, which is due to the presence of calcium bicarbonate and/or magnesium bicarbonate or calcium hydroxide (lime) can be removed by heating or boiling (thermal softening). The latter, due to calcium carbonate or magnesium carbonate or sulphate, must be removed by more elaborate methods. The degree of hardness of a water supply is usually given in mg calcium carbonate (or its chemical equivalent) in one litre of water.

Methods of water purification

Should the hardness of a given water supply be entirely of the temporary kind (which is unlikely), simple heating will be sufficient to remove it by thermal 'decalcification'. Heating such water from 20 °C/68 °F to 90 °C/194 °F precipitates more than half the dissolved solids. Heating it in a large boiler fitted with an exit filter is a cheap method, but not recommended for producing by itself water of a quality suitable for the cleaning of electronic assemblies. Chemical water softeners, which are based on an ion exchange mechanism using sodium chloride, are worse than useless for treating water for cleaning circuit boards: they replace calcium carbonate with strongly ionic sodium carbonate, a very unwelcome substance to find on a cleaned circuit board.[33]

In practice, water purity suitable for the cleaning of electronic assemblies can be achieved by one of two methods:

1. De-ionization with ion exchange resins gives good results. Mixed-bed de-ionization involves passing filtered water through one or more cartridges containing mixed anionic and cationic exchange resins. The dimensions and the number of the cartridges or columns of exchange resin needed for a given cleaning installation depend on the water consumption of the plant concerned. Regular and professional regeneration of the resins is essential. Normally, the cartridges are returned to the supplier for regeneration. Alternatively, some makers supply low-cost throw-away resins, where local regeneration facilities do not exist. The alternative to a mixed-bed system is separate-bed de-ionization, where anionic and cationic exchange resins are contained in separate columns, and the water passes through them in sequence. This system has a slightly lower efficiency than mixed bed de-ionization, but for all except the most demanding requirements it is good enough. Its advantage is its low running cost, because the resins can be regenerated by the user with hydrochloric acid.

2. Reverse osmosis (RO) is another option of water purification. In this method, the raw water passes under high pressure through a semipermeable membrane or 'permeator', which allows the water to pass through but which retains all dissolved substances, whether they are ionic or not. The principle involved is not a plain physical filtration. Put crudely, water is squeezed out of an electrolyte against the osmotic pressure of the solution. The purification ratio is between 10:1 and 20:1, depending on the design of the RO purifier. Any suspended solids in the raw water must be filtered out first, otherwise they clog the membrane of the RO filter. Excessive lime, giving a hardness $>8°$, and free Fe or Cl also render a feedwater unsuitable for RO. Thermal softening might be an answer to this problem, unless the initial contamination is high.

Degrees of water purity

The best yardstick for assessing the suitability of a given water supply for board cleaning is its conductivity, measured in micro-Siemens/cm ($\mu S/cm$) = megohm^{-1} cm^{-1} (S = Siemens = Ohm^{-1}). The various water qualities encountered in cleaning processes can be roughly classified as follows:

- *Normal mains water: 100–1000 $\mu S/cm$*
  Suitable for cleaning circuit boards for industrial applications with no excessive cleanliness requirements. However, after the last cleaning stage, the rinsing

water clinging to the boards must be blown off in a hot-air high-pressure drying stage (see Section 8.4.5).

- *Purified water: 10–100 µS/cm*
  For cleaning boards for domestic and industrial applications, with normal drying after washing. For more demanding applications, high-pressure drying is required.
- *High-purity water: 1–10 µS/cm*
  For military and avionic electronics.
- *Extreme-purity water: 0.8–1 µS/cm*
  For highly critical circuitry, carrying ICs with high lead counts.

8.4.3 Water recycling

The higher the purity requirements of the water, the more it costs to meet them. For this reason, the first stages of a water washing process are often carried out with mains water, provided it is of reasonable quality. Before the boards enter the last washing stage, adhering washing water is blown off, and only the last washing stage operates with purified water. Cleaning systems built on a module basis have the necessary flexibility to meet such situations.

The high-purity water of the finishing stage can be reconditioned by ion exchange or RO, either continuously or when a conductivity monitor indicates that the cleanliness has fallen below a preset value. Some soldering residues left on wavesoldered boards may cause problems here: certain watersoluble soldering oils are based on or contain non-ionic poly-alcohols which an ion exchange bed will not remove. RO filtering is suitable for dealing with these contaminants.

The quality requirements and the permissible concentration of contaminants in the effluent water, which is run to waste from a washing installation, vary not only from country to country, but also between different local authorities. Most organic constituents of soldering fluxes and their residues are bio-degradable, and since the amounts involved are small, their biological oxygen demand is modest.

However, because the action of every soldering flux is based on its ability to dissolve metal oxides, all flux residues contain metal salts. Therefore, washing water run to waste contains not only organic contaminants but also some metal compounds. The most likely metals to be found in defluxing waste water are Pb, Sn and Cu. The limits are set normally in mg/litre, for example, 0.5 mg Pb/litre and 1.5 mg Cu/litre and, in the USA, no limits for Sn.[34] One of the most demanding countries in this respect appears to be Switzerland, which sets a limit of 2 mg/litre for tin in the effluent water.

Should the waste water exceed one of these limits, the effluent is often diluted with mains water till the required limit is met. This practice is rightly beginning to be frowned upon, since what matters is not only the concentration of a pollutant in the effluent, but also its total tonnage over a given period. Thus it is very likely that an ion exchange treatment instead of simple dilution will be required by the various authorities at some future date with consequent increases in the cost of waterwashing.

The acceptable pH value of waste water lies normally in the range 6–9 and, therefore, some water-washing effluent may need pH adjustment before being discharged. Finally, most authorities object to foam and, if necessary, foam suppressants must be added before such water is run to waste.

8.4.4 Removal of residue from watersoluble fluxes

Residues from these fluxes consist mainly of their solids content, e.g. various amines or organic acids, together with thickeners, wetting agents, etc., partly in their original form and partly as pyrolytic decomposition products which have formed during soldering. Small amounts of metallic compounds are also present (see above). Most of these substances dissolve in hot water (in normal cleaning practice 60–80 °C/140–175 °F) reasonably quickly. Some Pb compounds may occasionally cause solubility problems, as can some soldering oils used in wavesoldering, which may polymerize and lose their solubility if they have been overheated or not renewed in good time (Section 8.2.1).

Though plain water without additives will remove all residues from watersoluble fluxes, for physical rather than chemical reasons the addition of 5–10% of high-purity (99.9%) isopropanol to the rinsing water of the last cleaning stage has been found to give several benefits. The isopropanol halves the surface tension of the water, and this assists rinsing. Also, on emerging from the rinsing stage, the boards are covered by a thin film of water instead of discrete drops. Therefore, they dry more readily, which saves thermal energy. Finally, the alcohol acts as a foam suppressant, and helps to dissolve remaining organics. In many cases, these advantages outweigh the cost of the isopropanol.[35]

8.4.5 Removal of residue from resinous fluxes

Resinous fluxes leave a heterogeneous residue: the rosin itself is non-polar and water-insoluble; the flux activators are polar and watersoluble; the metallic reaction products which are the result of the interaction between flux, metallic substrate and solder are partly watersoluble metal salts and partly water-insoluble metal resinates. This is true both for fluxes based on natural rosin, and on synthetic rosin substitutes.

To cope with this mixture, an additive is required. This may be one of the above-mentioned saponifiers, for example monoethanolamine, which converts the insoluble rosin into soluble resinous soaps. The reaction takes a certain time, especially if the rosin has polymerized because of overheating during the soldering process, or prolonged delay between soldering and cleaning. The resulting resin–soap solution is somewhat viscous, hence the need for high-pressure water jets. Alternatives which have recently become available contain surfactants (i.e. wetting agents) together with phosphates and silicates, or alternatively an alcohol which assists the solution of the rosin residue. The additives are claimed to be biodegradable within the definitions of the legal requirements of the EEC.

Resinous fluxes with low solids content, which have become popular in the last few years, are more amenable to water cleaning than the classic high-rosin fluxes: their residue contains a high proportion of the watersoluble activators, and therefore the washing water needs less saponifier to deal with them. Because of the preponderance of polar compounds in the residue from these fluxes, they are in fact more readily removed by water washing, with a surfactant (wetting agent) added to the water, than by solvent cleaning, should their removal be required.

The need to add a saponifier to the washing water has to be considered in the design of the washing plant. With batch cleaners, which work on the 'dishwasher' principle, the watertank has to be large enough to accommodate the saponifier addition. With in-line washing plants the saponifier addition must be metered so as to match the requirements of the boards in transit through the machine. With these machines, saponifier is added only to the first and the main wash, both of which must provide a transit time long enough to allow for complete saponification. The final wash operates with a separate clean-water circuit.

Saponified washing water has a pH of around 10. It should be buffered in order to prevent it from attacking light metals, such as certain component housings, or some board laminates or solder resists. Therefore, before using a given saponifier in a cleaning plant, it is advisable to test it at the maximum concentration which is likely to be encountered, and at the normal washing temperature, against all circuit board materials and components which will come in contact with the washing water.

Because of its high pH, saponifying washing water must be neutralized, mostly with dilute HCl, down to a pH acceptable by the local authority before it is discharged as waste. Most saponifiers are themselves biodegradable.

8.4.6 Water washing installations

Batch-cleaning systems

Batch-type water cleaning is sometimes known as 'dishwasher' cleaning. The name 'dishwasher' must not be allowed to mislead here: some makes of batch water cleaning machines are indeed made by well-known manufacturers of domestic and industrial washing and dish-washing equipment, but batch cleaners are dedicated machines, designed for the specific purpose of defluxing soldered circuit boards and, as a class, they have reached a high degree of technical sophistication. A batch cleaner needs provision for adding measured amounts of saponifier to the water of the first washing cycles, when the boards were soldered with a resinous flux, and for carrying out the last rinsing cycle with fresh, clean water.

With this type of equipment, the circuit boards are vertically arranged in fixtures or baskets, and placed at bench-level in a mostly front-loading washing chamber. They are then washed in an array of high pressure hot-waterjets, of a temperature of 60 °C/140 °F–70 °C/160 °F, which are projected from rotating or reciprocating spray arms. The cleaning sequence normally follows a program, which can be called up from a store, either supplied by the vendor or created by the user.

The water for the first washing cycle is pumped from a heated sump or tank. For the removal of rosin-free, fully watersoluble residue, a small amount of surfactant (wetting agent), sometimes together with a certain amount of alcohol, is often added to the water in order to speed up the solution of the residue. The removal of rosin-containing residue needs the addition of an alkaline 'saponifier', which converts the water-insoluble rosin into a watersoluble soap. A number of saponifiers are on offer from specialist vendors or from the machine vendor, and they raise the pH of the washing water up to 9–10. Some of them are in the form of concentrated solutions, which may contain a certain amount of alcohol.

With some machines, the water from the washing cycle is run to waste and replaced with the spent water from the rinsing cycle which follows. With others it is returned to the sump or tank, from which it is discharged to waste when it has reached a preset level of contamination. The tank may then be refilled with clean mains water and heated, or refilled with the run-off from a rinsing cycle.

The rinsing cycle normally operates with hot, demineralized and/or de-ionized water, of a quality governed by the cleanliness required from the finished board (see Section 8.4.2). Untreated mains water can sometimes be used if it is of high quality and if the cleanliness requirements of the product are modest. In all cases, the rinsing water is held in a separate tank, and often piped and sprayed through a separate plumbing system in order to avoid cross-contamination from the washing water. Some makers offer dishwasher-type machines operating with recirculated rinsing water which has passed through a de-ionizing column, and which is being monitored for its quality.

Zero-discharge batch washing machines are a relatively new concept. Here, spent rinsing water is used for the prewash, and then recycled through a set of activated charcoal filters, which remove non-ionic contaminants such as rosin residue, followed by a set of de-ionizing columns.

Having been washed and rinsed, the boards must be dried in a stream of hot air, sometimes created by an array of reciprocating hot-air knives. With some machines, the washing unit doubles as a drier, with a stream of hot air passing through the washing chamber. It is more economical, however, both in energy consumption and in time, to transfer the boards in their holding basket from the washing unit to a dedicated drying unit, in which the boards are subjected to a strong blast of hot air. It saves energy not to have to dry the interior of the washing unit and its spray-nozzles, as well as its load of boards, after every cycle. Time is saved and the capacity of the installation is doubled, because the next charge is being washed while the preceding one is dried.

A modern batch water washing system can handle boards with up to a total area of about 2.5 sq. m/22.5 sq. ft per batch. With two cycles per hour, a batch water washing system with separate washing and drying units can handle up to 5 sq. m/45 sq. ft board area per hour, sufficient to keep up with a small or medium sized soldering line.[36]

Dishwasher systems can give excellent cleanliness, at a capital cost and with space requirements which are both modest in comparison with in-line waterwashing plants. Their throughput, however, is limited and cannot be expected to keep pace with a continuously running large wavesoldering or reflowsoldering line.

Moreover, the boards must be placed by hand in their holding baskets, which means that batch washers cannot be integrated into a continuous soldering line.

In-line water washing systems

The output from one or more integrated full-sized soldering lines is best handled in an in-line washing plant, with the boards travelling on a continuous conveyor. For dealing with boards soldered with a watersoluble flux, the in-line cleaner need not be very long, at most three metres/nine feet, because of the ready solubility of the flux residue. The removal of resinous residues requires longer washing machines, because the rosin in the residue must be saponified and washed away before the subsequent rinsing stage. The nature of the process favours washing lines built up from modules. Modular design, offered by several makers, makes the process more flexible and allows the user to change from one soldering strategy to another without the capital investment for a complete new cleaning plant. A high–capacity modular water washing installation may be up to 9 m/27 ft in length.

Figure 8.15 shows the basic layout of an in-line water cleaner. Plants constructed from stainless steel have a long life, but are expensive and, being inherently more noisy, may need soundproofing. Polythene is cheaper and quite safe as a constructional material, provided the water temperature is reliably controlled so as not to exceed 70–80 °C/160–180 °F. At higher temperatures, polythene softens and may sag or crack under stress.

The boards travel on a horizontal, open-mesh stainless steel conveyor belt through successive arrays of high-pressure waterjets, which impinge on the boards from above and below. Because the boards lie loosely on the belt, the pressure and the flow rate of the upper and lower jet arrays must be precisely balanced, so as not

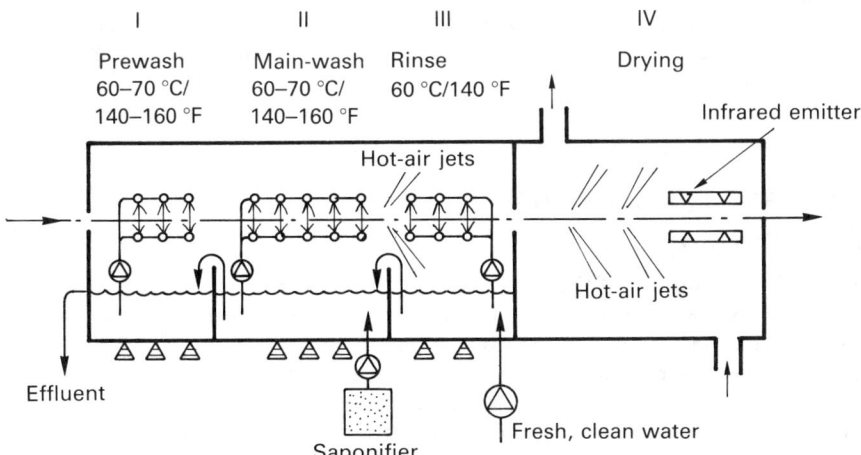

Figure 8.15 *In-line water washing machine*

to cause the boards to flip over. The kinetics of modern washing jets have been discussed already in Section 8.4.1.

With the system shown, washing proceeds in several stages. The prewash and the main-wash compartments are linked, with the feed-water entering the main-wash at its exit end, and cascading from tank to tank towards the inlet end of the machine, flowing against the direction of the conveyor belt. Saponifier, if needed, is added in metered amounts to the main-wash, either to the water tank or to the feed-water of the cleaning jets. The cleaning jets, which operate as a rule at about 2 bar (30 lb/sq. in), can be fed from the hot-water sump or with fresh, preheated water.

Both main-wash and prewash can be fed with mains water, provided this is of reasonable quality and free from suspended or organic matter. From the main-wash, the boards pass into the rinsing wash, through a curtain of high-pressure hot-air jets which blow the water from the wet boards back into the main-wash. This prevents contamination of the clean (and maybe expensive) rinsing water by carry-over of dirty washing water. Any plastic flaps which separate the two washing sections must be kept scrupulously clean so as not to redeposit dirt on the boards.

The water quality of the rinsing wash is chosen to suit the cleanliness and service requirements of the end product. Part or all of the rinsing water can be cascaded into the main-wash tank. Alternatively, the rinsing water can be recycled through an ion exchange column. Monitoring the water quality in the rinsing stage with conductivity meters can save water. Assuming the rinsing stage is subdivided into three modules, the conductivity can be set at, for example, $300\,\mu S/cm$ in the first tank, $30\,\mu S/cm$ in the next, and $3\,\mu S/cm$ in the exit tank. If these values are exceeded, either the water flow can be increased or the board conveyor can be slowed down.

From the rinsing stage the boards enter a drying chamber. Here they pass through one or more curtains of high-pressure hot-air jets, which blow away most of the adhering water. Finally a bank of infrared heaters warms the boards so that they lose any remaining surface moisture by natural drying after emerging from the machine.

8.5　Semi-aqueous cleaning

8.5.1　*Solvents*

The early phasing out of the non-flammable fully or partly halogenated solvents and the problems associated with the flammability of the lower alcohols have led to an intensive search for alternative solvents for the removal of not fully watersoluble flux residues. The solvents should have zero ODP and GWP and flashpoints above $37.8\,°C/100\,°F$. So far, a number of solvents which come under the broad classes of the terpenes, ketones, the modified alcohols and the so-called hydrocarbon mixtures have emerged (see Table 8.3). Since none of them contain halides, their ozone depletion potential is zero. Any global warming potential they may possess is claimed to be very low because of their low volatility. Being organic substances, they are combustible because their molecules contain carbon. They do not present a high fire-risk, but some precautions must be observed in their use, as will be discussed presently.

Their chemical constitution makes them good solvents for rosin and its synthetic equivalents. However, because their molecules are larger than those of the halogenated solvents and the lower alcohols, they are less mobile, i.e. more viscous. Once they have dissolved the flux residue, they are often, but not necessarily, rinsed off with water. A hydrocarbon mixture for example can be rinsed off with a similar, lower boiling one. What is then left on the board after that rinse can be allowed to dry in air.

In the last few years, a number of useful products of the terpene and the modified alcohol type have become available in commercial quantities. They are already being used in full-scale industrial installations for the cleaning of soldered assemblies, but the search for further such cleaning media is continuing.[37]

8.5.2 The concept

The solvent cleaning and water cleaning systems described so far employ one single cleaning medium, which fulfils two functions: first it dissolves the flux residue, and then it flushes away the resultant flux solution in the rinsing stage. The highly mobile halogenated solvents and the alcohols fulfil both functions very well, but they either endanger the environment or they are flammable. Water is slightly less mobile, but it needs additives to help it to dissolve the residue from rosin-based fluxes and oils.

The semi-aqueous cleaning process closely resembles the action of a cleansing cream, which is first applied to dirty hands and is then rinsed off with warm water: semi-aqueous cleaning systems separate the tasks of solubilizing and of rinsing. The task of dissolving the flux residue is taken by an organic, non-flammable, slightly viscous medium, which is miscible with or emulsifiable in water. Its principal function is to penetrate to all the places where flux residue lodges, and to interact with it. The subsequent rinsing task is taken over by water, very often warm or even hot, which flushes out and carries away the organic solubilizing medium together with the flux residue which it has absorbed. In the last few years, a variety of cleaning machines which are designed specifically for the semi-aqueous process have become commercially available. They are now established in a wide and diverse range of applications.

8.5.3 Terpenes and modified alcohols

The organic cleaning media which are available at the time of writing (1993) fall into two general classes (see Table 8.3). The terpenes together with the so-called 'hydrocarbon mixtures' are non-polar compounds; they are not miscible with water, but they readily emulsify in it, which means that a water rinse will remove them, together with the flux residue which they have dissolved. The 'modified alcohols' are a somewhat later development: being alcohols, they are polar compounds and miscible with water. Thus, a water rinsing stage will remove them together with the flux residue which they have absorbed.

Terpenes

Terpenes are a somewhat loosely defined, but very large, class of organic compounds, many of them liquids at room temperature. Most of them are derived from vegetable matter such as wood or fruit, and they have a distinctive, often not unpleasant, fruity smell. While over 10 000 naturally occurring terpenes are known, only a few of them which have a molecular structure similar to that of rosin are used for defluxing. Synthetic terpene-like cleaning media, known as hydrocarbon mixtures, with tailor-made, purpose-dedicated properties are also available.

Terpenes used for defluxing share several common features. Because of their molecular similarity to rosin they are very efficient rosin solvents, but their solvent power for inorganic or ionic flux activators is limited. Most commercially available defluxing terpenes contain an added surfactant (wetting agent) which assists them to penetrate into the recesses of the board and to be flushed out by the water in the rinsing stage.

Terpenes can hold up to 60% vol of rosin in solution before they become saturated. The rosin-loading of the terpene can be estimated by checking its specific gravity against a calibration curve, supplied by some vendors. Once it has become saturated with rosin and recovered from the defluxing line, it must either be disposed of, which is best entrusted to a specialist firm, or returned to the vendor if suitable arrangements exist.

Most terpenes are not miscible with water, so that terpene carried over into the water-rinsing stage must be separated out in a so called 'decanter' or settling tank. On the other hand, terpenes are soluble in alcohol, e.g. isopropanol. They are more viscous than water, but still reasonably mobile. A terpene can be separated from the rosin it has dissolved only by distillation. However, at normal atmospheric pressure, its boiling point is above its flashpoint, so that it must be vacuum-distilled at a pressure low enough to depress the boiling point to below the flashpoint. For this reason, reconditioning of flux-saturated terpene is best left to the vendor. Small amounts of terpene which enter the water of the final cleaning or 'polishing' stage which follows the rinsing are biodegradable.

Modified alcohols

The so-called modified alcohols are a class of water-miscible organic compounds, or a mixture of such compounds, which are derived from non-linear alcohols or glycols, sometimes with added ethers or amines. They have been formulated to possess good solvent power for both rosin and its synthetic equivalents, and for ionic or inorganic activators. Their flashpoints are in the region between 80 °C/180 °F and 100 °C/215 °F. They are mostly colourless fluids, and like the terpenes slightly viscous.

With most of them, the accumulated percentage of rosin in solution can be estimated by a calorimetric method, provided the rosin in the flux is not of the 'waterwhite' type or a synthetic resin (R. Wood, Dage (UK) Ltd, private communication). Vendors will provide information and equipment if required. Recovery by distillation is impractical for the average user for the same reason as outlined with the terpenes. The methods of disposal of spent modified alcohols depend on the country and the location of the user. Some vendors offer free collection of the spent solvent. Disposal by combustion is a possibility, and in fact modified alcohols can be considered as secondary industrial fuels. One vendor suggests the firing of cement kilns as a potential use for spent modified alcohol. Before any such disposal it is advisable, and it may sometimes be required, to establish the percentage of potentially harmful elements such as copper and lead in the spent solvent.

8.5.4 *Semi-aqueous washing installations for terpenes and hydrocarbon mixtures*

General features

Common to all semi-aqueous cleaning systems is a first washing stage which ensures thorough penetration of the organic solvent into all the spaces and crevices of the soldered boards. Though the wetting behaviour of both terpenes and modified alcohols towards the surfaces of boards, components and joints is good, and with some of them assisted by added surfactants, their viscosity means that they need some mechanical push to get to where they have to act. Section 8.2.1 describes the different methods which can be used to propel a solvent to where it is needed.

It is common practice to warm the cleaning medium in the washing stage, in order to reduce its viscosity and to speed up the solvent action. An operating temperature of 40 °C/100 °F to 50 °C/120 °F is normal, but in no case should this temperature approach the flashpoint of the medium closer than to within 20 °C/35 °F. Even when this precaution is observed, there is a problem when a terpene or modified alcohol is applied by a jet or spray. A mist or aerosol of a combustible fluid may spontaneously ignite and cause an explosion at temperatures below its flashpoint.[38] To guard against this danger, spray chambers of semi-aqueous washing machines must be filled with nitrogen or carbon dioxide.

This complication can be avoided if the boards, instead of being sprayed, are fully immersed in the cleaning medium, while the kinetic energy for propelling it into the crevices of the circuit board is provided by strong jets below the surface (jacuzzi effect) or by ultrasonic energy. An ultrasonic energy input into the liquid in the cleaning tank of about 20 W/litre (90 W/gal) is normal, where the energy is taken as the rated electrical power consumption of the ultrasonic transducers.

Conveyorized in-line systems

With an in-line system, the boards travel on a flat, straight open-mesh conveyor. They first enter a washing module, where they pass between successive arrays of solvent jets, which impinge on the boards with equal force from above and below, as with an in-line water washing machine. The explosion risk requires that the spray chamber is filled with nitrogen or carbon dioxide. If the boards travel fully immersed through an elongated jacuzzi tank on a conveyor which enters and leaves the bath at an angle, as is the case with some designs on the market, this complication is avoided. Before leaving the washing unit, an array of warm-air knives reduces the drag-out of solvent.

The boards then enter a rinsing or 'polishing' module via an intermediate conveyor, which prevents undue carry-over of solvent into the rinsing water. The rinsing stage is very similar in concept and operation to that of the rinsing stage of an in-line water washing system. The same considerations as to operational strategy, water purity and final drying apply (Figure 8.16).

As with all in-line systems, the boards travel through the machine in single, or at most double, file. This means that these machines are necessarily rather long. With washing and polishing modules having lengths of between 4 m/12 ft and 6 m/18 ft, and with an intermediate conveyor connecting them, overall lengths of 10 m/39 ft or more are not uncommon.

Batch systems

With the batch concept, the boards are placed in baskets or carriers, which are immersed in a succession of treatment tanks. With most installations on the market, the baskets are moved by mechanized, often automated and programm-

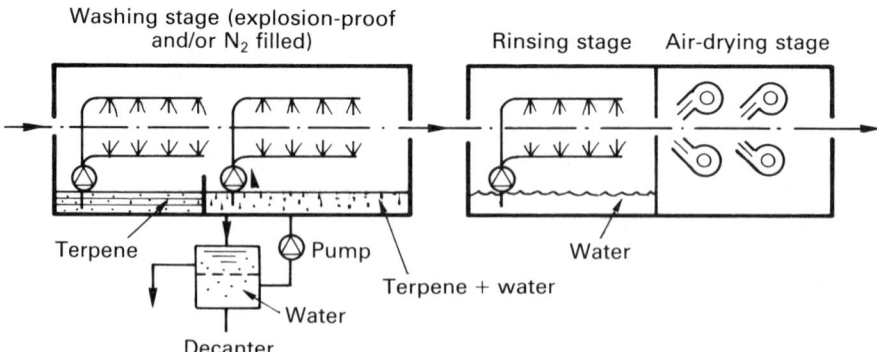

Figure 8.16 *In-line semi-aqueous cleaning (terpene)*

able, handling systems. The design details of a cleaning installation depend on whether a water-insoluble terpene or a watersoluble modified alcohol is used, but basically all batch cleaning systems are similar.

With a terpene solvent, the washing stage comprises two tanks. The first one contains warm solvent (40 °C/100 °F to 60 °C/140 °F, depending on its flashpoint). The solvent in the tank may be agitated ultrasonically by arrays of transducers which are so placed that their energy can reach the surface of all the boards in the holding fixture with adequate intensity. Alternatively, an array of strategically placed sub-surface jets of solvent provide the necessary kinetic energy in the solvent to penetrate into the recesses of the board assembly. With some installations, this tank is equipped with a heat pump, which extracts heat and recycles it, so that the temperature of the solvent may be maintained at its optimum, both for cleaning efficiency and for safety.

The dwell time of the boards in this tank depends on the complexity and the population density of their layout, on the type of flux used, on the soldering parameters, and on the time interval between soldering and cleaning. Normally it is no more than a few minutes.

The board is then transferred to the second tank of the washing module. This tank contains warm water (40 °C/104 °F to 60 °C/140 °F), with which the water-immiscible, rosin-loaded terpene forms an emulsion. Ultrasonic agitation or sub-surface jets flush this emulsion from the recesses of the board, removing the bulk of the flux residue. The surfactant in the terpene assists this rinsing process.

The terpene–water emulsion which forms in the cleaning tank is constantly pumped through a settling tank, called a 'decanter'. In the decanter, the water, which is heavier than the terpene, settles to the bottom, from where it is returned to the cleaning tank. The flux-loaded terpene rises to the top, where it is creamed off and transferred to containers for subsequent disposal by a specialist contractor, while the water is recycled.

The carry-over of terpene on the boards passing from the solution tank into the emulsion tank ensures that the concentration of rosin in the terpene in the solution tank remains roughly at the same level, normally about 50%. The carry-over is constantly made good by further terpene being added, which therefore must be considered as a consumable. This carry-over and the maintenance of a stable concentration of rosin in the solution tank are important: the solvent power of terpene for rosin depends on the rosin concentration in the terpene. It rises with rising rosin content and reaches a maximum at 40–50% rosin. At higher concentrations, the cleaning efficiency drops.[39]

After leaving the emulsion tank, the boards in their baskets enter the rinsing stages, where they pass through a succession of water-rinsing tanks. The successive washes follow the same pattern as applies to an in-line water washing installation (Section 8.4.6). Depending on the degree of cleanliness required, the boards then pass through several hot-water rinsing stages, of which the final ones may operate with de-ionized water if the product requires it. Finally, the baskets enter a hot-air drying chamber before they leave the machine. In principle, there is little difference between the back-end of a semi-aqueous cleaning system and that of a water cleaning one. Several processing options and configurations of this procedure, one of which is shown diagrammatically in Figure 8.17, are on offer to the market.

8.5.5 Semi-aqueous washing installations for modified alcohols

The main difference between the terpenes and the modified alcohols is the full miscibility of the latter with water and also with an alcohol like isopropanol. This allows much greater flexibility in devising a cleaning strategy, as shown in Figure 8.18.

The first immersion cleaning stage will always consist of either one tank or two tanks in sequence with either ultrasonic or sub-surface jet agitation. They operate at temperatures between 40 °C/100 °F and 60 °C/140 °F, and the dwell time in the tank or tanks depends, apart from the characteristics of the solvent, on the bath temperature and the intensity of the agitation. The dwell time is counted in minutes, normally well below five. Most solvent vendors recommend that the dissolved flux residue is stripped continuously from the solvent by circulating the contents of the tank or tanks through a carbon filter to remove the rosin and an ion exchange filter to remove the ionic flux activators. If a rosin–free watersoluble flux is used, only the latter is needed. The recycled clean solvent cascades from the second tank into the first. A number of options exist for the rest of the cleaning line.

The semi-aqueous option

If the normal semi-aqueous path is followed, the washing stage is followed by one or more water rinses, which in layout resemble a standard batch water cleaning sequence. For example, the first rinse may be fed with tap water, which is run to waste via a carbon absorption filter to strip any rosin carry-over (most modified alcohols are rated as biodegradable), or else is recycled in a closed loop.

Alternatively, the rinsing or polishing stage may be a cascade system, with recycled demineralized water entering the exit end of the last stage. The choice of

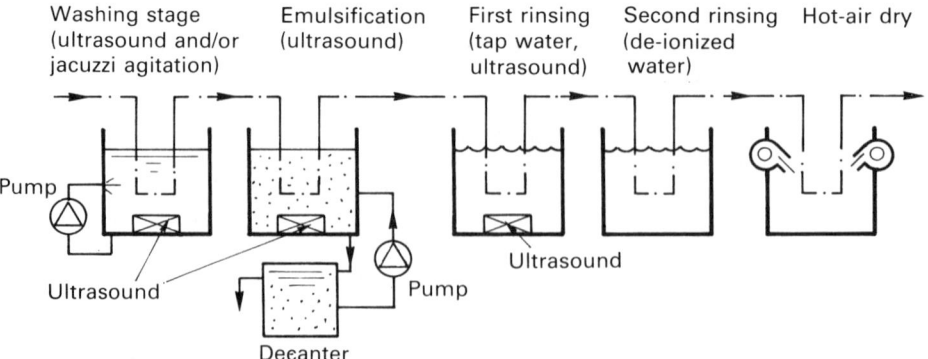

Figure 8.17 *Batch-type semi-aqueous cleaning (terpene)*

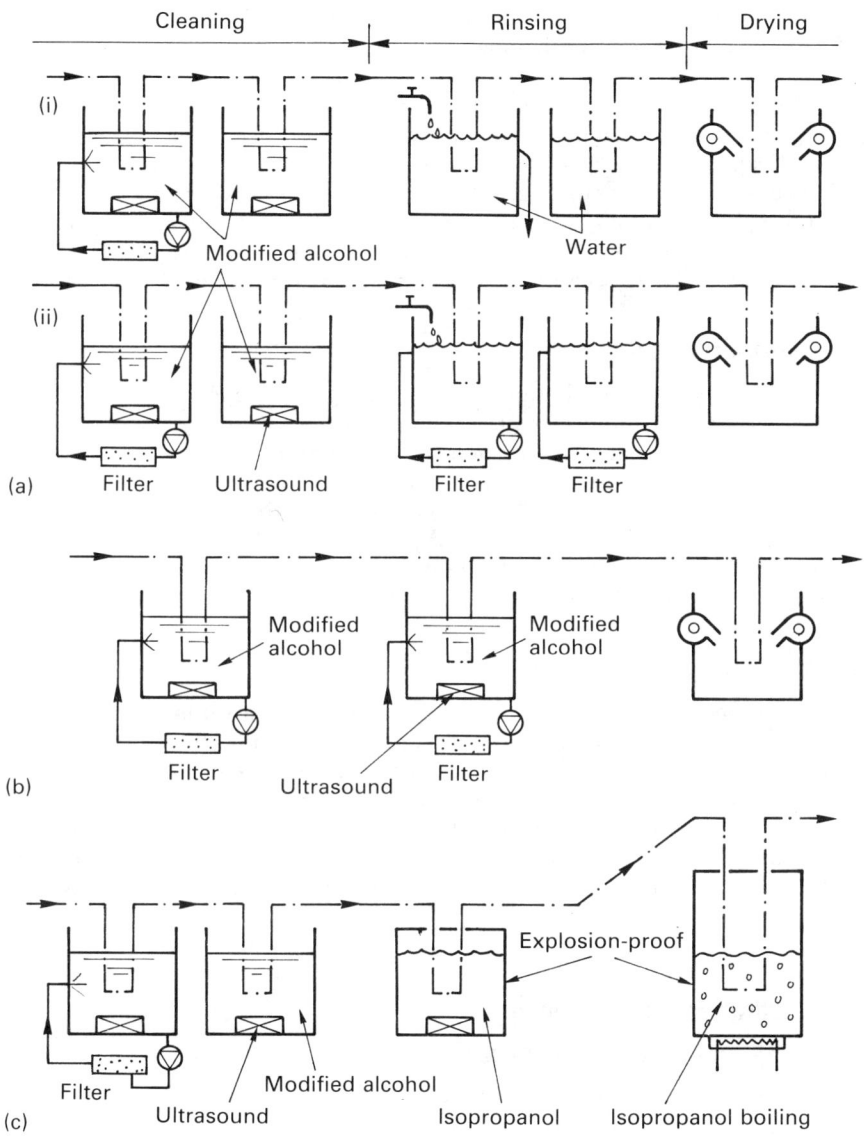

Figure 8.18 *Strategy options for cleaning with a modified alcohol. (a)(i) Semi-aqueous, (ii) semi-aqueous, zero effluent; (b) solvent only; (c) mixed solvents – rinsing and drying in isopropanol*

layout of the rinsing stage will be determined by the cleanliness which is required of the product. After leaving the final rinsing stage, the boards must be dried by forced circulation of hot air.

Depending on the dimensions of the plant, the hourly throughput of a semi-aqueous batch cleaning installation can be reckoned between 5 sq. m to 25 sq. m/45 sq. ft to 225 sq. ft of board area, occupying lengths from 5 m/15 ft to 6.5 m/20 ft. A simple, compact and low-cost but very effective installation for small throughputs, coping with up to 20 sq. m (180 sq. ft) board area per day, consists of just two adjacent tanks, one for cleaning and one for rinsing, which can be run concurrently. It occupies a floor area of 1.5 m × 0.75 m (60 in × 30 in).

The solvent-only option

With this option, both the cleaning and the rinsing stages operate with the same modified alcohol or similar solvent. With both stages, the solvent is recirculated through carbon and ion exchange filters to remove dissolved rosin and activator. The advantages are the convenience of operating a single-chemical system and the absence of an effluent problem. The disadvantages stem from the low volatility of the solvent, which means a longer drying time. On the other hand, energy consumption is low: though the boiling points of the modified alcohols are well above 100 °C/212 °F, their heat of evaporation is lower than that of water.

The solvent–alcohol option

Several equipment manufacturers offer a layout in which the rinsing stages operate with one of the lower alcohols, normally isopropanol, with which the modified alcohols are fully miscible. The first rinsing stage or stages operate with cold alcohol. The alcohol in the last stage boils, and the boards dwell for a short while in the vapour space in order to dry before they emerge. Naturally, the construction of such a plant must be explosion–proof, e.g. with Flutec©.[32] which raises the cost. On the other hand, there is no effluent problem and drying is rapid and consumes no extra energy.

8.6 Testing for cleanliness

8.6.1 The meaning of cleanliness

Before entering the complex field of testing circuit boards for cleanliness and of monitoring the effectiveness of a given cleaning treatment, a few concepts should perhaps be clarified. The first one is the meaning of 'clean' in the context of a soldered printed circuit board. Absolute cleanliness of an object or a surface cannot be achieved except in a hard vacuum or in outer space. Under normal circumstances, it can only be approached asymptotically. At which point on this asymptotic curve a given object is declared as clean depends on the context: for example, 'clean hands', when referring to a surgeon, means something quite different from when applied to a schoolboy. Neither cleanliness standard is arbitrary, both are purpose-related: the surgeon must not infect his patient, the schoolboy must not smudge his exercise book. The same is true for circuit boards; the link between the degree of cleanliness required from a circuit board and its function in service has been outlined in Section 8.1.1.

Testing or measuring the cleanliness of a circuit board involves a further concept. Here, cleanliness means compliance with a set of quantified requirements, which are defined in such a way that compliance with them can be objectively and accurately measured. As matters stand at present, two parameters have been chosen to serve as yardsticks for cleanliness.

The first one is the amount of ionizable matter per unit area which is present on a circuit board. This is a sensible choice, because under moist conditions ionizable deposits on a soldered board are liable to cause corrosion or they provide conductive paths between adjacent conductors. Leakage paths threaten the function of any electronic assembly, particularly when components with high impedance are involved. Detection of ionizable contamination is the object of the so-called MIL test, which will be discussed presently. The MIL-test method seeks out ionizable contamination only, disregarding non-ionic substances like drilling debris, oils or silicones. While these latter contaminants may not endanger the function of an affected circuit board, they may well prevent the adhesion of a conformal coating which is applied to the board after cleaning.

The second parameter chosen for testing the cleanliness of a circuit board is its surface insulation resistance (SIR) over a given distance, measured with the aid of one of the several standard interlacing-comb patterns. The SIR test requires separate, dedicated test boards: these can serve to assess the cleanability of a flux, the effectiveness of a cleaning solvent or a cleaning system, or the efficient working of a particular cleaning line in actual operation. Unless a production circuit board carries a test comb-pattern somewhere on its surface, for instance in a corner or under a component, its SIR cannot be directly determined.

8.6.2 Measuring ionic contamination (MIL test)

A test procedure, known by its abbreviated name as the MIL test, is contained in an American military specification, issued by Naval Electronic Systems Command, Washington DC. The full specification, with the title 'Printed Wiring Assemblies', last issued in 1981 under the number MIL-P-28809 A, deals with a large number of requirements which soldered and conformally coated circuit boards have to meet, one of them being the absence of ionic contamination. The specification describes a test method which involves rinsing both sides of the circuit board to be tested with a mixture of 75% high-purity isopropanol and 25% demineralized water which has been de-ionized until its conductivity is below $0.04 \mu S/cm$. Rinsing is carried out with the aid of a polythene wash-bottle for at least one minute, and the run-off is collected in a beaker. The conductivity of the run-off, which in total should amount to 10 ml for each sq. in of circuit board washed, is measured. From this, the amount of ionizable contamination which has been washed off the board is deduced. MIL-P-28809 requires that not more than $10.0 \mu g/sq.$ in ($1.56 \mu g/cm^2$) of ionizable matter, calculated as NaCl, should be found on a board, taking all board surfaces and an estimate of the component surfaces into account.

Returning to the image of the dirty hands, the MIL test is equivalent to judging the cleanliness of a pair of hands by measuring the amount of dirt in the washing water. On reflection it would seem that there is no reasonable or practicable alternative to this approach. The figure of 10.0 μg/sq. in is of course arbitrary. A provisional British Ministry of Defence Standard (Def. St. 34, Part 6, sect. 1, 1986) sets the soluble ionic impurity limit at $1.5\,\mu g/cm^2$ $(9.6\,\mu g/in^2)$ and adds the cut edges of the board to its calculated surface, together with a 10% allowance for the component surfaces.

A number of commercial instruments for the measurement of ionic contamination of printed circuit assemblies are on the market. They all share the same basic construction. The board to be tested is placed in a cuvette or cell where it is exposed to a stream of the propanol/water rinsing mixture under controlled conditions. The rinsing solvent circulates via a conductivity measuring sensor in a closed loop through the cell, and after the completion of a test cycle through a regenerating ion exchange column, to be stripped of the accumulated ionic contamination. The test cycle continues until the monitored rise in the conductivity of the rinsing solvent flattens out, indicating that all of the soluble contamination has been absorbed in the solvent (Figure 8.19). While the basic construction of most contamination meters follows this general outline, they differ in their testing and evaluation strategy. Most of them are controlled by microprocessors which automatically run the tests and evaluate and record the results.

The MIL test concept raises several issues:

1.   It assumes that cold rinsing solvent is capable of dissolving all of the resinous residue, rendering the ionic activators which are embedded in it accessible and

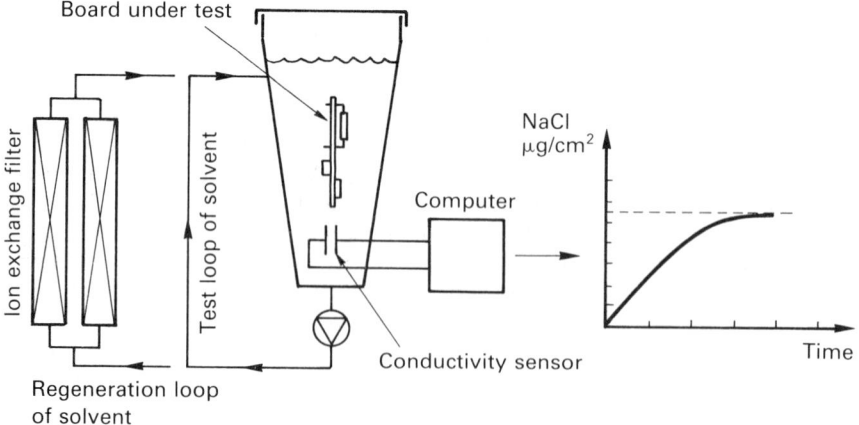

Figure 8.19   *Working principle of a MIL contamination meter*

soluble. This is not always the case, and therefore with some instruments the solvent can be heated up to 50 °C/120 °F.

2. Non-polar impurities like the rosin itself, oil residues and some of the constituents of soldercreams or of fluxes with low solids content remain 'invisible'.

3. Many locations on SMD-populated boards are not readily accessible to the rinsing solvent. This means that the conductivity of the solvent takes a long time to reach its asymptotic maximum. The standard mixture of isopropanol and demineralized water fails to remove any of the residue from a rosin-based solderpaste from a 0.15 mm/6 mil high gap, after 30 minutes immersion, at room temperature and without agitation. For this reason, several makers of MIL-test instruments have fitted the test cuvette with underwater jets, which produce a 'jacuzzi' effect.

    Since these modified contamination meters are more searching, but in different degrees, cleanliness requirements ought to specify the instrument with which compliance should be verified in order to be meaningful.

4. The limiting value of 10 μg/sq. in (1.56 μg/cm^2) is arbitrary and does not reflect the end use of the printed circuit assembly. A graded scale of contamination limits has therefore been proposed (Table 8.15).[40]

8.6.3 Measuring surface insulation resistance (SIR)

The determination of SIR requires a dedicated test-board which carries a standardized comb-pattern (Figure 8.20), unless a corner or a proportion of every production board is set aside for a test comb-pattern. The length of the tracks and

Table 8.15 *Proposal for grading MIL-test values*

End use	NaCl μg/cm^2	NaCl μg/in^2
Industrial electronics:		
no SMDs on board	2.0	13
with SMDs	1.5	10
Military, avionics, with high-impedance component:		
no SMDs on board	1.5	10
with SMDs	1.0	6.5
Ultracritical SMD assemblies, hybrids	0.5	3.5

Figure 8.20 *Comb pattern for a SIR test*

the spacing between them differs between the various specifications. The IPC-B-25 comb-pattern for instance prescribes spacings of 0.16, 0.25 or 0.31 mm (6.5, 10.0 or 12.5 mil). Further standards for such patterns, as well as for SIR testing in general, are for example MIL-STD-202, IPC-TM-650 or BELLCORE TR-TSY-000078.

SIR testing is normally carried out to evaluate the response of flux residues to a cleaning treatment, the need for cleaning in a given manufacturing situation, the effectiveness of a given cleaning treatment, or for monitoring the performance of a cleaning installation during a production run. A SIR test can take considerable time and, for that reason, it is unsuitable for giving a quick alarm if things should go wrong or are suspected to be going wrong in an ongoing production.

A number of commercial SIR testing instruments are on the market. Since the test programmes prescribed by the various standards require the exposure of the test boards to a variety of climatic environments, sometimes for prolonged periods, and repeated and intermittent testing, most testing instruments are provided with climatic test chambers, programmable and microprocessor controlled. Moreover, they must be capable of measuring high resistances, with orders of magnitude of up to 10^{12} ohm. Finally, test schedules are long and complex, involving a large number of measurements which must be evaluated and recorded. For all these reasons, SIR test equipment is of a high degree of sophistication, and expensive, as is SIR testing in itself.

8.7   The future of cleaning and of fluxing

The future of economic and environmentally acceptable cleaning depends on a rational and organized cooperation of several industries. Involved are principally the formulators and suppliers of solvents, and they have already come up with a number of useful and promising alternatives to the environmentally destructive and the flammable solvents. Second in line are the suppliers of cleaning plant, who have to design and supply installations for using these solvents, and they have already put a good choice of such equipment into the hands of the user.

Finally, and above all, the future of successful and economic cleaning, and perhaps the eventual abolition of cleaning, lies in the hands of the soldering industry, its customers and the suppliers of fluxes and soldering equipment:

1. Both the soldering industry and its customers must overcome their obsession with the need to use rosin-containing fluxes. Once rosin-free fluxes and solderpastes can be shown convincingly to leave truly innocuous residues which are acceptable to all except justifiably fastidious users, and once the electronics industry accepts them as such, there will be less cleaning, and such cleaning as must still be practised will be simpler and cheaper.
2. Flux manufacturers are striving and will continue to strive to produce fluxes which satisfy the no-clean requirements of all but the most fastidious classes of users of electronic equipment.

3. The makers of soldering machines are equally striving to perfect soldering equipment which permits soldering without using any flux at all.

8.8 References

1. Lea, C. (1992) After CFCs? Electrochemical Publications, Ayr, Scotland, p. 102 and (1993) Report to Soldering Science and Technology Club, 9 December, Teddington, England.
2. Lea, C. (1988) *A Scientific Guide to Surface Mount Technology*, Electrochemical Publications, Ayr, Scotland, p. 436.
3. Lea, C. (1988) *loc. cit.*, p. 438.
4. Lea, C. (1992) *loc. cit.*, p. 131.
5. Ellis, B. N. (1986) *Cleaning and Contamination of Electronics Components and Assemblies*. Electrochemical Publications, Ayr, Scotland, Appendix 1.
6. Lea, C. (1992) *loc. cit.*, p. 132.
7. Zado, M. F. (1983) Effects of Non-Ionic Watersoluble Flux Residues. *Western Electric Engineer*, **27**, 1, pp. 41–48.
8. US Patent 4.960.236.
9. Lea, C. (1992) *loc. cit.*, p. 88.
10. Richards, B. P., Prichard, D. J. and Footner, P. K. (1990) Assessing Cleaning Options. *Electronic Production, London*, Dec. 1990, **19**, 12, pp. 25–26.
11. Boswell, D. (1990) Surface Mount Components and the Ozone Layer Depletion Problem. *Proc. SMTCON, Atlantic City, NJ*, pp. 103–106.
12. Lea, C. (1988) *loc. cit.*, p. 449.
13. Lea, C. (1988) *loc. cit.*, p. 447.
14. Richard, B. P., Burton, P. and Footner, P. K. (1990) The Effects of Ultrasonic Cleaning on Device Degradation. *Circuit World*, **16**, 3, pp. 20–25.
15. Beine, H. (1993) *Entirely without Flux* (in German). *Productronic (Germany)*, 6, pp. 28–29.
16. Lea, C. (1992) *loc. cit.*, pp. 334–336.
17. Turbini, L. J., Eagle, J. G. and Stark, T. J. (1979) A comparison of removal of activated rosin flux by selected solvents. *Proc. IPC Fall Meeting, San Francisco, CA*, Techn. Paper IPC-TP-305.
18. Cabelka, T. D. and Archer, W. L. (1985) Cleaning, What Really Counts. *Proc. ISM Intern. Microel. Symposium, Anaheim, CA*, pp. 520–528.
19. Ellis, B. N. (1986) *loc. cit.*, p. 187.
20. Ellis, B. N. (1986) *loc. cit.*, p. 188.
21. Ellis, B. N. (1986) *loc. cit.*, pp. 170, 182.
22. Lea, C. (1992) *loc. cit.*, Chapters 1 and 2.
23. Seinfeld, J. H. (1986) *Atmospheric Chemistry and Physics of Air Pollution*, Wiley, New York.
24. UK Stratospheric Ozone Review Group (1988) *Stratospheric Ozone*, HMSO, London.
25. Polar Ozone Workshop (1988) *NASA Conference Publication 10014.*

26. Jones, R. R. (1987) Ozone Depletion and Cancer Risk. *Lancet*, pp. 443–446.
27. *New Scientist*, London, 22 Oct. 1988.
28. Lea, C. (1992) *loc. cit.*, Chapter 4.
29. Davidson, G. (1992) Icy Prospects for a Warmer World. *New Scientist*, 8 August 1992, pp. 23–26.
30. Murphy, K. P. (1990) The HCFC Alternative for CFC-113. *Proc. First CFC Alternatives Conference, San Francisco, CA.*
31. Tickell, O. Up in the Air. *New Scientist*, 20 October 1990, pp. 41–43.
32. Slinn, D. S. L. and Baxter, B. H. (1990) Alcohol Cleaning under a Non-flammable Perfluorocarbon Vapour Blanket. *Proc. Nepcon West '90*, Anaheim, CA, pp. 1810–1819.
33. Ellis, B. N. (1986) *loc. cit.*, pp. 205–206.
34. Andrus, J.J. (1991) PWA Aqueous and Semi-Aqueous Cleaning. *Proc. Nepcon West 91*, Anaheim, CA, pp. 281–291.
35. Ellis, B. N. (1986) *loc. cit.*, p. 211.
36. Ellis, B. N. (1986) *loc. cit.*, pp. 220–225.
37. Dishart, K. T. and Wolff, M. C. (1990) Advantages and Process Options of Hydrocarbon Based Formulations in Semi-Aqueous Cleaning. *Proc. NEPCON West '90*, Anaheim, CA, pp. 513–527.
38. Bodhurta (1980) *Industrial Explosion Prevention and Protection*, McGraw-Hill, New York.
39. Dishart, K. T. and Wolf, M. C. (1990) Circuit Assemblers can have a Reliable Future with New Cleaning Techniques. *Proc. Electronics Manufacturing and the Environment, Bournemouth, UK.*
40. Ellis, B. N. (1986) *loc. cit.*, p. 260.

9 Quality control and inspection

9.1 The meaning of 'quality'

The term 'quality', when applied to a technical product, can mean a number of different things, depending on the context in which the word is used. Some of them are listed below:

1. It can be used as a yardstick when comparing the same type of product made by different manufacturers, or a number of products in the same class. In this context, 'quality' expresses how closely the product approaches perfection of functioning, material specification, finish, appearance and so on.
2. It can be used for quantifying the working efficiency of a technical product, or the relationship between its cost and its performance.
3. It can be a measure of the capability of a technical product to meet a set of functional requirements, which may have been formulated by the maker, or which the purchaser can justifiably expect. In simple terms: Does it work as I expect it, or have been led to expect, and for how long will it work?

In this chapter, the word 'quality' will be used in this last sense.

9.1.1 Product quality and product reliability

The quality which a finished technical product must possess is only meaningful, and can only be achieved, if all the tasks which it will have to fulfil, the efficiency and perhaps the speed at which it must work, the circumstances under which it will have to function, and its reliability are precisely specified.

The formulation of these demands will depend on several factors:

1. Company-specific considerations like:

   Product or market philosophy
   Product or quality image
   What does the competition offer?

2. Application-specific considerations. Here, quality demands can be split into 'ability to function' and 'reliability', both of them in the environment in which the product will be used.

9.1.2   Classification according to reliability requirements

The following classification groups electronic products according to their reliability demands, and to some extent the consequences of a premature or unforeseen failure:

I   *Professional equipment, highest demands*
Space and satellite technology
Avionics, commercial and military
Telecom, submarine
Telecom, military and weapons systems
Medical, life support systems

II   *Professional and commercial equipment, very high demands*
Automotive, under hood
Telecom, commercial

III   *Professional equipment, high demands*
Industrial electronics, high quality
Computers
Telecom, private

IV   *Professional equipment, medium demands*
Industrial electronics, general
Medical electronics, general
Computer peripherals, medium price range

V   *Professional equipment, low demands*
Office equipment
Measuring instruments, general use

VI   *Semi-professional equipment*
Professional audio and video
Automobile, passenger compartment
Entertainment electronics, high-quality
Desktop and laptop electronics
Consumer electronics, high quality

VII   *Consumer electronics, general*
Entertainment electronics, domestic
Household equipment
Calculators
Toys

'Ability to function' in the context of defined quality requirements means that the equipment must be capable of working, not only at the time when it reaches the customer ('zero hour quality' or 'acceptance quality level') but also under its specified worst-case operating conditions.

'Reliability' in the context of quality can be defined as 'the ability of a product to function as expected for an expected period of time, without exceeding an expected failure level'.

A discussion of the reliability of an electronic assembly as such goes beyond the purpose of this book. The importance of defining, measuring and safeguarding the quality and reliability of technical products has been recognized in recent years, as witnessed by the large number of relevant national and international standards specifications (e.g. ISO 9001–9003, and MIL-Q-9858A). Quality and reliability in the context of industrial manufacture have become the subjects of academic teaching and research, and of a growing number of publications.[1]

What interests us here are the quality and reliability of the soldered joints of an electronic assembly. If just one of its joints is not soldered, or if one bridge between neighbouring conductors causes a short circuit, the whole assembly malfunctions or does not work at all. If one of the joints fails in service, the assembly begins to malfunction or stops working altogether. The reliability of soldered joints is dealt with in detail in Section 3.3.5. See also References 17–19 of Chapter 3.

9.2   Soldering success and soldering perfection

9.2.1   Soldering success and soldering faults

In soldering, two things matter before all others, as has been said several times already:

1. The solder must reach all the places where it is required, and it must stay there: there must be no open joints.
2. Solder must not remain anywhere where its presence causes a short circuit.

Soldering success shall be considered to have been achieved if both requirements have been met. If one or both remain unfulfilled, success has not been achieved and a 'soldering fault' has occurred: the circuit board cannot function.

In practical terms, a soldering fault is caused by one or more of the following.

Open joints

The electrical continuity of the circuit requires that no joint remains open. With wavesoldering, filling of every joint can be difficult, and it requires the optimization of both the board layout (Section 6.4) and of the flowpattern of the solderwave (Section 4.4.4) to persuade the solder to reach every joint. With reflowsoldering, every footprint must receive the correct solder depot and enough flux, be it as paste printdown or as a layer of solid solder. Also, every joint must have reached the required minimum temperature needed to melt the solder, and to connect the joint members. Briefly, there must not be a single skipped or open joint on any one board.

Thus, for example, a footprint without paste or solder depot must be considered a soldering fault even before the board is soldered: the solder never got to where it is required. If there was insufficient solderpaste on a footprint to reach the bent leg of a multilead device (lack of coplanarity), the joint will remain open. With a reflowed joint on a PLCC where wicking has occurred, the solder did not remain in its appointed place and the joint opens during soldering. In all cases, soldering success has not been achieved (Figure 9.1).

The correct position of every component belongs in the same category of soldering success. With SMDs, every component must sit and stay in its correct place. Seriously misaligned components, or components which have 'floated' or 'tombstoned', constitute soldering faults. It is convenient to include the placement of the components here, so that their correct identity, and in the case of active components their correct orientation, also counts under 'soldering success' (Figure 9.2).

Bridges

A circuit board cannot function if it contains a short circuit, i.e. a solder bridge. Wavesoldering without bridging demands special techniques, such as optimizing the configuration of the wave (Section 4.4.4) and the board layout (Section 6.4.1). Boards with a pitch below 1 mm/40 mil are difficult to wavesolder without faults, unless soldered in a nitrogen atmosphere.

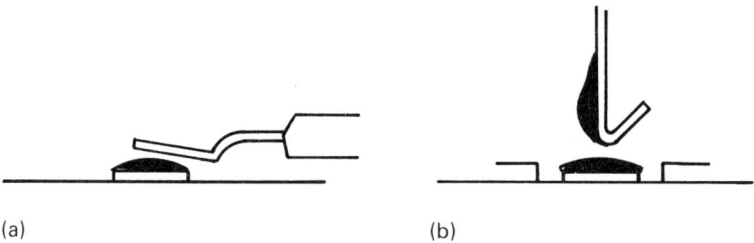

(a)   (b)

Figure 9.1 *Open joints caused by: (a) lack of coplanarity; (b) wicking*

Figure 9.2 *Displaced and misplaced components*

With reflowsoldering, especially of fine-pitch boards, the type of paste and its quality and the precision of the printing of it are key factors in achieving soldering success (Section 5.2.3).

Solderballs

Solderballs need not necessarily be classed as soldering faults. If a solderball sits between two neighbouring footprints on a fine-pitch board, it can constitute a shortcircuit, and prevents the board from functioning. Elsewhere, solderballs represent potential shortcircuits, and as such reduce the reliability to an extent which is difficult to quantify. How solderballs are to be regarded is very much a matter of individual company policy.

Even a single soldering fault on a board prevents it from functioning, and there are only two options: correct it or scrap the board. The choice between them depends on several factors, which will be discussed in Section 10.1. What must be stressed here is the following.

The existence of a soldering fault is an objective fact. A joint is either soldered or it is not soldered. A bridge is either there or else there is none. The soldering fault presents a 'yes/no situation' (Figure 9.3). To pronounce upon it is in the nature of a verdict upon an observed fact, and two or more inspectors must necessarily reach the same verdict.

Because of its objective 'yes/no' nature, the success/fault verdict can be entrusted to an automatic quality assessment system, which may be based on opto–electronic inspection or functional electronic testing (Section 9.5.5).

9.2.2 Soldering perfection and defects

Assessing soldering perfection presents an inspector with a fundamentally different situation: imperfect soldering does not prevent the affected circuit board from functioning, but it can be seen as endangering or reducing its reliability. It may also affect its saleability where the buyer has specified precise criteria.

Criteria for perfection may include the following features (Figure 9.4):

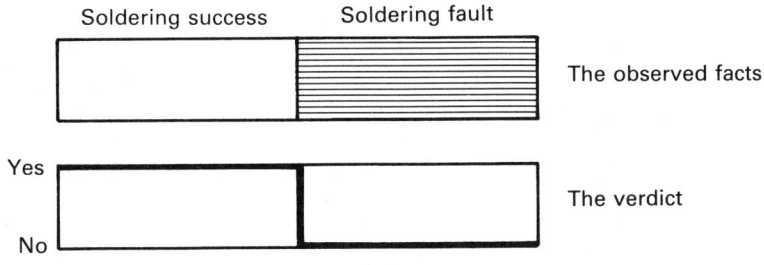

Figure 9.3 *The yes/no nature of soldering success*

Figure 9.4 *Some soldering imperfections or defects*

Wetting angle
Joint profile and amount of solder on a joint
Alignment or displacement of components

In the following, a soldering imperfection will be termed a 'soldering defect'.

If a defect disqualifies the product in the eyes of the customer, it becomes a soldering fault, because it makes the product unsaleable. A product which is unsaleable does not work as far as the vendor is concerned.

Being saleable is the first function any manufactured product must fulfil. A product which is not saleable in the market for which it has been made does not function from the point of view of its maker (unless it is still saleable elsewhere for less profit or at a loss). The offending feature must be corrected, or else the product must be scrapped.

In contrast to the unequivocal yes/no *verdict* upon the verifiable fact of soldering success or fault, a pronouncement upon the soldering perfection of a joint represents a *judgement*, which is necessarily subjective. The judgements arrived at by different inspectors represent points along a scale, which separate the 'perfect' or 'acceptable' from the 'imperfect' or 'non-acceptable' (Figure 9.5). On either side of the accept/reject divide are areas of doubtful acceptability and false alarm.

It has been found that only 44% of the quality judgements on the same set of soldered boards, made on two different days by the same inspector, agree with one another. The quality judgements of the same boards made by two different inspectors overlap by only 25%, while those made by three inspectors overlap by 14% (A. T. & T. Bell, Burlington, N. Carolina).

To sum up: deciding between soldering success and a soldering fault amounts to a verdict. Deciding whether soldering perfection has been approached sufficiently is a matter of judgement, and the making of this decision can be automated only with great difficulty.

The blowhole problem

Blowholes in wavesoldered throughplated joints, caused by 'gassing' of the walls of the hole, are a special form of defect (Figure 9.6).

The causes of gassing of throughplated holes, and the measures which are needed in order to avoid it, are by now well understood. Gassing can be prevented by ensuring the smoothness of the drilled holes and the continuity and adequate

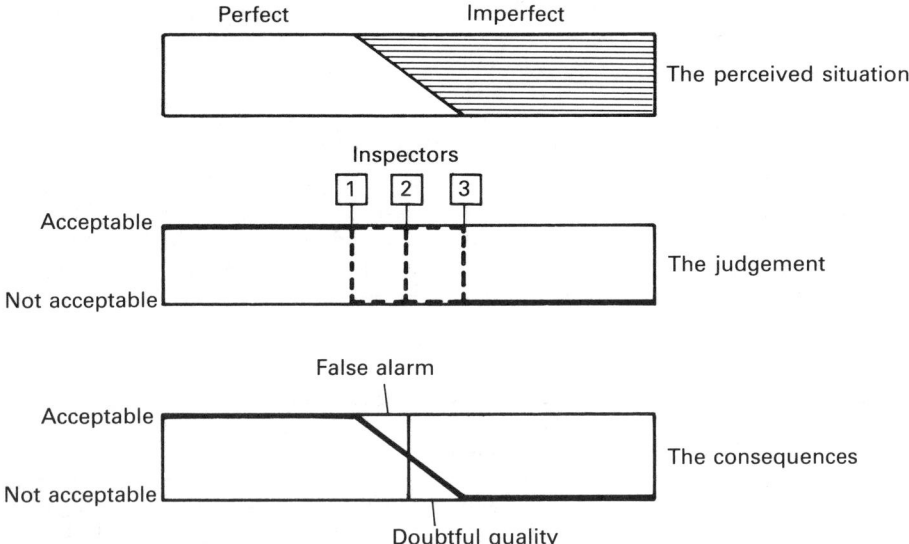

Figure 9.5 *The perfect/imperfect judgement scale*

thickness of the copper plating on their walls. It can be cured by a suitable heat treatment of boards which are liable to form blowholes by a suitable heat treatment before using them.[2]

Because blowholes do not interfere with the functioning of a circuit board, they are soldering defects rather than soldering faults, though their presence or absence is an unequivocal yes/no situation. Searching investigations have shown that they do not affect the life expectancy of joints or their reliability, in any way. Corrective

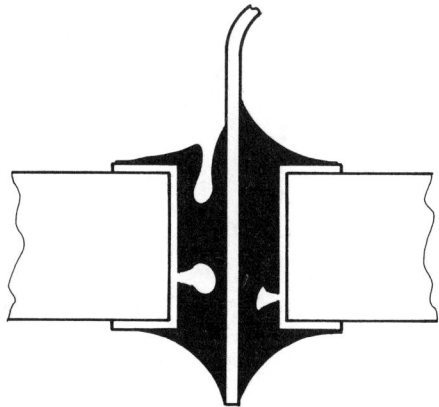

Figure 9.6 *Blowholes in a throughplated wavesoldered joint*

soldering can only mask, but not fill, a porous hole, and it is bound to shorten the life expectancy of the joint.[3]

9.3   Practical examples of soldering faults

The nature of a soldering fault means that a circuit board is faulty and cannot function until every single fault on it has been corrected. Therefore, the most important task of any quality-control system is to find every one of them. In Tables 9.1–9.3, the various types of soldering faults are listed and illustrated. For completeness' sake, faulty throughplated joints are included.

Table 9.1   *Soldering faults I: Open joints*

Soldering method	*Fault (C, if correction is possible)*	*Possible cause*
Wave	No solder on both footprint and wire or lead (C)	Solderwave did not reach the joint. (If the joint was not fluxed, it is covered by lumps of solder); or the solderwave did not overcome the shadow effect, because either layout or waveshape was unsuitable. Remedy: improve waveshape or layout, or change orientation of the board towards the direction of travel.
Reflow		No paste on footprint. Check screen or stencil for blockage
Wave	Solder on wire or lead, but not on land or footprint (C, but difficult and costly)	With wired components: land unsolderable through faulty solermask or misplaced marking
Reflow	Defect unlikely	With SMDs, as above or misplaced adhesive
Reflow	Wicking (C)	Gap between leg and footprint, due to lack of coplanarity of leads. Remedy: quality control before placement.
		Unsuitable soldering parameters. Remedy: put right, or choose paste with higher-melting solder

Table 9.2 *Soldering faults II: Bridging and solderballs*

Soldering method	Fault	Possible cause
Wave	Bridging between neighbouring lands or footprints (C)	Conveyor speed too high; unsuitable wave geometry. Flux too thin or too weak. Remedy: check (and change) flux, or soldering parameters; if layout unsuitable for wavesoldering, change orientation of board in carriage (does not always work)
	Ditto, with fine-pitch multileads	As above, or add solderthieves to layout. Change direction of travel by 90°
Reflow	As above (C)	Paste tends to form solderballs. Remedy: carry out solderballing test, and if necessary use fresh paste. Paste-printdown too thick: check and if necessary, correct
Wave	Solderballs near footprints (C)	Can happen with controlled atmosphere machines. Remedy: change to different or slightly thicker flux. Slow down conveyor
Reflow		Paste forms solderballs or spits. Remedy: if solderballing test confirms, use fresh paste
	Solderballs under melfs or chips (C: de-solder and resolder affected SMDs)	Printdown too large or too thick. Remedy: check and correct screen or stencil
Wave	Scattered solderballs, or 'spider's webs'. (C: pick up with tip of soldering iron)	Flux too thin, or insufficiently pre-dried. Remedy: check and, if necessary, adjust
Reflow	Scattered solderballs. (C: as above)	Paste spits because it has picked up moisture, or temperature profile too steep. Remedy: check and, if necessary, use fresh paste or adjust temperature profile of reflow oven *Note:* As pointed out above, solderballs do not necessarily constitute a soldering fault, unless they are loose, or are liable to become loose, and can roll about on the board

Table 9.3   *Soldering faults III: Displaced components*

Soldering method	Fault	Possible cause
Reflow	One or more leads or faces fail to connect with their footprints (C: desolder and resolder by hand)	Serious misplacement or floating. Remedies: check pick-and-place equipment; with chips or melfs, lack of solderability at one end, or lack of symmetry between footprints
	Floating or tombstoning (C): desolder and resolder	One end less solderable than the other. One end solders later than the other because of asymmetry of size or thermal behaviour of footprints. Remedy: check quality of components or correct layout fault

9.4   The ideal and the imperfect joint

The criteria of perfection in a soldered joint go back to the days of handsoldering. They have to do with two parameters: first, the wetting angle between the solder and the substrate; and secondly the amount of solder in or on the joint. Together, they determine the so-called joint profile. The ideal handsoldered joint has a 'lean' profile: the solder meniscus has a concave shape, so that the sharp wetting angle can be seen clearly. Also, the contours of the ends of the joint members must be visible, so that an inspector can be sure that, in the case of the leadwires of inserted components, the wires do in fact project through the hole and that all leads have been properly tinned (Figure 9.7).

The criteria of perfection of wavesoldered and reflowsoldered joints on circuit boards go back to these early days. They deal with surface contours, surface areas, the relationships between distances. It is possible to base judgements like good/bad, acceptable/unacceptable or beautiful/ugly on these criteria, provided every inspector can refer to a set of pictures or samples of 'perfect' and 'imperfect' joints. It is difficult to derive a clear yes/no verdict unless precise, time-consuming and therefore expensive measurements of individual joints are made. It is equally difficult, if not impossible, to base an automatic, opto-electronic inspection system on a 'good/bad' or 'beautiful/ugly' situation instead of a 'yes/no' one. Tables 9.4 and 9.5 illustrate practical examples of perfect and imperfect joints.

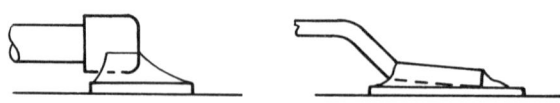

Figure 9.7   *The ideal handsoldered joint*

Table 9.4 *Too much or too little solder*

Soldering method	Defect		Judgement and possible remedy
Wave or Reflow	None $H_S = H_C + H_G$		The ideal joint
Wave	Too much solder $H_S > H_C + H_G$		Acceptability depends on the product category and the customer. Remedy: change machine parameters, lower the wave, speed up the conveyor
Wave Reflow	Too little solder $H_S < 30\%(H_C + H_G)$		Acceptability as above. Remedy: raise wave, slow down conveyor, more paste in case of reflow
Wave Reflow	Too little solder $W_S < 75\% W_C$		Acceptability as above. Remedy: check solderability of footprint and component
Reflow	Too little solder on PLCC J-leg $W_S < 50\% W_C$		Acceptability as above. Remedy: see 'wicking'

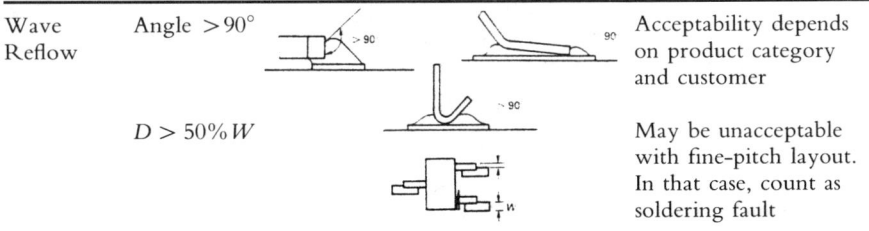

Table 9.5 *Unsatisfactory wetting angle; displaced components*

Wave Reflow	Angle $> 90°$		Acceptability depends on product category and customer
	$D > 50\% W$		May be unacceptable with fine-pitch layout. In that case, count as soldering fault

9.5 Inspection

No circuit board should leave its soldering stage without having been inspected. To inspect means to view or examine closely and critically. This implies that inspection is more than 'just having a look'. To be meaningful and cost effective, every inspection procedure must have a set of well-defined targets or criteria, preferably in the form of written, and sometimes illustrated, lists or explicit software.

The distinction between 'soldering success' and 'soldering perfection' (Section 9.2) simplifies the task of inspection, in the same way that it is easier to umpire a horse race than a beauty contest.[4] Without an umpire, both race and contest are pointless. Without inspection, a manufacturing process like the soldering of electronic circuit boards is incomplete.

There is a basic difference between the inspection of engineering products like a crankshaft and the inspection of a soldered circuit board. The dimensions of the former can be expressed numerically, and readily and automatically compared against prescribed standards, within a given set of tolerances. It is very difficult, and certainly expensive, to ascribe numerical values to a soldered joint. Hence the need for visual, optical or functional inspection of soldered circuit boards.

9.5.1   When to inspect

Inspecting every soldered board when it has reached the end of the production line is certainly necessary, but it is not enough. Unless intermediate inspections are carried out after various stages of production, errors or defects which are carried over into the next production stage can be very expensive to correct later.

The printdown of the adhesive which fixes SMDs to the board during wavesoldering must be checked for completeness and correct placement. Missing adhesive means a lost SMD; misplaced adhesive can make adjacent footprints unsolderable. Mistakes in adhesive printdown are very difficult and expensive to correct once the joints have been cured. Most adhesives are given conspicuous, sometimes luminescent, colouring to make visual or opto-electronic automatic inspection easier.

Similarly, the correct printdown of solderpaste must be checked before the SMDs are placed on the board. Faults are easily corrected at this stage. If detected after soldering in the form of empty joints or bridges, the cost of correction rises by at least one order of magnitude.

With hand-placed components, a final check before the soldering stage is advisable. Much mechanized pick-and-place equipment is equipped with integrated checks for correct identity, polarity and placement of the components (Section 7.4). If it is not, a final visual or optical check before soldering is worth while.

9.5.2   Visual inspection

It is useful to distinguish between two basic types of visual inspection, which one could call the 'general picture' and the 'detailed inspection'.

The general picture

In small-scale production, where boards are soldered individually, by hand or on benchtop equipment, the operator will naturally look at every single board before it leaves his workstation. If there is an obvious fault due to a malfunctioning of his equipment, or a defective board, he or she will put matters right before carrying on soldering.

With in-line soldering, the operator, or supervisor in charge of the soldering line, whether wave or reflow, should have a brief look at the boards leaving the line at regular intervals, perhaps at one board in every ten, in order to ensure that the line is running normally. Obvious major faults might be unfluxed areas or uneven soldering because of an unsteady solderwave, or reflowed boards which are scorched, or did not get hot enough for the paste to melt on all joints. Unless such disasters are spotted before many boards reach the next inspection station, the line may have been producing a good deal of expensive scrap.

The detailed inspection

The manner of the detailed visual inspection and the equipment used for it depend very much on the type and volume of production and the size of the boards. The type and specification of optical inspection equipment ranges from simple or illuminated magnifiers with a power of about five times, to sophisticated apparatus with zoom optics, binocular operation, stereoscopic vision, and facilities to look at the J-legs of PLCCs at an angle. In recent years, the advent of low-priced, small, readily manipulated video systems has added a new dimension to visual inspection.

As a general rule, optical systems where the operator has to look into a single eyepiece, or a binocular, which forces him to keep his head in a fixed position, are more fatiguing to operate than systems which show the object of observation on a screen. A good and flexible system of illumination is essential with all optical inspection methods. An easily operated handling system of the boards under test is equally important. With a number of systems, boards are mounted on a movable xy table, which allows for overall scanning, or indexing into preset positions where certain recurring faults tend to occur.

Recent studies pinpoint the problems of visual inspection: operators, often female, are under increasing stress, mental rather than physical, as boards get more complex and the pitch gets finer. They rate their stress factors in descending order as intense concentration, burden of responsibility and time pressure. Faulty ergonomics and noise can be additional problems. With fine-pitch layouts and components, the rate of inspection falls dramatically, and the stress is greater.

The solution is seen in systems where automated opto–electronic inspection precedes, and is linked with, inspection by a closely integrated team of about three operators, who visually inspect and manually correct faulty joints at the same time. The value of linking visual inspection with corrective soldering is increasingly recognized and practised (Section 10.1).

9.5.3 Automated opto-electronic inspection

Unless the distinction between 'soldering success' and 'soldering perfection' is made, automatic inspection must recognize both of them and be able to evaluate features like joint contours. This demands expensive systems of great complexity. If the judgement on soldering perfection is omitted, existing technology, which is still being refined, permits relatively straightforward practical solutions for automatically recognizing footprints without solderpaste, empty joints or the presence of bridges or solderballs. These systems are based on video scanning of a board surface, combined with an automatic comparison between the actual image and the ideal image of a faultless board. Equipment which operates fast enough to keep up with in-line soldering machines and reflow installations is commercially available.

Opto-electronic systems are able to recognize the following soldering failures:

Missing, misplaced or defective printdown of solderpaste
Missing or misplaced adhesive
Missing, misplaced or displaced components
Bridges, 'spider's webs' and solderballs

9.5.4 X-ray inspection

X-ray inspection represents an optical system, which operates at two levels. The board with its joints is scanned by penetrating X-ray radiation, which is absorbed most strongly by the lead-containing solder in either joints or paste printdown, less so by metallic conductors, ICs and other semiconductor devices, and least of all by organic substances like FR4 and ceramic or plastic component housings. The resulting X-ray image is converted to a monochrome image in the visible range, which can be evaluated visually by an operator, or processed photo-electronically as described above, and compared with the image of a faultless board.

X-ray images are shadowgraphs. In the last decade, X-ray sources have been developed with emitter-spots small enough to provide shadows of sufficient sharpness to allow even micrographic evaluation. By controlling the voltage applied to the X-ray tube, the penetrating characteristics of the radiation can be adjusted to suit the features which need to be examined.

X-rays show up the solder in the joint itself, which is a great advantage, but the image must be sensibly interpreted: as has been explained in Section 3.6.3 (Figure 3.19), almost all capillary joints contain voids. A distinction must be made here between two kinds of joint porosity. If the porosity appears as voids which are surrounded by solder, it is safe to conclude that the solder has wetted the joint surfaces with a satisfactory, sharp wetting angle. This type of porous joint will be as reliable as a completely filled one. If, however, the solder in the joint gap appears as an archipelago of separate islands, one of the joint surfaces is likely to have dewetted (Figure 9.8), and reliability of the joint is impaired. Image processing software might well be devised to be able to distinguish between these two types of

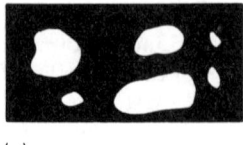
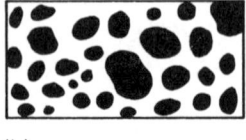

(a)   (b)

Figure 9.8 *X-ray images of capillary joints. (a) Innocuous cavities caused by trapped air or vapour; (b) pattern of unsoundness indicating a wetting or solderability problem.* ■ *Solder,* □ *void*

porosity, where the solder forms either a continuous or a discontinuous phase, and raise the alarm with dewetted capillary joints.

There is no reason why the adhesive for fixing SMDs prior to wavesoldering should not be made opaque to X-rays by the addition of a filler like a barium compound. Thus, X-ray scanning could be used for the intermediate scanning of boards for the correct presence of either adhesive or solderpaste before placing the components. X-rays are equally suitable for checking the correct placement of components before the boards enter their soldering stage. The virtues of such intermediate inspections between successive manufacturing stages are obvious.

9.5.5 Electronic inspection

Electronic equipment for identifying open joints between the legs of individual multilead components and their footprints is commercially available. The use of electronic methods to interrogate complex boards, populated with microprocessors and ASICs, for the presence and location of open joints and bridges is growing. Testability of this kind must be designed into a board from the start, and dedicated software must be written for the tests themselves. Expensive to develop and to install, electronic inspection of soldered boards becomes cost effective with growing complexity of both assemblies and components.

9.5.6 Thermographic inspection

For the sake of completeness, the so-called thermographic methods of sequentially checking all the joints on a board for their quality by measuring their thermal capacity must be mentioned here. The basic idea is to find out whether there is too much or too little solder on any given joint, or whether the joint is open, by measuring its thermal capacity and comparing it with an ideal value.

In order to do this, a short, accurately dosed pulse of Nd:YAG laser energy is focused onto the joint, and its temperature rise is picked up by a fast-response pyrometer and recorded.[5] Naturally, the reflective properties of the solder fillet on which the laser beam is focused also enter into the result and the evaluation of the test. The method requires a laser scan of great accuracy and, for a start, a board with every joint of an adequate degree of perfection – a so-called golden board – so that for every type of board, a library of ideal thermal capacities can be established.

The Vanzetti method was and still is used, especially with military and space electronics. Other, similar, thermographic testing systems have been developed in Europe,[6] but are not commercially available at the time of writing (1993).

9.6 References

1. Birolini, A. (1992) Guidelines for the development and design for quality, reliability and maintainability. *Report Z4*, 10 January 1992, Swiss Fed. Inst. Technology, Zurich.

2. Lea, C. *et al.* (1987) The Scientific Framework leading to the recommendations for the elimination of Blowholing in PTH Solder Fillets. *Circuit World*, **13**, No. 3, pp. 11–20.
3. Lea, C. (1990) The harmfulness of reworking cosmetically defective joints. *Soldering and SMT*, No. 5, pp. 4–9.
4. Strauss, R. (1992) The Difference between 'Soldering Success' and 'Soldering Quality'; Its Significance for Quality Control and Corrective Soldering. *Proc. 6th Intern. Conf. Interconnection Technol. in Electronics, Fellbach,* DVS Report 141, Duesseldorf, Germany (in German).
5. Vanzetti, R. (1984) Automatic Laser Inspection System for Solder-Joint Integrity Evaluation. *Proc. 3rd PC World Convention, Washington DC,* Paper WC III-44.
6. Geiler, H. D., Karge, H. and Kowalski, P. (1992) Possibilities and Limits of Modulated Thermographic Spectroscopy for the Nondestructive Testing of Joints. *6th International Conference on Interconnection Technology in Electronics, Fellbach.* D.V.S. Report 141, Duesseldorf, Germany, pp. 111–112.

10 Rework

10.1 The unavoidability of rework

10.1.1 Rework in the production process

In our imperfect world, zero-fault soldering does not exist. Soldering faults will occur, and because even one single fault makes a board unusable, each must be corrected by rework or corrective soldering.

It would be a mistake to regard rework as an unavoidable, tedious adjunct to electronic production. On the contrary, it is an essential link in the production chain. Unless it is taken seriously, properly organized, managed, monitored and integrated into production, reworking the soldered boards may well cost more than soldering them in the first place.

On the other hand, if rework is monitored systematically, so as to lead to a learning curve and a fault catalogue for each type of board, its cost can be reduced to its unavoidable minimum. At the same time, the fault catalogue will form a valuable tool, for use by management, designers, buying departments and quality managers. This applies equally to a large organization or manufacturing unit with many soldering lines as to a small manufacturer with a handful of employees.

Every rework operation must involve three steps:

1. *Diagnosis*   Having located a fault which must be put right, try to find out first why it has occurred. Don't start working on it, until you have satisfied yourself that you have found the answer, and have made a record of it. Otherwise, you may destroy vital evidence, which could have helped to prevent the fault recurring again.
2. *Remedy*   Put the fault right.
3. *Prevention*   Make sure that whoever in the organization could or should have prevented the fault from occurring, knows what you have found and done about it, and if possible that this information is recorded. Procedures and information technologies to take care of that are commercially available (Section 10.5.2). The rework rate can be regarded as the fever thermometer of a manufacturing line. If nobody cares to read it, the patient may well be moribund before anybody has noticed that he is sick.

10.1.2 Desoldering and resoldering

Rework itself often involves two closely linked operations: if the correction of a fault requires the removal of a faulty, misplaced or dislodged component and its replacement, the desoldering and resoldering operations which this implies should follow closely upon one another.

Removing bridges and, if necessary, solderballs, are simple desoldering operations. Desoldering leaded components from throughplated holes requires soldering irons which can suck the solder out of the hole. With SMDs, the removal of a defective component and its replacement with a new one can often be carried out with one and the same tool. With the desoldering of wavesoldered SMDs, separating the glued joint underneath the component is an additional operation. Open joints are normally filled with a small temperature-controlled handsoldering iron.

10.2 Basic considerations

10.2.1 Metallurgical and mechanical consequences of rework

A reworked joint is never as good as the first joint would have been had it not needed to be replaced or corrected. The reasons are found principally in the metallurgy of joint formation. Furthermore, the additional thermal, and sometimes mechanical, stresses of resoldering may easily weaken the bond between footprint and board. A lifted footprint is one of the most serious types of damage, caused by getting a joint too hot or heating it for too long during desoldering. Repairing a lifted footprint is possible, but it is expensive, takes time, and the board will never be as good again as it was in the first place.[1]

Industrial experience confirms the damaging effect of corrective soldering.[2] So did a cooperative research project carried out by a number of industrial companies in the UK under the auspices of the National Physical Laboratory.[3]

The latter counted the number of through-hole joints on standardized sample boards with 2000 joints each which showed visible cracks after they had been reworked with a soldering iron under controlled conditions and then exposed to up to 1000 temperature cycles between $-20\,°C/-4\,°F$ and $100\,°C/212\,°F$. For comparison, boards which had not been reworked were given the same thermal treatment. Some of the results are given in Table 10.1.

Two conclusions are clear:

1. The longer the rework contact time of the soldering-iron bit, the more soldered joints are visibly cracked after a given number of thermal fatigue cycles.
2. The higher the rework temperature, the more heat is supplied to the solder joint and the greater is the degradation of its performance under thermal fatigue, i.e. the loss of reliability.

Table 10.1   *The degradation of soldered jonts caused by rework*

Rework time at soldering temperature of 400°C/750°F	Number of joints out of 2000 cracked after 2000 cycles between −20°C/−4°F and 100°C/212°F
No rework	100
2 sec	160
6 sec	180

Rework temperature using rework time 4 sec	Number of cracked joints after cycles as above
350°C/660°F	130
400°C/750°F	170
450°C/840°F	200

The metallurgical reason for this degradation is the growth of the brittle intermetallic layer between solder and substrate. Its thickness is a function of the confrontation time between the molten solder and the substrate, and of its temperature.[4]

A further factor is the depletion of tin in the immediate neighbourhood of the solder/copper interface because it migrates to augment the intermetallic layer. This leaves a predominance of lead in the area next to that layer. Lead being weaker than solder, this circumstance further promotes cracking under thermal fatigue loads.

10.2.2   The cost of rework

Rework is a joint-by-joint procedure, and necessarily time-consuming and expensive. Reworking a single joint costs on average as much as did the first-time soldering of the complete circuit board on which it sits, and often considerably more. Depending on the type of product and its sale value, it may sometimes be cheaper to scrap a faulty circuit than to rework it. Some low-cost products like pocket calculators, cheap electronic watches and toys fall in that class.

In small-scale production in particular, where visual inspection and rework are carried out by the same operator or operators, it is important to keep this cost aspect in mind. It is tempting to say 'I might as well touch up this joint while I am looking at it.' Not only do joint quality and reliability suffer through this practice, but costs are liable to rise in an uncontrollable manner.

10.2.3   Lessons to be learned

The lessons to be drawn from these considerations are plain:

1. Whatever the method used for rework, choose only the best equipment available, and maintain it in top condition.
2. Complete every joint quickly. The solder should be molten for not longer than a few seconds.

3.   Keep the working temperature as low as is compatible with this requirement.
4.   Preheating the board, either locally or overall, before carrying out any rework
     on it brings two benefits. First, the specific heat of the FR4 substrate is almost
     four times that of copper or solder. This means that four times as much heat
     energy is needed to bring the board up to soldering temperature than to heat
     the joint itself. Preheating considerably shortens the heating time necessary to
     complete a joint, particularly with heavy multilayer boards.
         Secondly, with wavesoldered boards where the SMDs are glued to the
     board surface, preheating softens the adhesive joint and makes it easier to break
     when desoldering becomes necessary (Section 4.9.5). A preheating tempera-
     ture of about 60 °C/140 °F to 100 °C/210 °F is usually enough for this purpose.
         A stream of warm air directed against the underside of the board is the usual
     method of preheating. This avoids the danger of a sharply localized hot spot,
     which could distort the board.
5.   Rework for cosmetic reasons alone is an expensive and damaging luxury, and
     it should only be carried out if the customer or the market demands it and pays
     for it.

10.3   Rework equipment

10.3.1   Heat sources

Almost every heat source which is employed in production soldering finds its use
in rework: soldering irons in various forms, thermodes, solderwaves in miniature
form, infrared radiation, hot air or gas, and in recent years also the Nd: YAG laser.
With all of them, efficiency of the heat transfer from source to joint is of the
essence, together with precise temperature control.

Soldering irons

Molten solder is the ideal heat transfer medium: its heat conductivity is high and,
being molten, it adapts perfectly to the surface contours of the joint and the joint
members. For this reason, a soldering iron used for rework must have a well tinned
tip, preferably with a drop of molten solder on it, to establish instant and good
thermal contact with the joint. The best solder to use for rework, not only with an
iron but with any method, is the silver-containing 63% tin solder (Section 3.3.1).
    The size of the iron and the shape of the tip must suit the type and the
configuration of the joint, as will be shown in Section 10.3.2. Naturally, the
ergonomic aspects of the hand–held soldering iron must not be neglected.
    Precise control of the tip temperature, and a fast response to changes in the
amount of heat demanded from it, are very important. Recent years have seen
rapid advances in the technology of the soldering iron, which had remained static
through several decades. Above all, methods of heating the soldering tip, and of
shortening the path from the heat source to the end of the tip, have improved it

greatly. Many handsoldering work stations include a fast response temperature indicator of the soldering tip itself.

Bearing in mind what is at risk if rework is carried out badly, it is worth stressing the importance, not only of choosing a state-of-the-art soldering tool with antistatic protection and a quick response to sudden heat-demands or changes in thermostat setting, but also of maintaining it well and periodically checking the accuracy of its temperature control.

Heated tweezers and thermodes

Resistance-heated tweezers are a convenient tool for desoldering the two end-joints of a melf or chip simultaneously. PLCCs can be removed with a shaped tweezer, which grips the vertical shanks of the J-legs on all four sides of the components simultaneously. Tweezers are particularly useful for removing and resoldering these components on a crowded board, where the neighbours of the affected component must be protected from the soldering heat of the site of the repair (see below). Tweezers, like thermodes, are made from an untinnable metal like tungsten.

The thermodes used for desoldering are the same as those which are used for the impulse-soldering individual multilead SMDs (Section 5.7). Because their soldering surfaces are untinnable, they cannot be pretinned like a soldering iron, and molten solder cannot be used as a thermal link between the tool and the joints. A thin coating of a low-solids RMA flux, brushed over the row of leads on all four sides, will have to serve as a substitute for the molten solder in establishing a thermal link between the soldering tool and the joint. Once the thermode sits firmly on the flat leads, the heat transfer between the soldering tool and the joints takes place by conduction, as with a tinned soldering iron.

Miniature solderwaves

Miniature solderwaves were developed in the sixties for the desoldering and resoldering of inserted multipin devices, like DILs. With this method, the board is placed over the nozzle of a dedicated small wavemachine, so that the leads which have to be desoldered sit closely above it. As soon as the wave is switched on and touches the underside of the board, the solder in the joints melts and the DIL can be lifted off. A replacement DIL can then immediately be soldered into the empty holes, its legs having been fluxed first. Alternatively, the molten solder can be sucked out of the holes, an operation for which this type of machine is equipped, and the replacement part can be soldered in position, on the same wave, at a later stage.

Hot air or gas

Jets of hot air or gas, of a controlled temperature, are widely used for desoldering SMDs. Because they transfer their heat to the joints by convection, which is much less efficient than conduction through molten solder or contact with a thermode (Section 5.5), it takes longer to melt the solder in the joints, and this must be taken into account when deciding on the reworking procedure.

Single jets from hand-held nozzles can be used for desoldering and resoldering melfs, chips, SOs and multilead devices. For the latter, however, many vendors offer equipment with interchangeable arrays of jet nozzles, which direct the hot air or gas towards the joints on all four sides of the SMD at the same time. These jet arrays are useful for the desoldering and resoldering of PLCCs, where the joints are not directly accessible to a plain thermode.

The hot air stream from a hand-held nozzle is usully controlled by the operator via a pedal, while multijet soldering heads are operated by a timer, the correct duration of heating having been determined empirically for a given board and component. With most hot-air desoldering machines, the board is preheated locally from underneath, for reasons already mentioned. Preheating is especially important with heavy multilayer boards.

Infrared radiation

Beams of infrared radiation, focused on the joints of a multilead device, are another option for supplying the heat for desoldering and resoldering. Similar to the jet nozzles, the IR desoldering heads are interchangeable to suit specific component dimensions. Heating by radiation represents a non-equilibrium system (Section 5.8) where the temperature of the heat source is much higher than the target temperature of the joint. For this reason, such systems demand precise dosage of the radiation input, and ideally a feedback from the joint temperature to the heatsource.

Infrared desoldering heads normally operate with medium wavelength emitters, operating in the $300\,°C/550\,°F$ to $500\,°C/950\,°F$ range.

10.3.2   Rework stations

The workplace or work station for carrying out desoldering and resoldering must be user-friendly. Reworking can be a stressful task, and requires constant concentration and decision making. The workplace must therefore be well lit, provided with an exhaust which removes the flux fumes, and the equipment must be ergonomically designed so that it can be operated with a minimum of fatigue.

For manual rework, adjustable inclined frames which hold the board are standard. Wide-field magnifiers with a power which need not exceed five, or at most ten, times are also standard. Often, provision is made for a set of soldering irons of different size to be readily at hand. In addition to soldering irons, workstations may provide sets of hot air or gas jets, vacuum pipettes for handling components, and heated tweezers for desoldering and resoldering them. Small mechanical tools like rotary drills and reamers to clean or remove footprints are also often part of a rework station.

More elaborate manual stations are fitted with an overhead illuminator, which direct a pencil of light in sequence to the several joints or components which require rework. This sequence can be created by the inspection station and stored on software in cases where visual inspection and manual rework are carried out separately from one another.

From simple manual workstations, the market offers a gradual succession of increasingly complex and sophisticated equipment for rework, particularly for use with multilead SDMs. Most of them use sets of quickly interchangeable hot-air nozzles or thermodes, together with placement systems for aligning the component to be dealt with accurately with the soldering tool, visually or by opto–electronic means.

Notwithstanding these electromechanical aids, the accuracy of feedback between the human eye and hand must not be underestimated. It has been stated that a trained operator, aided by a low-power magnifier, can place a multilead device with a 0.5 mm/20 mil pitch on its footprints with sufficient accuracy by hand. This means a lateral accuracy of ± 0.125 mm/5 mil, since the maximum permissible lateral misplacement of fine-pitch components is one quarter of the pitch distance (M. Cannon, Pace Europe Ltd, verbal communication).

10.4 Rework tasks and procedures

Most of the published information and know-how in this field can be found in the voluminous vendors' literature. A few comprehensive accounts have appeared in the technical press.[5–7]

10.4.1 Removing bridges and solderballs

A solder bridge is removed by touching it briefly, for not more than a few seconds, with the well-tinned tip of a small soldering iron, with a rated output of about 40 W, and set to a temperature of about 250 °C/480 °F. For de-bridging fine-pitch leads, a conical 20 W soldering bit is best. All that is necessary is to disrupt the bridge. Do not try to 'tidy up' the joints on either side, and do not use flux. A solder wick (see below) should not be used either, because it is likely to suck out too much solder from the joints themselves and thus weaken them.

Solderballs are removed in the same way. As soon as the tip of the iron touches one, it will disappear. The same will happen with 'spiders' webs'. Because a web covers more ground than a solderball, a chisel-shaped soldering tip is more convenient for removing them.

Excess solder is removed from the tip of a soldering iron by wiping it with a piece of linen or cotton, but never with fabric made from synthetic or mixed fibre. Many handsoldering stations are equipped with a heat resistant sponge for cleaning the soldering tip.

10.4.2   Desoldering SMDs

Any of the following circumstances makes desoldering of an SMD obligatory:

- Melfs and chips, if one or both of their metallized faces are seen to be badly solderable or unsolderable, or if the metallization has been leached off by the molten solder. The desoldered components are discarded.
- Chips which have 'tombstoned'. The desoldered components are discarded, because their solderability is doubtful.
- Melfs, chips and SOs which have floated away from their footprints. The desoldered components are discarded, unless there is a desperate need for their re-use.
- Any SMD which is in the wrong place, or wrongly orientated. If carefully desoldered, the component can be reused.

Desoldering melfs and chips with a solder-wick

Melfs and chips are joined to their footprints by butt joints. The joints of the larger ones contain too much solder to be sucked up by the tip of the hot soldering iron. Instead, a device called a 'solder-wick' is used to remove it. A solder-wick is a tape of closely braided fine copper wire, a few millimetres wide, which is impregnated with an RMA flux. Recently, solder-wick impregnated with a no-clean flux has become available. The wick acts towards molten solder like blotting paper against ink: for desoldering, it is pressed into the corner between the endface of the component and the footprint (Figure 10.1) with the end of a chisel-shaped soldering tip held at a temperature of about 250 °C/480 °F–300 °C/570 °F. As soon as the heat has travelled from the tip through the wick of the solder, the latter melts and is sucked by capillarity into the fluxed wire braid. Tip and wick are lifted off, and the solder-filled end of the wick is discarded. This operation should take no longer than a few seconds, but the wick should be pressed into the corner for long enough to let the heat travel into the gap underneath the component so that all the solder is sucked out of it.

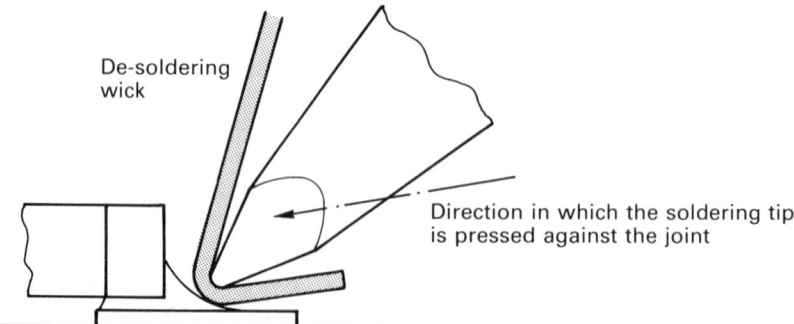

Figure 10.1   *Desoldering a melf or chip with a solder-wick*

With wavesoldered components, the glued joint between component and board must now be broken. To make this possible without undue force, the adhesive must be above its 'glass-transition temperature', which is normally about 60 °C/140 °F to 80 °C/180 °F. Above that temperature, the adhesive changes from a rigid body to a soft pliable substance (Section 8.5). This makes it possible to break the glued joint by twisting the component (Figure 10.2).

If the temperature is not high enough and the adhesive still too stiff, the board might get damaged. If there are any conductor tracks running underneath the component, between its footprints, the glued joint sits on top of the soldermask which covers them: twisting the joint apart with too much force can damage the soldermask and expose the tracks underneath. Local preheating of the board will raise the temperature enough to soften the adhesive. The heat of the desoldering will do the rest. If this is not enough, further heat from a hand–held hot-air nozzle will have to be applied to the component.

Desoldering melfs, chips and PLCCs with heated tweezers

A simpler method of removal employs a pair of tweezers, heated in the same manner as a thermode to about 220 °C/400 °F. In this way, the joints at both ends of a component are desoldered simultaneously. The jaws of the tweezers are pressed against the ends of the component until the solder can be seen to melt. As soon as this happens, the component is lifted off or, in the case of a wavesoldered one, twisted from its adhesive joint (Figure 10.3).

The same principle, using tweezers with interchangeable heated jaws which enclose the sides of a PLCC, is a useful option. The jaws press against the sides of the J-legs, and the component is lifted or twisted off as soon as the solder melts.

Desoldering SOs, SOICs and multilead components

With these components, the leads cannot be gripped with a sideways pressure without damaging them, which would be undesirable where removed compo-

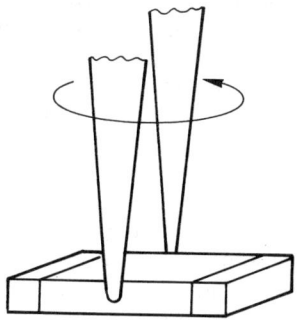

Figure 10.2 *Twisting a glued joint apart*

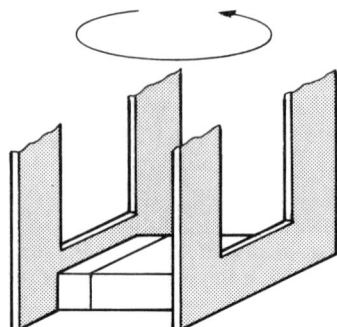

Figure 10.3 *Desoldering with heated tweezers*

nents may have to be re-used. Here, a heated tool of a shape which conforms to the pattern of the footprints, in other words a thermode as it is used in impulse soldering (Section 5.7), is pressed downwards against the horizontal ends of the gullwing legs (see Section 5.8.3).

The hot jaws only heat the joints, but do not grip the leads. Lift-off is effected by a vacuum pipette, which is located between the jaws and seats against the top of the component housing. It is activated by the operator when the joints are seen to have melted and the component is ready to be lifted.

With thermode desoldering, a thin coating of RMA flux brushed over the component leads improves the thermal contact between them and the face of the thermode. The presence of flux also reduces to some extent the sharp spikes of solder which are likely to remain on the footprints after the component has been lifted clear.

Failing desoldering equipment which allows the use of a thermode, multilead components can be desoldered joint by joint, with a temperature-controlled soldering iron (about 220 °C/430 °F). The well-tinned tip of the chisel-shaped iron is pressed against the joint: as soon as the solder is seen to melt, the iron is lifted off and the gullwing leg is lifted clear of the footprint with a pair of tweezers. Another way is to thread a thin stainless steel wire underneath the component: as soon as the joint has started to melt, the legs are pulled upwards with the wire, one after the other (Figure 10.4). After desoldering operations such as these, the component is mostly a write-off because of its bent legs. Also, unless skilfully executed, these methods put the footprints in danger of lifting off the board because of overheating. For all these reasons, it is best to invest in a thermode desoldering tool.

Cleaning the empty footprints

Before the replacement component is put into place, the footprints must be cleared of remaining blobs or spikes of solder. This is done by lightly fluxing them again and then passing the tip of a soldering iron, preferably chisel-shaped, across all of them. After that, they are fluxed once more, and the replacement SMD is soldered into place as soon as possible.

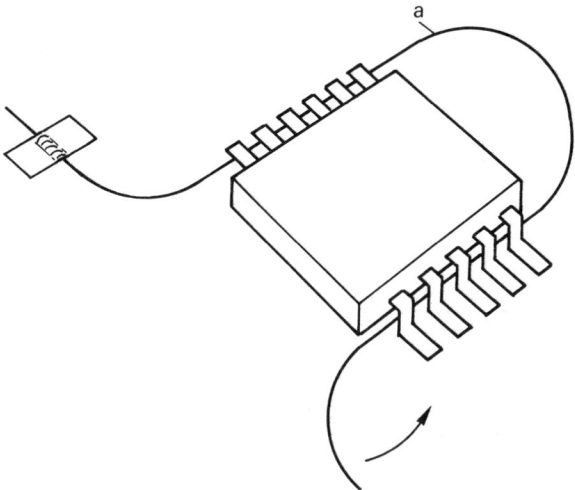

Figure 10.4   *Desoldering a SOIC leg by leg (courtesy Edsyn Corp.)*

Caution!

Any footprint cleared of solder must be given a coating of flux as soon as possible, for the following reason: if the soldercoating is removed from a footprint, or for that matter from any soldered copper surface, the top layer of the diffusion zone formed by the intermetallic compound 'eta' (μ, Cu_6Sn_5) (Section 3.1) is exposed to the air. This compound will immediately begin to oxidize and become unsolderable, even with a strong flux, unless it is quickly protected from the atmosphere by a thin coating of flux. All the same, a replacement component should be soldered to its footprints as soon as possible.

10.4.3 Filling empty joints

Finding the empty joint

An empty butt joint or open joint is easily recognized provided it is fully visible. PLCCs are the exception, because they are half hidden under the component edge. With a little practice and the right optical aid this problem is easily overcome.

With empty capillary joints, only the top surface of the component lead is directly visible. A bent lead or a footprint without solder is immediately apparent. If the footprint is covered with solder, but the solder along the edge of the lead looks doubtful or if there is none at all, the joint can be tested by lifting the end of the lead with a sharp probe. If it lifts, the joint must be corrected. If it stays firm, it is best left alone. For non-visual methods of inspection, see Section 9.4.3.

Finding the cause of the defect

Before a joint which on inspection has been found to be empty is repaired and filled with solder, it is essential to find out why the fault happened, and that the finding be recorded. If this is not done, and the reason remains undetected, and the repair unrecorded, the fault may recur again and again throughout the complete batch of boards.

If a joint at the end of a melf or chip has remained empty because the metallization was unsolderable, has dewetted or leached away (as the evidence will clearly show) it is useless to try resoldering. The component must be desoldered, discarded and replaced. If there are several similar open joints, it is advisable to query the whole batch of components, and meanwhile to continue production with a fresh batch, having checked its solderability.

If the footprint did not accept the solder, the reason for it must be discovered before an attempt is made to patch up the joint.

The soldermask may have been faulty, and have partially or completely covered the footprint. Again, first check whether this is a single occurrence or whether several boards show the same defect in the same place. If the latter, production management must be informed at once. If the former, a decision must be made whether to desolder the component and repair the affected footprint, or whether to scrap the board. Removing cured soldermask from a footprint is a specialized and time-consuming task, which requires either a special solvent or mechanical removal with a rotating miniature burr or milling cutter, similar to a dentist's drill (many vendors of rework equipment supply such equipment).

Alternatively, the printed marking ink on the board may have partially covered the footprint. Misplaced ink can be removed with a suitable solvent, but it is as well to check whether this is a recurrent fault. If it is, action must be taken as above.

If the board was wavesoldered, adhesive which has spread across one or both footprints may be responsible for an open joint. This can happen if too much adhesive has been put down between the footprints, and was squeezed out as the component was placed on the board. The closely spaced footprints of small melfs and chips are especially at risk from this cause. The fault can be readily spotted by the distinctive colour of the adhesive. If the board is to be saved, the components must be desoldered and the adhesive joint must be parted. The cured adhesive can be removed from the pad only by mechanical means, with one of the rotating tools described above. This should only be done by trained operators, because at this stage the board carries a high investment of production effort, which a mistake at this stage puts at risk.

On wavesoldered boards, an unsuitable layout, or an unfavourable orientation of the board towards the direction of travel over the wave, may have made it difficult for the solder to reach some joints. The result will be a 'skipped joint'. What has happened will be obvious from the evidence which such joints present. It is not difficult to fill them with a fine soldering iron with a chisel tip, having lightly fluxed them first (RMA, no-clean or watersoluble flux), and using a thin, flux-cored solder wire of silver-containing eutectic solder and filled with an RMA flux. Solderwire filled with no-clean or watersoluble flux has recently become available (Section 3.4.3).

The occasional empty joint on a multilead component on a reflowed board is often due to the gullwing leg having been bent upwards for some reason. This fault is called 'lack of coplanarity' (Section 1.4). It can be caused by unsuitable packaging or handling. Coplanarity should normally be within ± 0.1 mm/4 mil, but even within these limits a thin paste printdown may not be able to reach the occasional lifted leg.

Filling the open joint

Soldering iron

The correct way of soldering open joints with a soldering iron is shown in Figure 10.5. In all cases, the chisel-shaped soldering tip is placed against the joint in such a way that the transfer of heat from tip to joint is as efficient as possible. The end of the solderwire is always placed close to the tip of the iron, but never on it: if that were done, contact with the hot iron would have killed the flux by the time it reached the joint. Placing the solderwire close to both tip and joint lets the heat travel to it, and it will begin to melt as soon as the joint has reached the correct temperature. As soon as the solder is seen to have flowed into the joint, both the solderwire and the iron are lifted off.

The temperature of the soldering iron is set at about 210 °C/410 °F to 220 °C/440 °F. Joints on thick multilayer boards may need a higher soldering temperature. In any case, the operation should be completed in not much more than three seconds.

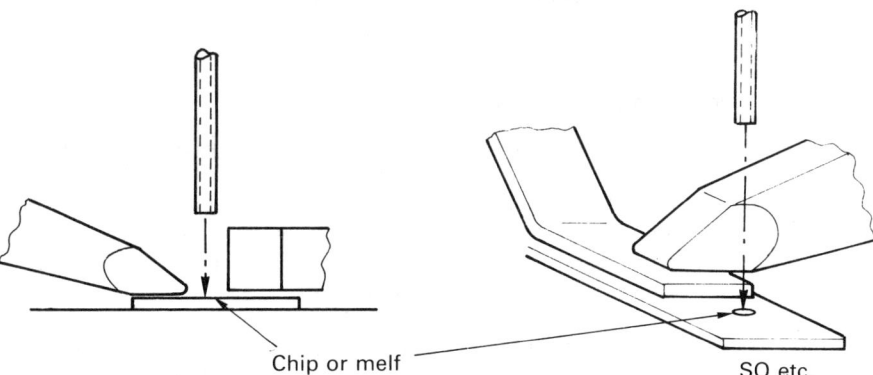

Chip or melf   SO etc.

Figure 10.5 *Filling an open joint with a soldering iron. Place the end of the solderwire on the footprint in spot (1). Then place the flat side of the soldering iron against the footprint or component leg as shown*

Solderpaste can of course be used instead of solderwire for the filling of such joints. A small spot of paste is placed into, or next to, the joint before heating it. Generally it is found that handsoldering with paste instead of wire leaves more flux residue on the soldering iron and on the joint, and needs more cleaning up after soldering.

Hot air or gas

Being a convection method of heating, filling an empty joint with a hot air or gas jet (about 300 °C/570 °F) may take more time than soldering with an iron, maybe from five up to ten seconds. It may be advisable to use local preheat from below, to speed things up. On the other hand, handling a hot air jet requires less skill than the simultaneous handling of a soldering iron and solder wire; also, the progress of the soldering operation is easier to observe. With hot air or gas soldering, a small spot of solderpaste is preplaced into or near the joint, and filling the joint becomes a one-hand operation.

Thermode

Filling an open joint with a thermode is not often practised because it involves all joints on a multipin component for the sake of correcting one single joint.

10.4.4   Resoldering SMDs

Resoldering in this context means replacing a component which has been desoldered for one of the reasons discussed above, mostly by a new one. Soldering a component which has been desoldered back into its place should only be done for the compelling reason of its value or irreplaceability, and after its correct functioning has been checked. Desoldered passive components like melfs and chips, particularly ceramic condensers, should always be discarded if the situation allows it.

Handling and placing the component

Here, the first rule is never to touch a component with the fingers. However clean the operators' hands are (and they should be clean of course), some sweat or contamination is bound to transfer to the soldering surfaces and reduce their solderability. Components are handled with tweezers or preferably a vacuum pipette. The latter have reached a high level of ergonomic perfection and all of them allow the component to be rotated into its correct orientation after having been picked up.

Most operators find it more convenient to manipulate components directly with a hand-held pipette, instead of via 'joystick' like manipulator, which is manoeuvrable in the x, y and z directions, and rotatable round the vertical z axis. As has been said already, the coordination between the human eye and the human hand is remarkably precise, to within 0.125 mm/5 mil.

Handsoldering with an iron

Using the conductive method of heating, handsoldering is fast. Melfs, chips and SOs are placed and held in position so that their leads sit correctly on their footprints. A little RMA flux is then applied to the joints with a small brush, and the joints are touched with the tip of the soldering iron, which is preferably chisel shaped and held at a temperature of about 220 °C/430 °F. Before this, the tip of the iron has been touched briefly with the end of a thin flux-coated solder wire, so that it carries a small drop of molten solder at its end.

A 'half-chisel' shape of soldering iron tip has been specially developed for the handsoldering of SMDs:[8] the endface of the cylindrical bit is flat and set at an angle so that it has an oval surface (Figure 10.6). This enables it to hold a fair amount of solder, which can be run into the joint very quickly while presenting a sharp, horizontal edge to the joint.

The butt joints of melfs or chips are touched at the corner where the metallization and the footprint meet; capillary joints are touched at the joint entry. The whole operation should not take longer than a few seconds – three at most.

With multilead components with gullwing legs, the component is placed carefully on its prefluxed footprints by hand. That this is perfectly feasible even with fine-pitch components as has already been explained. Once in place, the component is tacked to the board by touching one joint each at two diagonally opposed corners. After that, the component need not be held down any longer, and the rest of the joints can be soldered by running the tip with its reservoir of solder along each row of joints, letting the solder run into them.

The J-legs of PLCCs are dealt with similarly. The edge of the tip is run along the line where the legs sit on their footprints.

Resoldering with a hot air/gas jet

Resoldering with a hand-held jet of hot air or gas needs only simple equipment, but it requires skill and experience. The temperature of the air or gas is about 300 °C/540 °F to 350 °C/630 °F, and the force of the jet can be controlled with a

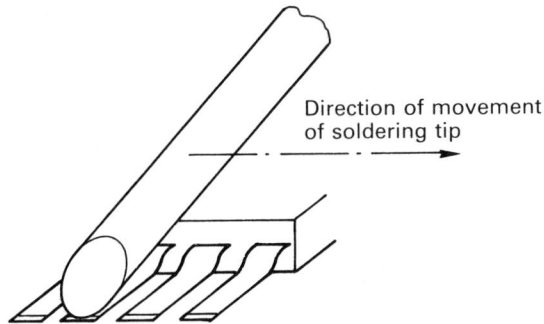

Direction of movement
of soldering tip

Figure 10.6 *Soldering bit with an oblique endface (courtesy Pace Inc.)*

footpedal. The footprints are thinly covered with solderpaste, which can be put down with a hand- or air-operated syringe.

The method demands skill because the heating rate is slower than a soldering iron, and therefore soldering takes longer. With densely populated boards, particular care must be taken to avoid getting neighbouring joints too hot, so that they start to melt. Nevertheless, this method is very useful for boards with a simple layout, and it requires little investment.

10.4.5 Cleaning after rework

Cleaning after rework is often neglected or forgotten. There are several tell-tale marks which indicate, or betray, where a joint has been reworked: the usual sign is a conspicuous flux residue on an otherwise clean board; less frequently, and perhaps less excusably, an uneven or maybe clumsy dot of solder. If rework is carried out before the boards are cleaned, all flux traces of any rework will of course disappear. If there is no cleaning after soldering, or if rework is carried out after cleaning, the question of cleaning up after rework must be addressed. Unless it is the deliberate policy of quality inspection to show where correction has taken place, traces of rework should be eliminated for the sake of appearance and customer satisfaction. If the finished board is to be lacquered or has to receive a conformal coating, there is a technical reason for a clean-up after rework.

Solderwire with a core of no-clean flux is commercially available, and there may therefore be no technical reason for its removal. The residue left by some of these wires is very inconspicuous, and whether it can be left as it is must be a matter of individual judgement. If the flux residue is plainly visible, it ought to be removed.

Any cleaning method must of course be environmentally acceptable. Spray dispensers of non-flammable flux solvent are commercially available. A sprayed-on solvent will disperse the flux residue and make it less conspicuous, but it will not remove it. The best method of local flux removal is dabbing with a small swab of linen or cotton, soaked in isopropanol and held with a tweezer. With some practise, this method gives perfectly adequate results.

Here are a few 'don'ts':

- Don't use a very volatile solvent such as methylene chloride (Section 8.3.1), which is often the basis of nail-varnish remover or of the fluid used for removing adhesive plaster: it dries so quickly that it does not remove the flux residue but only spreads it around.
- Don't use cotton wool, or natural wool, for a swab. Both will leave fluff behind, which is as bad if not worse than the flux residue you want to remove.
- Don't use synthetic fibre fabric for a swab, because it does not soak up liquids very well.

10.4.6 Semi-mechanized work stations

A variety of semi-mechanized, multi-purpose workstations are commercially available. These can be used not only for rework such as desoldering and resoldering multilead components, but also for soldering them individually to a circuit board during normal production. Some work with thermodes, but most of them use hot air or gas as the heating medium. Their design spans a wide range of technical sophistication:

1. Most of them embody means for optically checking the correct positioning of the component on its footprints. This may be done visually through stereoscopic or vertical sighting, or through video observation on a screen.
2. A set of hot air/gas nozzles of various sizes to match the component to be soldered is normally carried. On some stations, the nozzles are carried on a rotating turret. The design of the nozzles confines the stream of heating gas to the joints of the component, and prevents it from affecting neighbouring components.
3. The correct heating cycle for a given component is empirically determined, and can be stored in a databank.
4. Postcooling with room–temperature air, before the pipette which holds down the component is lifted off, is often provided.
5. Depending on the type of reworkstation, a complex range of data storage and processing can be incorporated. So-called 'paperless rework stations' can fulfil the following tasks:

   - identify (mostly with the aid of a bar code) and record the type of board that is being reworked, the identity of the reworked component and the type of fault that caused the rework
   - record the rework parameters and the identity of the operator
   - process the recorded data, so as to integrate them into a CIM system and to make them available for statistical process control (see also Section 10.5.1).

10.5 Integrating rework into the production process

Rework, together with inspection, must be seen not only as a process in itself, but also as an essential part of the whole manufacturing activity. The links between rework and product design, materials selection and control, and the manufacturing parameters in general, must be understood, established and maintained, and above all acted upon. Unless this is the case, inspection and rework, while preventing faulty products from leaving the factory and reaching the customer, will be unproductive and unnecessarily expensive.

10.5.1 Rework personnel

Rework is entirely a personal matter. The decision whether rework is necessary may have been made by a quality inspector, or by automatic inspection equipment, but the rework itself is done by a person, not a machine. As has been shown in the preceding sections, the question whether to rework or not is very often the responsibility, and the result of an on-the-spot decision, of the same person.

Inspectors and reworking personnel, like policemen, lawyers and doctors, only go into action when things have gone wrong. It is natural that they are tempted to take a jaundiced view of the work of their colleagues further back along the production line, from the designers and the purchasing departments down through all the manufacturing sections.

It is important that this problem is appreciated by management, and that every effort is made to prevent the resulting attitude from taking root and hardening. Inspectors and the reworking teams must be made familiar not only with the manufacturing steps upstream of their own place in the chain, but they must also be given the opportunity to establish some personal contacts with the people who work there. It is important not to leave these links entirely to some bureaucracy or an impersonal CIM system.

Thus, the training of rework personnel must include familiarizing visits to these departments, and occasional renewal of contacts, as well as being kept informed of what goes on and what is new.

10.5.2 Closed-loop soldering

The term 'closed-loop soldering' is an apt formulation of the concept of which the above integration of inspection and rework into the entire manufacturing chain is only a part.[9] In a closed-loop production, inspection and rework have two functions: first, they must ensure product integrity; secondly, they must identify design problems and provide the background for material and process control.

Practising closed-loop soldering without computerized data processing is possible, but with medium and large organizations it involves much paperwork and administrative effort. A number of individuals and companies have begun to develop and market both software and hardware which are designed to take over where the human operator would be swamped with recording and evaluating what he is doing.

Inspection verdicts and rework operations can be entered into data stores, where they are classified and evaluated. Paperless rework stations are able to log their operations automatically. Programs are already in existence which can be interrogated to relate fault rates of a specific board or a component to suppliers, storage time or any of the manufacturing parameters relating to the assembly being repaired.[10] Statistical process control can be added to such a program, so that warnings are sounded as soon as the rates of faults, rejects or rework exceed an empirically established standard fluctuation. To use an image introduced earlier, closed-loop soldering automatically checks the fever thermometer, which signals the state of health of the manufacturing unit, and it sounds a warning as soon as it begins to drift beyond its normal reading.

10.6 References

1. Strauss, R. (1990) The Metallurgical Aspects of Corrective Soldering of Printed Circuit Boards. *Proc. 5th Intern. Conference Interconnection Technology in Electronics*, DVS Report 129, Duesseldorf, Germany, pp. 155–157 (in

German).

2. Keller, J. (1989) Solder Joint Failure Detection: Does Inspection ensure Reliability? *Soldering and SMT*, No. 3, October 1989, pp. 56–57.

3. Lea, C. (1990) The Harmfulness of Touching up of Cosmetically Defective Solderjoints. *Soldering and SMT*, No. 5, June 1990, pp. 4–9.

4. London, J. and Ashall, D. W. (1986) Compound Growth and Fracture at Copper/Tin–Silver Solder Interfaces. *Brazing and Soldering (UK)*, No. 11, pp. 49–55.

5. Verguld, M. M. F. and Leenaerts, M. H. W. (1988) Repair of Printed Circuit Boards carrying SMDs. *Circ. World (UK)*, No. 2, January 1988, pp. 11–15.

6. Klein Wassink, R. J. (1989) *Soldering in Electronics, 2nd ed.*, Electrochem. Publ., Ayr, Scotland, pp. 631–641.

7. Morris, B. (1993) Rework and Repair of SMD Assemblies. *Electronic Mats. & Packages (Elsevier)*, March/April, pp. 27–32.

8. Abbagnaro, L. (1993) Manual Operations take the Heat out of Rework, *Electronic Manuf. (UK)*, June, pp. 8–10.

9. Davy, J. G. (1990) Closed-Loop Soldering, *Proc. 14th Electronics Manufacturing Seminar, Naval Weapons Centre, China Lake, CAL*, February 1990.

10. Habenicht, G., Janker, A. and Wuermseher, H. (1992) Automatic Registration of Quality Data, *Proc. 6th Intern. Conf., Interconnection Technol. in Electronics, Fellbach (Germany)*, DVS Report 141, Duesseldorf, Germany (in German).

Index